CHEMISTRY FOR HEALTH SCIENCES

CHEMISTRY FOR HEALTH SCIENCES

M. SATAKE
Y. MIDO

Consulting Editors.
M. S. SETHI
S.A. IQBAL

DISCOVERY PUBLISHING HOUSE
NEW DELHI—110 002

First Published—1995

Second print–2003

ISBN 81–7141–284–X

Published by:
DISCOVERY PUBLISHING HOUSE
4831/24, Ansari Road, Prahlad Street,
Darya Ganj, New Delhi-110 002 *(INDIA)*
Phone: 3279245
Fax: 91-11-3253475

Laser Typesetting by :
Allied Computers,
Karnal

Printed at

Tarun Offset Printers
Phone : 2260794

Preface

The teaching of Chemistry at the introductory stage becomes each day a more challenging task as the subject matter becomes more diverse and more complex. These challenges have evoked a series of responses—the present set of introductory chemistry monographs is one such. The teaching of chemistry recognises a number of problems that confront those who select text books. In order to overcome these problems, this volume "**Chemistry for Health Sciences**"—one of about fifty in the Chemistry Monograph Series—is introduced. Each volume is independent of the others deals with one of Chemistry topics and constitutes and complete entity. Each volume is more comprehensive than can be possible in a single volume text. It is intended to provide a range of topics to cover most undergraduate and chemistry main courses of study. These volumes can be used to enrich the more conventional courses of study.

Suggestions for improvement are welcome and shall be gratefully acknowledged.

Authors

Contents

1

Some Basic Concepts

Advances in health sciences have been achieved through the use of discoveries in chemistry. Chemical research in the pharmaceutical industry has led to the synthesis of new drugs. The majority of pharmaceuticals prescribed today were not available 20 years ago. New chemical processes have yielded synthetic biomaterials for blood vessel replacements, cardiac components, hip joints, artificial kidneys, and supplemental lenses. Techniques have been developed through which the chemcial bases of disease at the molecular level have been elucidated for several specific cases.

All of these advances have been made based on a fundamental understanding of chemistry, the study of the changes that matter undergoes and the energy relationships accompanying these changes. This chapter deals with some fundamental concepts of chemistry.

MATTER

Matter is *anything that has mass and occupies space.* Our entire universe is composed of matter. Familiar examples include wood, glass, paper, sugar and air.

States of Matter

Three physical states are possible—solid, liquid, and gas. The melting of ice and evaporation of perfume indicates that the physical state of a material depends on conditions such as the temperature and pressure under which the observation is made.

The three states of matter—solid, liquid, and gas—have physical characteristics that are detectable by the human senses. Solids and liquids can usually be identified by sight and touch. A piece of iron has a definite shape and is hard, whereas liquid water takes the shape of its container and is soft to the touch. Gases are not so easily identified unless they are colored. Thus, reddish brown nitrogen dioxide (a common air pollutant from automobile exhaust) fills a transparent bottle and is easily detected in the bottle by its color. Air, however, being colorless, may be detected by trying to put other matter in its place and obtaining evidence that the space is already occupied.

Terms describing changes between the solid, liquid, and gas state are illustrated for water.

$$\underset{\text{(Solid)}}{\text{Ice}} \underset{\text{freeze}}{\overset{\text{melt}}{\rightleftharpoons}} \underset{\text{(Liquid)}}{\text{Water}} \underset{\text{condense}}{\overset{\text{evaporate}}{\rightleftharpoons}} \underset{\text{(Gas)}}{\text{Water vapor}}$$

Nature of Matter

Democritus, an ancient Greek philosopher, proposed by 400 B.C. that matter was composed of small indivisible particles called atoms. He had no experimental means of testing his theory. Experimental data was eventually forthcoming that led John Dalton (a British chemist) to a restatement in 1808 that *all matter is composed of tiny, discrete particles called atoms.* The particle nature of matter is fundamental to chemistry.

Extensive observations over many centuries have led to the concept that a sample of matter may be categorised as being either a mixture or a pure substance (Fig.1.1). Mixtures contain two or more components that are present in variable amounts; air is an example. Samples of air taken from a urban area, from a rural site,

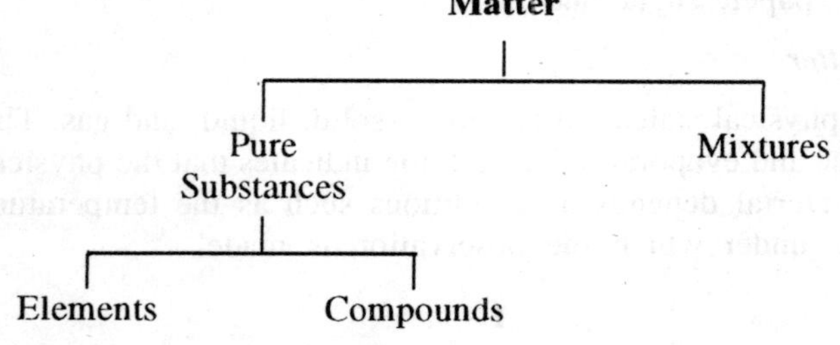

Fig. 1.1 : Classification of matter

and over the ocean would differ in the type and amount of their constituents.

Elements

A pure substance may be further categorised as being either an element or a compound. *An element is a pure substance that cannot be broken down into simpler substances by ordinary means such as heat or light.* To date, the existence of 105 elements has been verified; 89 are naturally occurring on earth, while 16 have been made by physicists and chemists using specialized techniques.

Mercury in a thermometer is an **element** because it is *a substance that is composed of only one kind of atom.* The various chemical elements are often indicated by a shorthand notation called **chemical symbols**, decided upon by international agreement. These symbols (see Table 1.1) consist of a capitalized initial letter, sometimes followed by a lower case second letter to distinguish between elements with the same initial letter. Thus, the symbols for beryllium, boron, bromine, barium, bismuth, and berkelium are Be, B, Br, Ba, Bi, and Bk, respectively. Others, derived from the Latin name for the element (Ag for silver from *argentum*) or the German name for the element

Table 1.1 Symbols of Some Common Elements

		Percent by Mass				*Percent by Mass*	
Element	*Symbol*	*Natural*	*Human*	*Element*	*Symbol*	*Natural*	*Human*
Aluminium	Al	7.5		Lead	Pb		
Argon	**Ar**			Magnesium	Mg	1.9	
Arsenic	**As**			Mercury	Hg		
Barium	Ba			Nickel	Ni		
Bromine	**Br**			**Nitrogen**	**N**		**3.1**
Calcium	Ca	3.4	2.0	**Oxygen**	**O**	**49.2**	**64.6**
Carbon	**C**		**18.1**	**Phosphorus**	**P**		**1.1**
Chlorine	**Cl**		**0.45**	Potassium	K	2.4	0.37
Chromium	Cr			**Silicon**	**Si**	**25.7**	
Copper	Cu			Silver	Ag		
Fluorine	**F**			Sodium	Na	2.6	0.11
Gold	Au			**Sulfur**	**S**		**0.25**
Hydrogen	**H**	**0.9**	**10.0**	Tin	Sn		
Iodine	**I**			Titanium	Ti	0.6	
Iron	Fe	4.7		Zinc	Zn		

(W for tungsten from *Wolfram*), have no correspondence with their current names. However, extensive use in the past has led to their continued use.

The natural distribution by mass of the ten most common elements in the earth's crust, oceans, and atmosphere is also shown in Table 1.1. The elements are distributed very unequally. Oxygen, the most abundant element, comprises nearly 50% of the total; oxygen and silicon combined constitute nearly 75% of the total; and the ten most abundant elements combined make up nearly 99% of the total.

The chemical elements are distributed in the human body in very different proportions from their natural abundances, as shown by the data in Table 1.1. Oxygen, the most abundant element naturally, comprises nearly 65% by mass of the human body; oxygen and carbon combined constitute nearly 84% of the total; and oxygen, carbon, hydrogen, and nitrogen make up about 96% of the total. Each of these four elements is present in a much higher percent in living organisms than in nature.

Elements are often classified further as metals or nonmetals. *Metals* such as iron, aluminium, magnesium, and copper are solids at room temperature (only mercury is an exception), shiny, good conductors of heat and electricity, and are easily hammered into sheets or drawn into wire. Metals are used in such things as cooking ware, tools, cars, and bridge and building frameworks. *Nonmetals* such as carbon, sulphur, and iodine are brittle solids; bromine is a liquid; and some nonmetals are gases at room temperature. All are poor conductors of heat and electricity.

Compounds

A **compound** is *a pure substance that is formed when two or more atoms of the same element or of different elements combine.* Diatomic oxygen is a compound formed from two atoms of the *same* element. The elements in a compound are combined chemically in a fixed relationship by weight so that a compound has a definite, uniform composition. The chemical union of elements to form a compound gives the compound properties distinctly different from those of its component elements. Consider, for example, that the element sodium is a soft metalic lustrous solid that vigorously reacts with water; the element chlorine is a pale green, poisonous gas at normal temperature and pressure. When these two elements react, they produce a white solid that does not react vigorously with water.

The white solid is sodium chloride (common table salt), quite different from the elements of which it was formed.

A compound is represented by a chemical formula that indicates the elements present in the compound. The formula for water (H_2O) indicates that water consists of the elements hydrogen and oxygen combined in a definite relationship.

DALTON'S ATOMIC THEORY

The main points of Dalton atomic theory are

1. Matter consists of small particles called atoms.
2. Atoms are not destroyed during chemical reactions but simply become involved in new kinds of chemical combinations (law of conservation of matter).
3. An element's atoms are all alike.
4. Elements differ because of the weight and nature of their component atoms.
5. When forming a chemical compound, atoms of the reacting elements combine in a definite ratio.

Dalton's proposals were based upon conclusions drawn from experimental data gathered by him and several of his contemporaries. Dalton's theory caused a flurry of scientific investigations seeking new elements, a refinement of atomic weight values, and an understanding of the forces and energies associated with chemical reactions. Other than describing an atom as a small, indivisible particle, however, Dalton's theory leaves some questions unanswered such as Why do atoms combine? Why do atoms combine in certain ratios when forming compounds? Why do energy changes accompany chemical changes? There has been a continuing search for answers to these questions since Dalton's time.

Many compounds are composed of small distinguishable units called molecules. A **molecule** is *an electrically neutral collection of atoms bonded together strongly enough to be considered a recognizable unit*. Two nitrogen atoms bind together to form a molecule of nitrogen. Two hydrogen atoms and two oxygen atoms bind together to form a molecule of hydrogen peroxide (a common antiseptic). One sulphur atom and two oxygen atoms bind together to form a molecule of sulphur dioxide. Fifty-seven carbon atoms, 110 hydrogen atoms, and 6 oxygen atoms bind together to form a molecule of stearin (the main component of beef fat and a precursor in the preparation of soaps). A molecule exhibits distinctive characteristics that may not be related in any obvious fashion to the properties of the atoms comprising the molecule.

Just as chemical symbols are used to signify elements, chemical formulas are used as abbreviations for compounds. A **formula** *conveys information about the composition and structure of a compound.* An empirical formula is the simplest kind of formula. It gives only a minimum of information about the composition of a substance and says nothing about its structure. An **empirical formula** states only *the* ***relative*** *numbers of atoms in a compound.* By convention an empirical formula includes symbols of the elements making up a compound and subscripts affixed to the symbols of the lements making up a compound and subscripts affixed to the symbols to designate the relative numbers of atoms. Thus, the empirical formula of hydrogen peroxide is HO, meaning that there is one atom of hydrogen for each atom of oxygen in the compound.

A **molecular formula** states *the* ***actual*** *numbers of atoms in a molecule,* but says nothing about its strucuture. A molecular formula includes symbols of the elements making up a molecule with subscripts affixed to the symbols to designate the actual numbers of atoms. The molecular formula of hydrogen peroxide is H_2O_2 in contrast to its empirical formula, HO. The molecular formula of sulphur dioxide is SO_2 and of stearin is $C_{57}H_{110}O_6$, the same as their respective empirical formulas.

A **structural formula** gives a maximum amount of information about a molecule. It gives *the actual numbers of atoms in a molecule, indicates which atoms are bonded to each other, and may indicate the three-dimensional configuration of the molecule.*

Mass is *a measure of the amount of matter.* The mass of a body is a fixed property of that body. It is *constant* everywhere, independent of other objects. The mass of a sample of moon rock is the same on the moon, in space, or at any altitude on the earth's surface. Mass is the property that gives matter the tendency to stay at rest when already at rest or to stay in motion when already in motion.

Weight is *a measure of the pull of gravity on an object.* The magnitude of the gravitational pull is directly proportional to the mass of the object and the mass of the attracting body (usually the earth). Thus, the weight or pull of gravity on a moon rock is less the smaller the mass of the moon rock. Likewise, the weight of a given moon rock determined on the moon is smaller than the weight of the same moon rock determined in your laboratory because the less massive moon exerts less gravitational force on the rock than does our more massive earth. Moreover, the gravitational pull is inversely proportional to the distance between the centers of the attracting bodies; that is, the gravitational pull is larger when the bodies are closer together and is smaller when the bodies are farther apart. Thus, the gravitational pull and the weight vary depending upon where the measurement is made. The weight of the moon rock is smaller when determined at a higher altitude on earth or when determined eighty miles out in space than when determined in your laboratory. The distinction between mass and weight may be understood in terms of an astronaut in space. The astronaut is weightless whenever he is so far away from the earth or other planetary bodies as to be unaffected by their force of gravity, but the astronaut is never massless since mass is constant everywhere.

Mass and weight can be distinguished by the method of their determination. The mass of an object is determined by balancing it against standard masses. The chemical balances use this principle. *By convention the mass of an object equals the weight of the same object only at sea level on the earth.* The weight of an object is frequently determined by means of a spring balance. No standard masses are involved and the weight will depend only on the force of gravity at the point of measurement. Bathroom scales and grocer's scales use this principle. Unfortunately, determining mass and determining weight are both commonly called weighing.

It is important to recognize the difference between mass and weight and realize that we are using mass in laboratory experiments and calculations when we compare *amounts of matter.* Many texts use mass and weight interchangeably, in fact, mass will be the appropriate term in almost all cases.

ENERGY

Energy (Greek, *en*, in; *ergon*, work) is *the capacity to do work* or to move an object through a distance against a force. Energy changes always accompany chemical changes; thus, the concept of energy is fundamental in chemistry. When your leg muscles act as you pedal a bicycle up a hill against both frictional forces and the force of gravity, energy is expended. Your leg muscles possess energy and are able to do work causing the bicycle pedals to move; the pedals move the crank and chain, causing the wheels to move; the wheels possess energy and are able to do work causing the bicycle to move; and the bicycle possesses energy and is able to do work causing your body to move.

Kinetic Energy

When a bicycle is in motion, it possesses energy that it does not have when it is standing still. *Energy of motion* is called **kinetic energy.** Any mass in motion possesses kinetic energy. Particularly in relation to equipment with moving parts, kinetic energy is often called *mechanical energy.*

Potential Energy

When an automobile is at rest near the top of a hill, it has no kinetic energy, but it does possess energy of position that it does not have at the bottom of the hill. This energy is converted easily to kinetic energy by releasing the brake. When a bow is held motionless in its bent form, it has no kinetic energy, but it possesses energy of configuration that it does not have in its normal shape. Releasing the string imparts kinetic energy to an arrow. A mixture of hydrogen gas and oxygen gas possesses energy because, under appropriate conditions, these gases react explosively to produce sound, heat, light, and mechanical energy. These examples are manifestations of potential energy. **Potential energy** is *that energy which matter has stored in it by reason of its relative position, configuration, or chemical composition.*

Chemical energy and nuclear energy result when there is a difference in the potential energy of reactants and products in a chemical reaction or in a nuclear reaction. When a electric current is passed through water, hydrogen gas and oxygen gas are liberated in a chemical process called electrolysis. When calcium hydroxide is dehydrated by heating, water and calcium oxide are formed. These are endotghermic processes. An **endothermic** reaction is a *reaction that absorbs energy* or one in which the potential energy of the reactants is less than that of the products. Many endothermic reactions are responsible for the building of tissues in the human body. Through an appropriate reaction a lightening bug emits light. If charcoal is burned in air, carbon dioxide is formed and heat is liberated. These are exothermic processes. An **exothermic** reaction is a *reaction that releases energy* or one in which the potential energy of the reactants is greater than the potential energy of the products.

The importance of the various forms of energy in everyday life arises primarily from the relatively easy transformation of one form of energy into another. For example, the potential energy of water held behind a dam is converted to the mechanical energy of rushing water. The rushing water moves blades in a generator and allows conversion of mechanical energy to electrical energy. Electric current is passed through a coil in a toaster to produce heat, through a filament in an incandescent bulb to produce light, through an electrolytic cell to produce chemical energy in the electrolysis of water, through an electric motor to convert the electrical energy back to mechanical energy, or through a circuit.

The various forms of energy are interconverted such that no **net** gain or loss of energy occurs. In the operation of a flashlight, the appearance of light as radiant energy must be accompanied by an equivalent loss of stored energy from the flashlight's battery. In ordinary chemical processes, neither mass nor energy is lost. In nuclear reactions, a small but significant amount of mass is lost, manifesting itself as an equivalent amount of energy. In 1905, Albert Einstein proposed an equation expressing this mass-energy equivalency: $E = mc^2$. This relationship along with the interconversion between various forms of energies leads to *the law of conservation of matter and energy : The total nergy of the universe is unchanging.*

The Joule is an energy unit smaller than the calorie and is also used in describing changes in energy.

$$1 \text{ cal} = 4.18 \text{ Joules}$$

Kinetic energy is manifested in a number of forms that differ only in the nature of the particles in motion.

Electrical energy. If the matter in motion is a stream of electrons flowing through a conductor and giving rise to an electric current, the energy is called electrical energy. Electrical energy causes some chemical reactions to occur, and electrical energy can be produced by other chemical reactions. Electrical energy can be converted easily to light as in the case of the incandescent bulb. The electrical nature of nerve impulses is of particular interest to biochemists.

Heat. If the movement involves random thermal motion of atoms or groups of atoms in a substance, the energy is called heat. Heat increases as matter becomes hotter and its atoms move more rapidly. Heat and temperature are related, but should be distinguished clearly. **Temperature** is *a measure of the* ***intesity*** *of heat,* but specifies nothing about quantity of heat. Thus, a burning match and a bonfire may both be at the same temperature, but the bonfire certainly gives off more heat.

Temperatures near room temperature are usually measured by following the expansion of mercury with increased temperature. The mercury is contained in a glass thermometer consisting of a terminal bulb surmounted by a calibrated capillary tube from which the temperature is read.

Temperature is measured in units called *degrees* using one of three temperature scales. The *Fahrenheit* scale (°F), used commonly in the United States, is based upon 0°F as the freezing point of a solution containing a maximal amount of table salt dissolved in water and is based upon 100°F as body temperature. On this scale, the freezing point of water is 32°F and the boiling point of water is 212°F. Each Fahrenheit degree is $\frac{1}{180}$ of the temperature interval between these two points. The *Celsius* scale (°C), used commonly in chemistry, is based upon the freezing point of water as 0°C and the boiling point of water as 100°C. Each Celsius degree is $\frac{1}{100}$ of the temperature interval between these two points. It follows that a Celsius degree is 1.80 times as large as a Fahrenheit degree and that the Celsius freezing point of water is 32° lower than the Fahrenheit value. Thus, Fahrenheit temperatures (°F) and Celsius temperatures (°C) can be converted to each other using the following equations.

$$°F = 1.80°C + 32° \qquad °C = 0.556\ (°F - 32°)$$

Certain chemical calculations, particularly those involving gases, require the use of *absolute* temperatures (suffixed with a K for Kelvin that does *not* require a superscript degree sign before it). The size of each degree on the absolute scale is identical to a degree on the Celsius scale, but each absolute temperature is 273.16° higher than its corresponding Celsius temperature. Thus, Celsius temperatures (°C) and absolute temperatures (K) can be converted to each other using the following equations.

$$K = °C + 273.16° \qquad °C = K - 273.16°$$

The relationship of the three temepratute scales is shown in Fig. 1.2b. Unless stated otherwise, Celsius temperatures will be used in this book and frequently the C will be omitted.

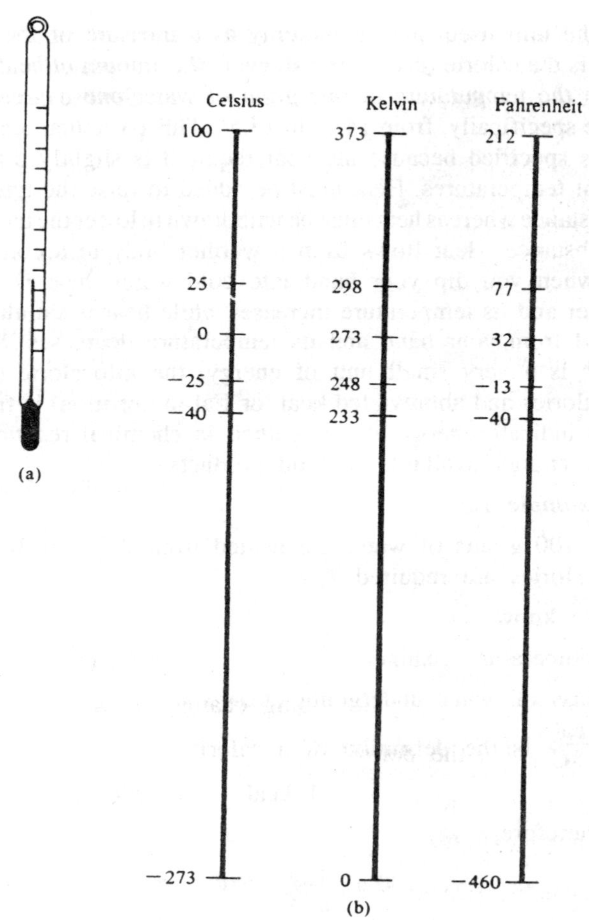

Fig. 1.2 : (a) Mercury thermometer (–10° to 110°, Celsius scale). (b) Relationship of three temperature scales.

Example 1.1 :

Calculate the Celsius and absolute temperatures that are equivalent to 64°F.

First convert °F to °C

$$°C = 0.556\ (°F - 32°) = 0.556\ (64° - 32°) = 18°$$

Then convert °C to K.

$$K = °C + 273.16° = 18° + 273.16° = 291\ K$$

The unit used most commonly as a measure of the *quantity* of heat is the calorie (cal). One **calorie** is *the amount of heat required to raise the temperature of one gram of water one degree Celsius,* or more specifically, from 14.5° to 15.5°. This particular temperature range is specified because the heat required is slightly different at different temperatures. Heat must be added to raise the temperature of a substance whereas heat must be withdrawn to lower the temperature of a substance. Heat flows from a warmer body to a colder body. Thus, when you dip your hand into cold water, heat is added to the water and its temperature increases while heat is simultaneously removed from your hand and its temperature decreases. Since the calorie* is a very small unit of energy, the kilocalorie (equal to 1000 calories and abbreviated kcal, or Cal in nutrition) is frequently used to indicate energy lost or gained in chemical reactions or to indicate energy available in food products.

Example 1.2

100 grams of water are heated from 25°C to 30°C. How many calories are required ?

We know

Temperature change = 5°C

Mass of water undergoing change = 100 g

$\frac{1 \text{ cal}}{\text{g °C}}$ is the definition of a calorie

1 kcal = 1000 cal

Therefore,

$$100 \not{g} \times \frac{1 \text{ cal}}{\not{g} \text{ °}\not{C}} \times 5\text{°}\not{C}$$

Radiant energy. If the movement involves photons, or packets of electromagnetic radiation, the energy is called radiant energy. If the temeperature of the heating coils in an oven is increased gradually, invisible radiation called *infrared radiation* (of lower nergy than visible light) is emitted. As the temperature is increased further, visible rays called *light* are emitted. At very high temperatures, as in the sun, invisible *ultraviolet radiation* (of higher energy than visible light) is emitted. Infrared radiation, light, and ultraviolet radiation are forms of radiant energy.

---

* The Joule is an energy unit in the SI system, is smaller than the calorie and is also used in describing chemical changes
1 calorie = 4.18 J

Mixtures

Many of the materials that we encounter are **mixtures** of susbtances; that is, they *contain two or more elements and/or compounds* not bound to each other and not in any fixed proportions. Air is a mixture consisting primarily of nitrogen gas and oxygen gas, but also containing small amounts of water vapor, carbon dioxide, and other substances. The composition of a mixture may be variable, sometimes over a wide range. The amount of water vapor in a air changes over a relatively narrow range, wheras the composition of a mixture of sand and table salt can be varied continuously in any ratio desired to produce an infinite number of compositions.

A given mixture can be studied in great detail and can be described further as a solution or as a heterogeneous mixture. Any given **solution** has *a constant composition, structure, and physical and chemical behavior.* Given quantities of solid sugar dissolved in water, liquid ethyl alcohol dissolved in water, or hydrogen chloride gas dissolved in water form three different solutions of constant composition.

Portions of **heterogeneous mixtures** *exhibit different compositions, structures, and physical and chemical properties, so that abrupt discontinuities or boundaries may be observed.* Thus, a glass filled with iced tea and ice cubes (Figure 1.3a) contains a heterogeneous mixture since there is an observable boundary between the liquid solution and solid ice. There may also be undissolved solid sugar in the bottom of the glass (Figure 1.3b). Heterogeneous mixtures may be formed by composites of elements and/or compounds. Thus, a mixture of magnesium turnings and powdered copper (Figure 1.3c) is heterogeneous and the distinct fragments of each metal can be perceived easily. Likewise, a sealed flask containing sand, water, and air (Figure 1.3d) holds a heterogeneous solid-liquid-gas mixture.

The term **phase** is often applied to *a mechanically separate, homogeneous portion of matter. Any solution is,* therefore, *a one-phase system.* Natural gas, a solution of a number of compounds containing carbon and hydrogen, is a one-phase system. All gas mixtures are homogeneous. A *heterogeneous mixture contains two or more phases.* Thus, a sealed wine bottle contains two phases : the wine which is itself a solution, and the air above the wine which is also a solution.

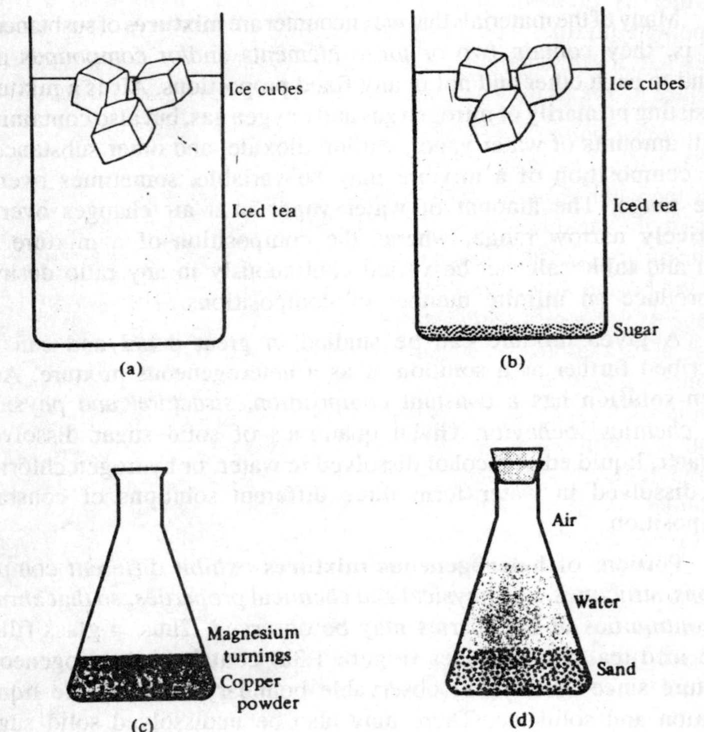

Fig. 1.3 : Examples of heterogeneous mixtures. (a) Ice cubes and iced tea (two phases); (b) ice cubes, iced tea, and solid sugar (three phases); (c) magnesium turnings and powdered copper (two phases); (d) air, water and sand (three phases).

Separation of Substances

When carrying out a chemical reaction in the laboratory, one frequently encounters a mixture of two or more substances. The task then is to separate the mixture into its components, hopefully without converting the substances into new ones and losing the desired product.

Two-component mixtures are easiest to resolve into their components but the same principles and techniques apply to mixtures with more than two components. If a two-component mixture is heterogeneous, that is, has two phases, the separation may be relatively easy. One of the components can be simply removed physically from the other by methods such as decantation and filtration. If a two-component mixture is homogeneous, that is, has only one phase, somewhat more sophisticated techniques such as crystallization, distillation, sublimation, extraction, or chromatography are required.

Decantation. When one cleans his fish tank, he first removes the fish, then carefully pours the water from the tank, leaving behind most of the sand, shells, and plants. When you drink your coffee, you sip more carefully near the bottom of the cup, leaving the small residue of coffee grounds behind in the cup. These are examples of decantation. **Decantation** is *a gentle pouring off of a liquid without disturbing a solid sediment.* Decantation is useful for separating a liquid from a solid that has settled to the bottom of a container. A special form of decantation is utilized to separate two immiscible liquids (for example, motor oil and water) using a separatory funnel (Figure 1.4). After the separatory funnel has been left standing and the two immiscible liquids have separated, the lower layer (the liquid having the greater density or mass per unit volume) is drawn off through a stopcock into another container, thus providing a separation.

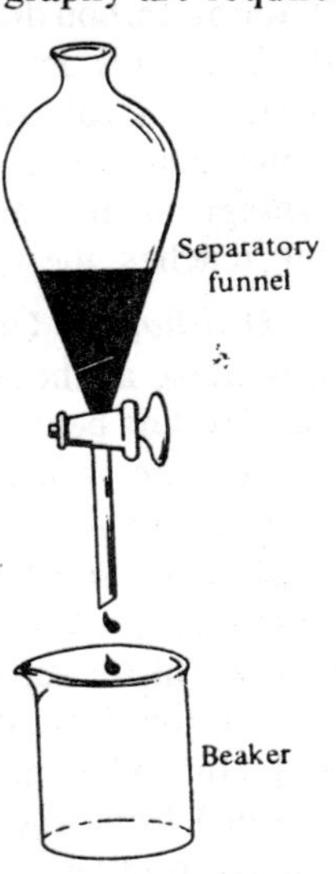

Fig. 1.4 : Separating two immiscible liquids. The more dense liquid forms the lower layer and can be drawn off through the stopcock.

Filtration. Primary treatment in a sewage disposal plant involves passage of sewage through coarse screens to remove pieces of wood, metal, rags, garbage, and other sizable materials that would otherwise damage pumps in the sewage disposal plant. In some coffeepots boiling water is forced up a hollow stem after which it percolates down through ground coffee contained in a perforated pan and then returns to the pot below. In many vacuum cleaners air is pumped through a porous bag that acts as a filter and collects dust. These are examples of filtration. **Filtration** is *the passage of a liquid or gas through a porous substance for the purpose of removing suspended solids.* Filtration is useful for separating solids from liquids as in the first two cases above or solids from gases as in the third case above.

Crystallization. Many minerals in the earth's crust were probably formed by crystallization under appropriate conditions from liquid mixtures. Calcite (calcium carbonate) occurs as limestone, a common sedimentary rock, and as marble, a common metamorphic rock. Calcite is relatively insoluble in water, but is appreciably soluble in water containing carbon dioxide that is picked up as water percolates through soil. Stalactites and stalagmites in limestone caves form due to the deposition of calcite from ground water as carbon dioxide escapes into the atmosphere. This is an example of crystallization, and takes advantage of the change in carbon dioxide concentration as ground water reaches the earth's surface.

Distillation. Ordinary tap water should not be used in a steam iron because as the water is heated and converted to steam, deposits of salt are left behind. These salts react chemically with the interior, corroding the iron and eventually closing the steam holes so that the iron no longer has any steaming capability. This is an *undesirable* separation of substances in a solution by distillation. **Distillation** *involves evaporation and subsequent condensation of a substance,* and is useful for separating two substances (gases, liquids, or solids) in a solution that have quite different boiling points. Steam iron manufacturers suggest purchasing distilled water that has already been freed from trace quantities of dissolved salts as a way to avoid the above problem. The basic apparatus and techniques utilised for distilling water are illustrated in Figure 1.5. The water is heated and boiled, the vapor is then condensed to the liquid state by passing it through a condenser in which cold water is circulated, and the distilled water is collected in a receiver flask. Meanwhile, the nonvolatile salts originally in the natural water remain in the distillation flask.

Nature provides a cycle involving distillation that is very important to us. Water evaporates from the oceans into the air. When moist air colls, the water vapor condenses. Fog or stratus clouds (high fog sheets) form if condensation occurs close to the earth. Altostratus or altocumulus clouds form if condensation occurs higher in the atmosphere. Cirrus clouds composed of ice crystals fonn if condensation occurs high enough that the temperature is below freezing. The cycle is completed when moisture returns to the oceans as dew, rain, or snow.

Sublimation. If dry ice (solid carbon dioxide) is left exposed at room temperature, it seems to disappear. Actually it has been

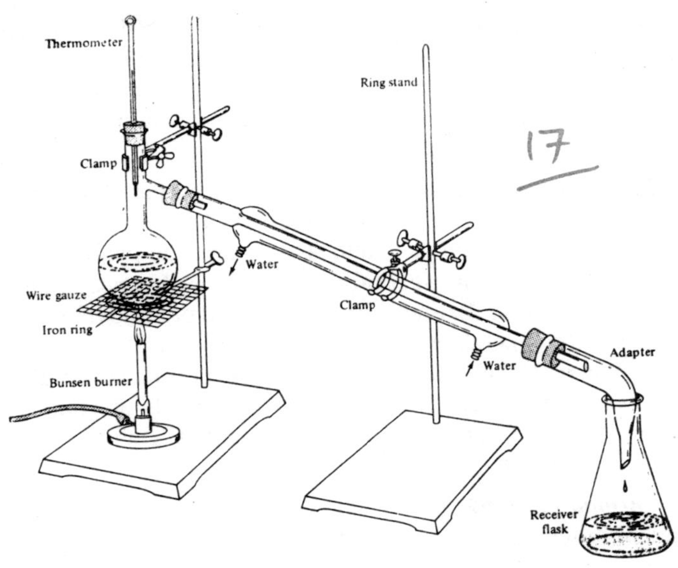

Fig. 1.5 : The distillation of water. Water in the round bottom flask is heated and boiled; water vapor is condensed to liquid water by passing it through a condenser in which cold water is circulated; and the distilled water is collected in a receiver flask.

converted directly from the solid to the gas state. If a few purple crystals of iodine are placed in an evacuated bottle at room temperature, the purple color of iodine vapor soon forms above the crystals and eventually fills the bottle. These examples illustrate **sublimation,** which is *the direct conversion of a solid to a gas without passing through the liquid state.* If an ice cube is held against the side of the evacuated bottle containing iodine crystals, the purple crystals in the bottom of the bottle begin to disappear and purple crystals form on the side of the bottle at the ice cube. This illustrates sublimation followed by crystallization. This is a common method for separating a volatile solid from a nonvolatile solid, for example, separating a mixture of iodine and table salt. The mixture is placed in an evacuated bottle and an ice cube is held against the side of the bottle. Iodine crystals will sublime and crystallize on the side of the bottle whereas the less volatile table salt will remain on the bottom of the bottle. Thus, a separation is achieved.

Extraction. Cookbooks recommend adding a pared and cut potato to a stew that is too salty. The potato absorbs salt from the stew and makes the dish more palatable. This is an example of extraction. **Extraction** is *the withdrawal or forcing out of a substance by chemical or mechanical action.* Extraction is one method of removing perfumes from roses. Rose petals are immersed in hot petroleum ether. Heat ruptures the petal cells and the discharged perfumes are absorbed by the petroleum ether. The remains of the petals are removed by filtration; then the petroleum either is distilled leaving behind the perfume components.

Chromatography. If a solution of color pigments from leaves is passed through a vertical glass tube containing activated charcoal, and if more solvent is pured slowly through the column, a separation of the pigments into bands, called a chromatogram, is obtained by the time they pass from the bottom of the column. This is an example of liquid *chromatography.* Interpretation of the results of such an experiment in 1903 allowed Michael Tswett, a Russian botanist, to elucidate the basic principles of adsorption chromatography. A solution is poured into a chromatographic column containing a packing that has a greater affinity for one or more components in the solution. This packing material is referred to as the stationary phase. The mixture is adsorbed by the stationary phase and held at the top of the column. When a second moving phase is passed slowly through the column, the components of the mixture are subjected repeatedly to extraction by the mobile phase and adsorption on the stationary phase. The more strongly a component of the mixture are subjected repeatedly to extraction by the mobile phase and adsorption on the stationary phase and the less soluble it is in the mobile phase, the slower it moves through the column. Thus, a separation is achieved and pure zones of substances are collected at the bottom of the column. Various types of chromatography have been developed: adsorption, gas or vapor phase (VPC), ion exchange, partition or paper, and thin-layer (TLC). All utilize the same basic principles, but differ in the nature of their stationary and mobile phases and in the kinds of substances they can separate most effectively. Both gas and thin-layer chromatography require only micrograms of a sample (1 microgram = 0.000001 gram, or about $^{1}/_{100}$ of the mass of a grain of sugar). Gas chromatography is more capable of separating similar substances than any other method presently known.

Various methods for separating substances from mixtures have been discussed individually. In fact, these methods are often used in conjunction with each other as illustrated in the basic procedures for production of sugar from sugar beets. The beets are first placed in a perforated, tumbling washer that selectively *filters* dirt and small stones from the larger beets. The beets are then sliced into chips and placed in a heated tank where hot water *extracts* sugar from the beet chips. The sugar solution is purified by the addition of lime which precipitates impurities so that they can be separated by *filtration*. Sulphur dioxide is added to bleach residual coloring matter and to precipitate excess lime so that it can be seprated by *filtration*. Water is *distilled* under vacuum from the sugar solution so that the sugar slowly *crystallizes*. The sugar crystals are then *filtered* in porous centrifugal baskets. Even at this point adsorption *chromatography* indicates the presence of a number of sugars (with sucrose being the dominant one) as well as small amounts of other impurities.

Measurements

The basic principles of chemistry have been deduced from the interpretation of results from experiments carried out under well-defined conditions. These results became meaningful only as they become reproducible. Results can be reproduced only when the experimental conditions have been described in sufficient detail so that they can be repeated exactly. Experimental conditions and results require both qualitative and quantitative description, that is, the questions as to both *what* and *how much* must be answered.

Accurate and precise measurements as well as a standard numerical form and system of units are required to describe the properties of matter quantitatively.

Fractions

A fraction is a ratio of two numbers, the numerator and denominator, e.g.,

$$\frac{240}{6} \begin{array}{l}\leftarrow \text{numerator}\\ \leftarrow \text{denominator}\end{array}$$

The value of a fraction is determined by dividing the denominator into the numerator. The answer is called the *quotient.* The meaning of the fraction above can be expressed

$$\begin{array}{r}\text{numerator} \rightarrow\\ \text{denominator} \rightarrow\end{array} \frac{240}{6} = 240 \div 6 = 6\overline{)240} = 40 \leftarrow \text{quotient}$$

When the numerator is greater than the denominator, e.g.,

$$\frac{75}{25} = 3$$

the quotient is greater than 1. When the numerator equals the denominator, the quotient is 1 :

$$\frac{135}{135} = 1 \qquad \frac{8642}{8642} = 1$$

When the numerator is less than the denominator, the quotient is less than 1 :

$$\frac{75}{150} = 0.5 \qquad \frac{5}{20} = 0.25$$

Complex Fractions

A complex fraction is a combination of several simpler fractions. An example of a complex fraction is

$$\frac{4 \times 12 \times 6}{3 \times 8 \times 7}$$

This expression is the product of three simpler reactions :

$$\frac{4}{3} \times \frac{12}{8} \times \frac{6}{7} = \frac{4 \times 12 \times 6}{3 \times 8 \times 7}$$

In the general case,

$$a \times \frac{b}{c} \times \frac{d}{e} = \frac{\cdots}{\cdots} = \frac{a(b)(d)(\cdots)}{c(e)(\cdots)}$$

Thus, rather than performing the operations of each of the simpler fractions, the numerator terms are multiplied to give a single product; the denominator terms are multiplied to give a single terms; the denominator is then divided into the numerator. From the example above,

$$\frac{4 \times 12 \times 6}{3 \times 8 \times 7} = \frac{288}{168} = 1.71$$

Example 1.4

$$6 \times \frac{28}{16} \times \frac{4}{50} = \frac{6 \times 28 \times 4}{16 \times 15} = \frac{672}{800} = 0.84$$

EXPONENTS AND EXPONENTIAL NOTATION

Exponents and exponentials notation are condensed ways to express large and small numbers. An exponent is a superscript that follows another number. The exponent indicates how many times the preceding number is to be mutliplied by itself.

10^2 (read ten squared) = 10×10 = 100

10^4 (read ten to the fourth power) $= 10 \times 10 \times 10 \times 10 = 10{,}000$

A number expressed in standard exponential form has only one number to the left of the decimal point. Exponential form is also called *scientific notation.* Since exponential notation depends on understanding values of powers of ten, the following table summarizes these relations :

Number	*Exponentially Express*	
1,000,000	1.0×10^6	decreasing magnitude ↓
10,000	1.0×10^4	
1000	1.0×10^3	
100	1.0×10^2	
10	1.0×10^1	
1	1.0×10^0	
0.1	1.0×10^{-1}	
0.01	1.0×10^{-2}	
0.001	1.0×10^{-3}	
0.0001	1.0×10^{-4}	
0.0000001	1.0×10^{-7}	

Note that for positive exponents a *higher-numbered exponent* indicates and expression of *larger value*; for *negative exponents,* as the *exponent increases,* the expression's *numerical value decreases.* For example, $1 \times 10^4 > 1 \times 10^3$ but $1 \times 10^{-4} < 1 \times 10^{-3}$ (> represents *greater than*; < represents *less than*).

Numbers expressed exponentially can be considered as the product of two numbers, one of which is a power of ten. The other number is written to the left of the power of ten term. When written in proper exponential form, this number has only one figure to the left of the decimal point.

$$300 = 3 \times 10 \times 10 = 3.0 \times 10^2$$
$$4000 = 4 \times 10 \times 10 \times 10 = 4.0 \times 10^3$$
$$800{,}000 = 8 \times 10 \times 10 \times 10 \times 10 \times 10 = 8.0 \times 10^5$$

300 is converted to exponential form by moving the decimal point two places to the left; 3 0 0. The original number is 300, not 3.0. To make the exponential form value equal to the original value, 3.0 is multiplied by 10 raised to the correct exponent. The exponent is determined by counting the number of decimal places that the decimal point was moved. If the decimal was moved to the left, the exponent is a positive number; if the decimal was moved to the right, the exponent is a negative number :

$$4500 = 4\ 5\ 0\ 0. = 4.5 \times 10^3$$
$$71{,}000 = 7.1 \times 10^4$$
$$275 = 2.75 \times 10^2$$
$$0.041 = 0.0\ 4\ 1 = 4.1 \times 10^{-2}$$
$$0.00067 = 6.7 \times 10^{-4}$$
$$0.82 = 8.2 \times 10^{-1}$$

There are times when a number expressed in standard exponential form has to be converted to another exponential form. This is done by moving the decimal point and correspondingly changing the exponent.

The number 2000 can be expressed several ways :

$$2000 = 2000. \times 10^0$$
$$200.0 \times 10^1$$
$$20.0 \times 10^2$$
$$2.0 \times 10^3$$

as this decreases, the power of 10 increases

For numbers less than 1, a similar technique is used, but with negative exponents the larger the exponent, the smaller its numerical value; i.e., $10^{-5} < 10^{-4}$ and $10^{-3} < 10^{-2}$. The number 0.0052 can be expressed

$$0.0052 = 0.52 \times 10{-}2$$
$$= 5.2 \times 10{-}3$$
$$= 52 \times 10{-}4$$

as this decreases this increases

$$6.4 \times 10^{-5} = 64 \times 10^{-2}$$

What is the exponent required? Changing 6.4 to 64 corresponds to multiplying by 10; thus, the exponent of 10 must be decreased by 1. Therefore, 10^{-5} is reduced to 10^{-6} and $6.4 \times 10^{-5} = 64 \times 10^{-6}$. Note in the previous example that $5.2 \times 10^{-3} = 52 \times 10^{-4}$.

USING SIGNED NUMBERS; MULTIPLICATION AND DIVISION OF EXPONENTS

A signed number is one that is either positive (+3, +7) or negative (—3. —8). In chemistry, the most common use of signed numbers is in operations involving exponential values. The exponents are signed numbers.

The Law of Exponents states

1. When multiplying exponential numbers, the exponents are *added algebraically.*
2. When dividing exponential numbers, the exponent of the divisor is *algebraically subtracted from* the exponent of the numerator.

Algebraically adding or subtracting signed numbers involves the use of the Law of Signs applied to exponents :

1. Adding two positive exponents gives a positive exponent.

Example 1.5

$$(1 \times 10^5) \times (1 \times 10^3) = 1 \times 10^{5+3} = 1 \times 10^8$$
$$(2 \times 10^4) \times (3 \times 10^9) = 6 \times 10^{4+9} = 1 \times 10^{13}$$
$$(6 \times 10^3) \times (4 \times 10^4) = 24 \times 10^{3+4} = 24 \times 10^7 = 2.4 \times 10^8$$

2. Adding two negative exponents yields a negative exponent.

Example 1.6

$$(1 \times 10^{-2}) \times (2 \times 10^{-4}) = 2 \times 10^{(-2)+(-4)} = 2 \times 10^{-6}$$
$$(1 \times 10^{-5}) \times (1 \times 10^{-3}) = 1 \times 10^{(-5)+(-3)} = 1 \times 10^{-8}$$
$$(6 \times 10^{-6}) \times (3 \times 10^{-8}) = 12 \times 10^{(-6)+(-8)} = 12\times10^{-14} = 1.2\times10^{-13}$$

Continuing with the Law of Signs applied to exponents:

3. Adding a positive exponent and a negative exponent gives a new exponent whose sign is that of the numerically larger *exponent.*

Example 1.7

$(1 \times 10^4) \times (2 \times 10^{-3}) = 2 \times 10^{(4)+(-3)} = 2 \times 10^1$

$(1 \times 10^3) \times (2 \times 10^{-4}) = 2 \times 10^{(3)+(-4)} = 2 \times 10^{-1}$

$(3 \times 10^8) \times (3 \times 10^{-3}) = 9 \times 10^{(8)+(-3)} = 9 \times 10^5$

$(3 \times 10^3) \times (3 \times 10^{-8}) = 9 \times 10^{(3)+(-8)} = 9 \times 10^{-5}$

$(6 \times 10^{-11}) \times (5 \times 10^8) = 30 \times 10^{(-11)+(8)} = 30 \times 10^{-3} = 3.0 \times 10^{-2}$

4. Subtracting a positive exponent from a larger positive exponent yields a positive exponent.

Example 1.8

$$\frac{1 \times 10^8}{1 \times 10^6} = 1 \times 10^{8-6} = 1 \times 10^2$$

$$\frac{4 \times 10^9}{2 \times 10^4} = 2 \times 10^{9-4} = 2 \times 10^5$$

5. Subtracting a positive exponent from a smaller positive exponent yields a negative exponent.

Example 1.9

$$\frac{1 \times 10^8}{1 \times 10^{10}} = 1 \times 10^{8-10} = 1 \times 10^{-2}$$

$$\frac{6 \times 10^4}{3 \times 10^9} = 2 \times 10^{4-9} = 2 \times 10^{4-9} = 2 \times 10^{-5}$$

$$\frac{8 \times 10^3}{2 \times 10^6} = 4 \times 10^{3-6} = 4 \times 10^{-3}$$

$$\frac{4 \times 10^2}{8 \times 10^8} = 0.5 \times 10^{2-8} = 0.5 \times 10^{-6} = 5 \times 10^{-7}$$

6. Subtracting a negative exponent from a larger negative exponent yields a negative exponent.

Example 1.10

$$\frac{1 \times 10^{-3}}{1 \times 10^{-2}} = 1 \times 10^{(-3)-(-2)} = 1 \times 10^{-3+2} = 1 \times 10^{-1}$$

$$\frac{1 \times 10^{-8}}{1 \times 10^{-3}} = 1 \times 10^{(-8)-(-3)} = 1 \times 10^{-8+3} = 1 \times 10^{-5}$$

$$\frac{3 \times 10^{-12}}{6 \times 10^{-4}} = 0.5 \times 10^{(-12)-(-4)} = 0.5 \times 10^{-12+4} = 0.5 \times 10^{-8} = 5 \times 10^{-9}$$

7. Subtracting a negative exponent from a smaller negative exponent yields a positive exponent.

Example 1.11

$$\frac{1\times10^{-9}}{1\times10^{-15}}=1\times10^{(-9)-(-15)}=1\times10^{-9+15}=1\times10^{6}$$

$$\frac{6\times10^{-7}}{3\times10^{-12}}=2\times10^{(-7)-(-12)}=2\times10^{-7+12}=2\times10^{5}$$

8. Subtracting positive and negative exponents is done as follows:

   (a) Subtracting a negative exponent from a positive exponent gives a positive exponent.

Example 1.12

$$\frac{1\times10^{3}}{1\times10^{-2}}=1\times10^{(3)-(-2)}=1\times10^{3+2}=1\times10^{5}$$

$$\frac{9\times10^{25}}{4\times10^{-18}}=2.25\times10^{(25)-(-18)}=2.25\times10^{25+18}=2.25\times10^{43}$$

$$\frac{3\times10^{10}}{6\times10^{-3}}=0.5\times10^{(10)-(-3)}=0.5\times10^{10+3}=0.5\times10^{13}=5\times10^{12}$$

(b) Subracting a positive exponent from a negative exponent yields a negative exponent.

Example 1.13

$$\frac{1\times10^{-3}}{1\times10^{2}}=1\times10^{(-3)-(2)}=1\times10^{-5}$$

$$\frac{7.0\times10^{-12}}{3.5\times10^{6}}=2\times10^{(-12)-(6)}=2\times10^{-18}$$

$$\frac{4.8\times10^{-22}}{1.2\times10^{20}}=4\times10^{(-22)-(20)}=4\times10^{-42}$$

If the exponents are equal in value but of opposite signs, the resulting exponent is zero.

Example 1.14

$$(1\times10^{3})(1\times10^{-3})=1\times10^{(3)-(-3)}=1\times10^{0}=1$$

$$\frac{5\times10^{-3}}{1\times10^{-3}}=5\times10^{(-3)-(-3)}=5\times10^{(-3)+3}=5\times10^{0}=5$$

Metric System

The metric system is used commonly in many other countries, and has become the standard system for scientists. In the metric system the common units of length, volume, and mass are the meter, the litre, and the gram, respectively. The main advantage of the metric system is that subdivisions or multiples of its basic units are in factors of ten and are expressed using suitable prefixes attached to the basic unit. Common prefixes are

mega-	1×10^{6}
kilo-	1×10^{3}
centi-	1×10^{-2}
milli-	1×10^{-3}
micro-	1×10^{-6}
nano-	1×10^{-9}
pico-	1×10^{-12}

Length. The fundamental unit of length in the metric system is the meter (m). The meter was defined originally as the length between two marks on an inert platinum-iridium reference stick kept at Sevres, France. A meter is slightly longer than three feet. Conversion factors between various units of length in the metric system and the British system are given below.

1 kilometer (km)	= 1000 meters (m)	= 0.6214 mile
	1 meter	= 39.37 inches
1 centimeter (cm)	= 1×10^{-2} meter	= 0.3937 inch
1 millimeter (mm)	= 1×10^{-3} meter	= 0.03937 inch
1 micrometer (μm)	= 1×10^{-6} meter	
1 nanometer (nm)	= 1×10^{-9} meter	
1 Angstrom (A)	= 1×10^{-10} meter	

Another commonly used conversion factor is 2.54 cm = 1 in. Lengths commonly measured in the laboratory using a meter stick are expressed in millimeters or centimeters.

Volume. The common unit of volume in the metric system is the litre. The millilitre (ml), equal to 0.001 litre is used for small volumes, especially for liquids and gases. The cubic centimeter (cm^3), exactly equal to 1 ml, is also used frequently, particularly for solids.

Volumes of sizable solids, depending on their shapes, can be determined by measuring lengths or radii, or by displacement of measured volumes of liquid. Volumes of liquids are measured in graduated cylinders, burets, transfer pipets, measuring pipets, or volumetric flasks (Figure 1.6). These vessels are designed to contain or deliver known, calibrated volumes. Volumes of gases are measured in known, calibrated, closed volumes.

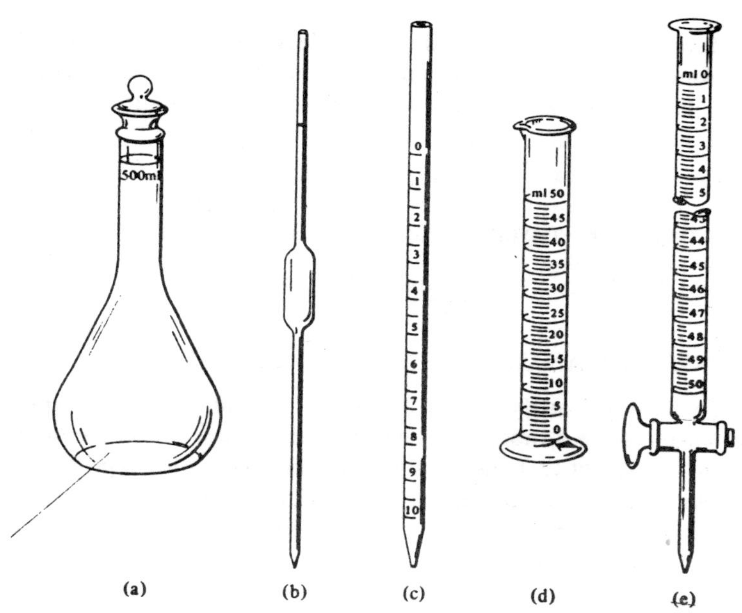

Fig. 1.6 : Vessels for measuring volumes of liquids. (a) Volumetric flask; (b) transfer pipet; (c) measuring pipet; (d) graduated cylinder; (e) buret.

Mass. The standard unit of mass in the metric system is the kilogram; however, the *gram* (g) is used more commonly. The gram is defined as one thousandth of the mass of an inert, platinum-iridium, standard kilogram block kept at Sevres. A gram is slightly less than half the mass of a dime. The common conversion factor between the British and metric systems is 1 pound = 453.6 grams. Mass is always used to specify the amount of a solid because the volumes of solids vary markedly with shape, state of subdivision, and packing of the particles.

Table 1 : Basic metric units of mass, length, and volume

	Mass	Length	Volume
Basic Unit	*Gram (g)*	*Meter (m)*	*Litre (L)*
Derived Units	nanogram (ng)(1×10^{-9}g)	nanometer (nm)($1/10^{-9}$m)	
	microgram (μg)(1×10^{-6}g)	micrometer (μm)($1/10^{-6}$m)	
	milligram (mg)(1×10^{-3}g)	millimeter (mm)($1/10^{-3}$m)	milliliter (mℓ)(1×10^{-3}L)
	centigram (cg)(1×10^{-2}g)	centimeter (cm) ($1/10^{-2}$m)	centilitre (cℓ) (1×10^{-2}L)
	decigram (dg)(1×10^{-1}g)	decimeter (dm) ($1/10^{-1}$m)	deciliter (dℓ) (1×10^{-1}L)
	kilogram (kg) (1000g)	kilometer (km) (1000 m)	kilolitre (kℓ) (1000 L)

Table 1 indicates the relationships between basic and derived units. An equality between two measurements is called a *conversion factor,* which can be expressed as a ratio. For example,

$$1\text{ g} = 1000\text{ mg} \frac{1\text{ g}}{1000\text{ mg}} = \frac{1\text{ g}}{1\text{ g}} = 1$$

Alternatively, the conversion factor can be written as

$$\frac{1000\text{ mg}}{1\text{ g}} = \frac{1\text{ g}}{1\text{ g}} = 1$$

DIMENSIONAL ANALYSIS (*FACTOR-LABEL METHOD*)

Dimensional analysis is a technique used in problem solving in which units are arranged so that all units cancel except the unit of the desired quantity. In using conversion factors to solve a problem, first determine what quantities are given and what units are associated with them. Then, correctly arrange the appropriate conversion factors so that all units cancel except that of the desired answer.

Example 1.15

Suppose a patient to receive 250 mg of a particular medication. How many grams does this represents ?

The problem is asking, "Is there any relationship between milligrams and grams?" The relationship we know is

$$1000\text{ mg} = 1\text{ g} \frac{1\text{ g}}{1000\text{ mg}};\ \frac{1000\text{ mg}}{1\text{ g}}$$

We want to eliminate milligram units and keep grams as our final unit. Setting up the problem so that milligram units cancel will provide the answer:

$$250\ \cancel{\text{mg}} \times \frac{1\text{ g}}{1000\ \cancel{\text{mg}}} = \frac{250}{1000}\text{ g} = 0.250\text{ g}$$

Several conversion factors can be used together.

Example 1.16

A total of 0.300 L of medication was administered equally to ten patients. How many milliliters of medication did each patient receive?

$$\frac{0.300\ \cancel{\text{L}}}{10\text{ patients}} \times \frac{1000\text{ m L}}{1\ \cancel{\text{L}}} = 30.0 \frac{\text{m L}}{\text{patient}}$$

TEMPERATURE

The Fahrenheit and Celsius scales are two widely used temperature scales. Conversions between the two scales can be done using the relationship °F = 1.8 °C + 32. For example, a body temperature of 98.6 °F equals 37 °C.

$$98.6\ °F = 1.8\ °C + 32$$
$$98.6\ °F - 32 = 66.6\ °F = 1.8\ °C$$
$$\frac{66.6}{1.8°C} = 37°C$$

DENSITY

The ratio of a substance's mass to its volume is called *density.*

$$\text{density} = \frac{\text{mass}}{\text{volume}}$$

Density is commonly expressed as grams per milliliter or grams per centimenter.

If two of the three values in the density expression are known, the third can be calculated.

Example 1.17

(a) A urine sample has a density of 1.030 g/mℓ. What is the mass of a 25.0-mℓ sample?

$$25.0\ \cancel{m\ell} \times \frac{1.030\ g}{1.0\ \cancel{m\ell}} = 25.75\ g$$

(b) What is the volume of a 22.6-g urine sample having a density of 1.027g mL ?

The density term can be written as 1.027g 1.0 mL. It can be rearranged to 1.0 mL/1.027 g. Using this,

$$22.6\ g \times \frac{1.0\ mL}{1.027g} = 22.0\ mL$$

Note that by rearranging the density expression, grams cancel leaving the desired unit of volume, milliliters.

SPECIFIC GRAVITY

Specific gravity is a comparision of the density of a substance to the density of water measured at the same temperature.

$$\text{specific gravity} = \frac{\text{density of substance}}{\text{density of water}}$$

Both densities are expressed in the *same* units and the units cancel. Thus, specific gravity is a unitless quantity.

A convenient way to measure a liquid's specific gravity is with a hydrometer (see Fig. 1.7). The depth to which the hydrometer sinks in the liquid depends on the specific gravity of the liquid, the higher the specific gravity, the higher the hydrometer floats; conversely, it will sink to a lower depth in liquids of lesser specific gravity. The hydrometer stem marking at the liquid level is calibrated to correspond accurately to the actual specific value of the test liquid.

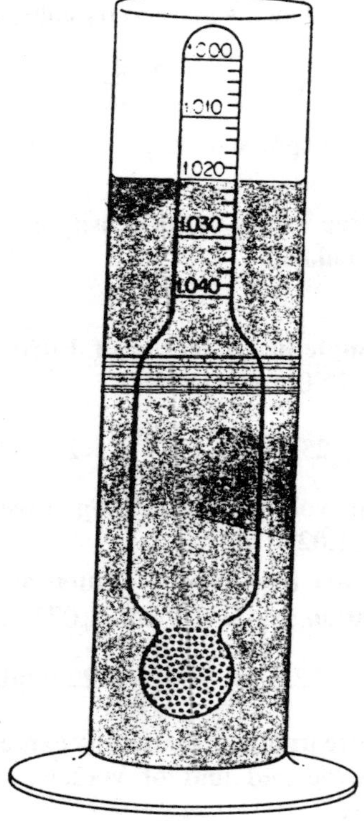

Fig. 1.7 : A hydrometer for measuring specific gravity of liquids having values between 1.000 and 1.040. During measurement, the calibrated stem marking at the liquid surface is the specific gravity of the liquid at a particular temperature. In this figure, the specific gravity reading is 1.020.

2

Atomic Theory and Laws of Chemical Combination

The Law of Conservation of Matter (or mass)

This law states: *In a chemical reaction the total mass of the products equals the total mass of the reactants.* First put forward by the French scientist Antoine Lavoisier in 1774 it seems to have been assumed by some scientists before that date. Lavoisier's great contributions to chemistry are due in large part to his careful quantitative experiments, i.e., he carefully weighed and measured things. Consequently he performed a number of experiments to verify the law. Although they were as accurate as Lavoisier could make them they would be considered very inadequate by present standards, and it was not until a century later that a series of really accurate experiments was performed by Landolt. His apparatus consisted of an H-tube, as shown in Fig. 2.1. The regents were placed in different limbs of the tube and, after sealing each limb to ensure that nothing could enter or leave the apparatus, it was weighed. Alter tilting the tube to permit chemical reaction the apparatus was reweighed. Using a very accurate chemical balance, capable of accuracy to one part in ten million, Landolt found no change in mass.

Landolt was careful to choose reactions with which there was no evolution of gas and little evolution of heat. Such reactions occur on mixing solutions of silver nitrate and sodium chloride, or of sodium sulphate and barium chloride.

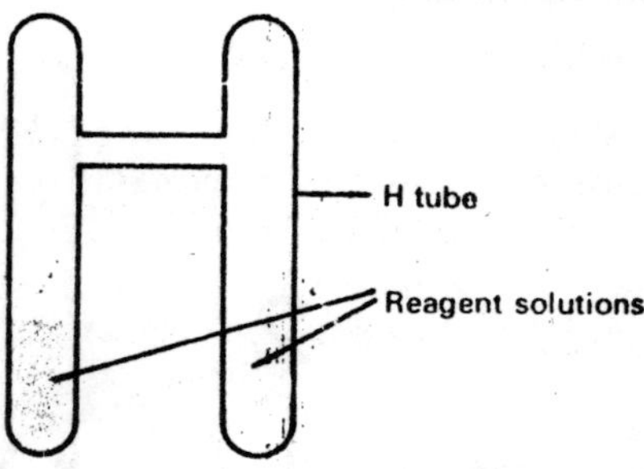

Fig. 2.1 : Landolt's H-tube.

Silver nitrate +Sodium chloride → Silver chloride +Sodium nitrate

Barium chloride +Sodium sulphate → Barium sulphate+Sodium chloride

The choice of such reactions was necessary, because a heavy apparatus would have been required to withstand the pressure of an evolved gas. Further, heat causes the evaporation of thin film of moisture normally present on the surface of the glass. This means that if the apparatus is weighed before it has time to regain this moisture there is an apparent loss of mass. Because there is no change in mass in the comparatively small number of reactions investigated, it is assumed that the law holds true for all reactions.

The law can be verified in the school laboratory by a simple adaptation of Landolt's apparatus. Using a flask and test-tube as shown in Fig 2.2 one of the reagents is placed in the bottom of the flask, the other in the small test-tube suspended by a thread. To mix the reagents the flask has to be tilted.

You may feel that there are several obvious objections to the law. Certain substances—for example, paper and coal—seem to disappear almost completely when they are burnt. The formation of invisible gaseous products is the cause of this apparent destruction. Thus, paper (a compound of carbon, hydrogen and oxygen) gives the gas carbon dioxide, and water vapour, If all the reactants, including the air which provides a proportion of the oxygen, and all the products were weighed the law would be found to be true in this case.

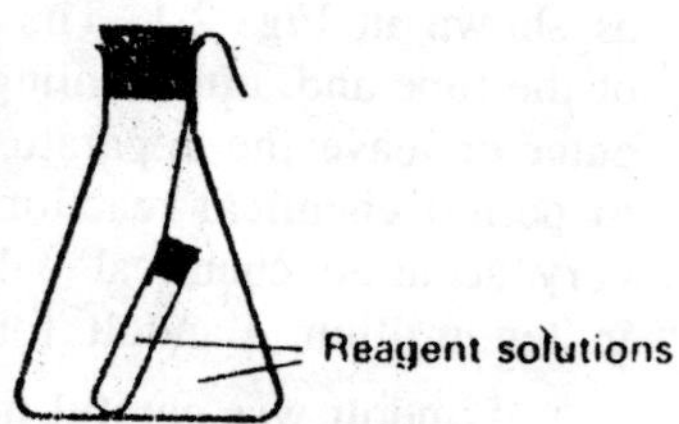

Fig. 2.2 : A laboratory version of Landolt's tube

The Law of Constant Composition

This law states: *No matter how a given chemical compound is prepared, it always contains the same elements in the same proportions by mass.* It was first stated by Proust in 1799, but was assumed by lavoisier and other scientists before that date. In common with the Law of Conservation of Mass, it cannot possibly be proved in every case, but the truth of it is accepted on the basis of those compounds which have been investigated.

Experiment—Verification of the Law of Constant Composition

To verify the law it is necessary to obtain samples of given compound from several different sources, and then to analyze each sample. Copper (II) oxide is suitable for use in the school laboratory, as it can be prepared readily be a variety of means. Use the following two methods; others we will learn later.

(a) *From copper (II) carbonate.* Place a few grams of the pale green copper (II) carbonate in a crucible and heat strongly Fig 2.3 The colour will change from green to black as the carbonate is converted into oxide.

Copper (II) carbonate → Copper (II) oxide+Carbon dioxide

Heat to constant mass to ensure complete decomposition.

As copper (II) oxide absorbs moisture from damp air, keep the product in a desiccator until ready for use.

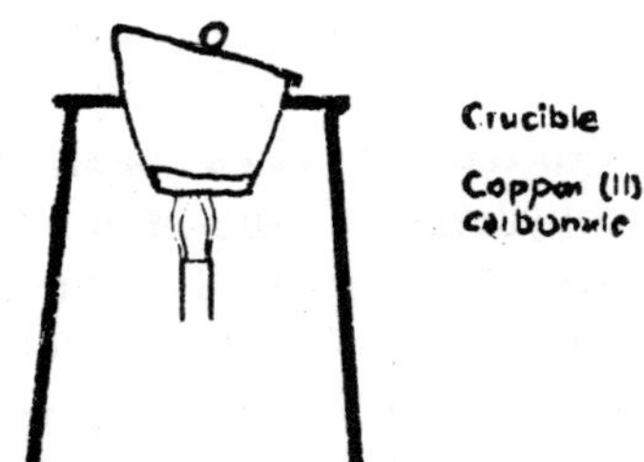

Fig. 2.3 : Preparation of Copper (II) Oxide.

(b) *From copper (II) nitrate.* Similarly, heat a few grams of copper (II) nitrate to constant mass. This time the solid will change from blue to black. Brown fumes of nitrogen dioxide are evolved during this chemical change.

Copper (II) nitrate → Copper (II) oxide+Nitrogen dioxide+Oxygen

N.B. It should be noted that copper (II) oxide cannot be prepared satisfactorily by heating copper in air or oxygen. Unless the copper is finely powdered and the heating prolonged, the copper (II) oxide forms only on the surface of the copper metal.

To analyze the samples of copper (II) oxide obtained, heat known masses in hydrogen or town gas, when the following reaction occurs:

Copper (II) oxide + Hydrogen → Copper + Water.

Then if the masses of copper formed are known it is a simple matter to calculate the composition of copper (II) oxide. Proceed as follows:

First weigh two porcelain boats empty and then containing the samples of copper (II) oxide. Place them in a hard glass tube (Fig. 2.4) and pass hydrogen through, until a test-tube full of the gas collected at the other end burns calmly. This shows that the hydrogen is not mixed with air—a necessary precaution, as mixture of an inflammable gas with air are explosive.

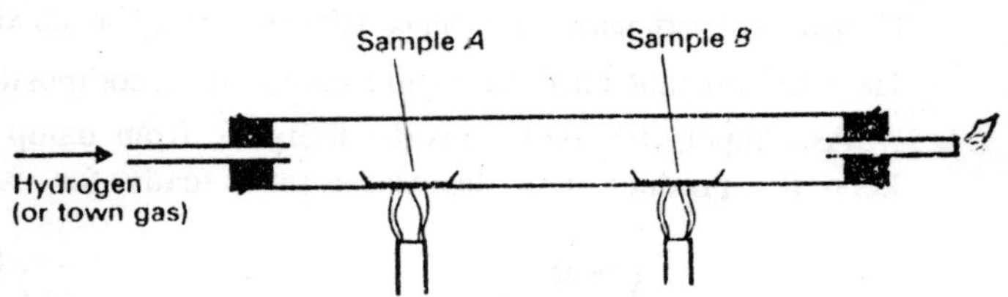

Fig. 2.4 : Verification of the Law of Constant Composition.

Then light the hydrogen as shown and heat both samples. A glow will pass through the copper (II) oxide and clouds of steam replace the hydrogen, extinguishing the flame. When the reaction is complete no more steam will be formed. Then relight the hydrogen and allow the apparatus to cool with hydrogen still passing. Do not remove the copper whilst it is still hot, because it will combine to some extent with oxygen in the air, re-forming copper (II) oxide. List your data in tabular form and calculate as indicated (Table 2.1) *i.e.* within the limits of experimental error, you should find your results to be in accordance with the Law of Constant composition.

This method can be used with any other oxide capable of being reduced by hydrogen, *e.g.* lead (II) oxide.

Table 2.1

	Sample A	*Sample B*
Mass of boat + copper (II) oxide	3.69 g	3.89 g
Mass of boat	2.01 g	1.85 g
Mass of copper (II) oxide	1.68 g	2.04 g
Mass of boat + copper	3.35 g	3.47 g
Mass of boat	2.01 g	1.85 g
Mass of copper	1.34 g	1.62 g
Percentage of copper in copper (II) oxide	$\frac{1.34 \times 100}{1.68}$ = 79.8	$\frac{1.62 - 100}{2.04}$ =79.4
Percentage of oxygen in copper (II) oxide	(100 – 79.8) = 20.2	(100 – 79.4) = 20.6

LAW OF MULTIPLE PROPORTIONS

This law, which was probably deduced by Dalton and only later subjected to experimental proof, states:

If two elements combine together to form more than one compound, then the masses of one of these elements combining with a fixed mass of the other are in a simple ratio to one another.

For example, carbon forms two common compounds with oxygen: carbon monoxide and carbon dioxide.

With carbon monoxide. 1.33 g of oxygen are combined with 1 g of carbon.

With carbon dioxide, 2.67 g of oxygen are combined with 1 g of carbon. That is, the masses of oxygen combining with a fixed mass of carbon are in the ratio 1:2. To verify the law in the school laboratory, examples that use solid substances are more suitable *e.g.* the oxides of lead.

Experiment—Verification of law of multiple proportions

Weigh three porcelain boats, first empty and then containing samples of lead (II) oxide, red lead oxide and tead (IV) oxide. Heat these oxides in a stream of hydrogen or town gas, taking the precautions outlined in the previous experiment. Each oxide is thereby converted into lead. Reweigh the boats containing the metal.

Here are some typical results:

Table 2.2

	Lead (II) oxide	*Red lead oxide*	*Lead (IV) oxide*
Mass of boat	4.55 g	4.70 g	5.10 g
Mass of boat + oxide	11.50 g	11.32 g	12.14 g
Mass of boat + lead	11.00 g	10.71 g	11.19 g
Mass of lead (b)	6.45 g	6.01 g	6.09 g
Mass of oxide (a)	6.95 g	6.62 g	7.04 g
Mass of oxygen (a – b)	0.50 g	0.61 g	0.95 g

With lead (II) oxide:

0.50 g of oxygen is combined with 6.45 g of lead.

1.00 g of oxygen is combined with $\frac{6 \cdot 45 - 1 \cdot 00}{0 \cdot 50}$ g of lead

= 12.90g of lead

With red lead oxide:

0.61 g of oxygen is combined with 6.01 g of lead.

1.00 g of oxygen is combined with $\frac{6 \cdot 01 - 1 \cdot 00}{0 \cdot 61}$ g of lead

= 9.85g of lead.

With lead (IV) oxide:

0.95 g of oxygen is combined with 6.09 g of lead.

1.00 g of oxygen is combined with $\frac{6 \cdot 09 \times 1 \cdot 00}{0 \cdot 95}$ g of lead

= 6·41g of lead.

We see, therefore, that the masses of lead combining with a fixed mass of oxygen (1 g) are in the ratio of 12·90:9·85:6·41, i.e. 4:3:2 (within the limits of experimental error). The results obtained in a laboratory experiment are usually much poorer than this.

Example 2.1

A metal forms two chlorides containing respectively 65.6 and 55.9 per cent of chlorine. Show that these figures are in accordance with the Law of Multiple Proportions.

The compositions of the chlorides are:

	%chlorine	% metal
Chloride *A*	65.6	34.4
Chloride *B*	55.9	44.1

In chloride *A*,
65·6 g of chlorine combine with 34·4 g of metal.

$$1\text{ g of chlorine combines with } \frac{34\cdot 4}{65\cdot 6}\text{ g of metal}$$

$$= 0\cdot 525\text{ g of metal}$$

In chloride *B*,
55·9 g of chlorine combine with 44·1 g of metal

$$1\text{ g of chlorine combines with } \frac{44\cdot 1}{55\cdot 9}\text{ g of metal}$$

$$= 0\cdot 789\text{ g of metal}$$

i.e. the masses of metal combining with a fixed mass of chlorine are in the ratio 0·789: 0·525 = 3:2. These figures are in accordance with the Law of Multiple proportions.

It should be noted that it is immaterial which element, or what value for its fixed mass, is chosen. If with this same problem, 34·4 g of metal is taken as the fixed mass, we get:

In chloride *A*,
34·4 of metal combine with 65·6g of chlorine.

In chloride *B*,
44·1 g of metal combine with 55·9 g of chlorine.

$$34.4\text{ of metal combine with} \frac{55\cdot 9\times 34\cdot 4}{44\cdot 1}\text{g of chlorine}$$

$$= 43\cdot 6\text{ g of chlorine}$$

i.e. the masses of chlorine combining with a fixed mass of metal are in the ratio 65·6 : 43·6 + 3 : 2

DALTON'S ATOMIC THEORY

We indicated earlier that there have been theories about atoms since very early times. Credit must go, however, to John Dalton for placing the atomic theory on a firm basis. His own account of his early history in interesting.

"The writer of this was born at the village of Eaglesfield, about two miles west of Cockermouth, Cumberland. Attended the village schools, there and in the neighbourhood, till eleven years of age, at which period he had gone through a course of mensuration, surveying, navigation, etc.; began about twelve years to teach the village school and continued it about two years; afterwards was occasionally employed in husbandry for a year or more; removed to Kendal at fifteen years of age as assistant in a boarding school; remained in that capacity for three or four years, then undertook the same school as principal and continued it for eight years; whilst at Kendal employed his leisure in studying Latin, Greek, French and the mathematics, with natural philosophy; removed thence to Manchester in 1793 as tutor in mathematics and natural philosophy in the new College; was six years in that engagement and after was employed as private and public teacher of mathematics and chemistry in Manchester, but occasionally by invitation in London, Edinburgh, Glasgow, Birmingham and Leeds."

John Dalton was a Quaker, and, even later in life when he had achieved fame, remained completely unassuming, continuing, for example, to coach local boys and girls in their arithmetic.

His atomic theory states:

(a) matter is made up of small, indivisible particles called atoms.

(b) Atoms can neither be created not destroyed.

To use some of Dalton's own words: "We might as well attempt to introduce a new planet into the solar system, or to annihilate one already in existence, as to create or destroy one particle of hydrogen".

(c) The atoms of a particular element are alike in every respect, including mass. They differ from the atom of any other element.

(d) When atoms combine they do so in small whole number to form "compound atoms". "Compound atoms" formed by such combination are alike in every respect (These "compound atoms" are now called MOLECULES.*)*

This theory leads us to a common definition of an atom:

An atom is the smallest particle of an element that can take part in a chemical change.

Both the words "law" and "theory" have now been used. It is essential to understand what is implied by these terms. A *law* is a general statement describing a way in which substances behave; it summarizes a number of isolated facts, which can be directly verified by experiment. A theory provides a generally accepted explanation of some way or other in which substances behave. A *theory*, once formulated, often leads to the discovery of fresh facts; in time, facts are found which are not in accordance with a particular theory. It then needs modifying, or may have to be abandoned altogether.

These points are well illustrated by reference to Dalton's theory. It provides satisfactory explanations for the laws mentioned earlier in the chapter.

(a) Law of Conservation of Mass

By Dalton's theory atoms can neither be created nor destroyed. It follows that a chemical reaction can consist only of a rearrangement of atoms, with no possible change in mass, i.e. the Law of conservation of Mass will be obeyed.

(b) Law of Constant Composition

By Dalton's theory every molecule of a particular compound contains the same number of each type of atom present. Each atom has a fixed mass. It is therefore obvious that all samples of the compound should have the same percentage composition.

From the sort of result obtained from the experiment described on page 33 some idea of *relative atomic masses* may be obtained. Roughly 4 g of copper combine with 1 g of oxygen. If it is assumed (as Dalton did) that copper (II) oxide is composed of atoms with one atom of copper for every atom of oxygen, this means that a copper atom is four times as heavy as an oxygen atom. Atoms do not, however, always combine in the sample ratio of 1:1.

(c) Law of Multiple Proportions

If atoms combine together in *small whole numbers*, as suggested by Dalton, then the Law of Multiple Proportions must be true. For example, suppose two elements X and Y form two compounds.

Assume that in one compound, one atom of X combines with one atom of Y and that in the other compound, one atom of X combines with two atoms of Y.

As Dalton assumed all atoms to be alike in every respect, including mass, then it is obvious that the mass of Y combining with a fixed mass of X are in the ratio 1 : 2.

Whatever *small* number of atoms of X and Y are assumed to combine with one another, a simple ratio will exist for the masses of one of the elements combining with a fixed mass of the other.

3

Atom, Mole and Chemical Arithematic

In this chapter, the concepts such as atomic mass, mole, molar mass and simple chemical calculations are described. We will see how chemists measure amounts of compounds, and decide what amount of a substance in required to react with a given amount of another substance to produce the desired reaction products in desired quantities:

ATOM

An **atom** (Greek, *atomos*, indivisible) is *the smallest particle of a chemical element that possesses all the inherent characteristics of that element and that remains unchanged in an ordinary chemical reaction.* Atoms can be further subdivided in nuclear processes, but in such processes the original chemical identity of the atom is destroyed. Atoms have a nuclear region made up of protons and neutrons (section and an extranuclear region containing electrons. The masses of protons (1.00728 amu) and neutrons (1.00867 amu) are comparable. However, the mass of an electron is only 1/835 that of a proton, and mass changes due to gain or loss of electrons in chemical reactions are assumed to be negligible.

ATOMIC MASS UNIT

Two units of mass are convenient for chemical calculations, the *gram* and the *atomic mass unit.* The gram is the most commonly used unit of mass in the metric system and is defined as 0.001 kilogram.

Until 1961 the mass of the hydrogen atom was taken as a standard and relative atomic mass defined as the number of times one atom of a particular element was heavier than one atom of hydrogen. In 1961, the definition was changed to: *The relative atomic mass of an element is the number of times one atom of that element is heavier than one twelfth of an atom of the main isotope of carbon (known as Carbon-12).* Using this standard the gram is equal to $^1/_{12}$ of the mass of 1 mole of ^{12}C. Using the same standard the atomic mass unit, abbreviated amu, is defined as $^1/_{12}$ of the mass of *one atom* of ^{12}C. Since an atom is extremely small and has very little mass, it follows that the atomic mass unit is extraordinarily small compared to the gram.

If this second definition is used the relative atomic mass of hydrogen changes form 1·000 to 1·008, so that in fact, whichever definition is used the relative atomic masses are almost are same.

Nowadays the relative atomic masses are obtained using a large expensive instrument known as a *mass-spectrometer.* By shooting the atoms through some complicated electric and magnetic fields, this instrument is able to sort atoms according to their mass, and to measure every accurately the masses of the atom. For this method it does not matter at all in what ratios atoms combine.

ATOMIC MASS

The **atomic mass** of an element is *the average mass for an aggregate of atoms of that element compared to the mass of a ^{12}C atom having a mass of* 12.000 *amu.* The atomic mass of helium is 4.00260 amu; thus, the average mass of a helium atom is about one third of the mass of a ^{12}C atom. The atomic mass of titanium is 47.90 amu; thus, the average mass of a titanium atom is about four times the mass of a ^{12}C atom.

The average atomic mass for an aggregate of atoms of the same kind can be determined by taking a weighted average of the stable isotopes of that atom.

Example 3.1

If boron has naturally occurring isotopes with mass numbers 10 and 11 amu with relative percentage abundances of 19.6 and 80.4, respectively, calculate the average atomic mass of boron.

The relative contribution of each isotope must be considered to obtain the average atomic mass. Each contribution is determined by multiplying the mass number by the percent abundance changed to a decimal.

$$\begin{aligned}(10\ \text{amu})(0.196) &= 1.96\ \text{amu}\\ (11\ \text{amu})(0.804) &= 8.84\ \text{amu}\\ \hline &\quad 10.80\ \text{amu}\end{aligned}$$

MOLECULE

A **molecule** is *an electrically neutral collection of atoms bonded together strongly enough to be considered as a recognizable unit.* Two carbon atoms and four hydrogen atoms bind together to form a molecule called ethylene, a reactive gaseous compound. Six carbon atoms and twelve hydrogen atoms bind together to form a molecule called cyclohexane a liquid in which many organic compound dissolve readily. A molecule exhibits distinctive characteristics that may not be related in any obvious fashion to the properties of the atoms that comprise the molecule.

ION

An ion is *an atom or distinguishable group of atoms bearing one or more positive or negative charges due to the loss or gain of one or more electrons.* Sodium chloride, is composed of Na^+ and Cl^- ions. These ions are formed from single atoms. Ammonium sulphate consists of Nh_4^+ and SO_4^{2-} ions, both of which are formed from distinguishable groups of atoms. *A positive ion* such as Na^+ or NH_4^+ is called a **cation**. *A negative ion* such as Cl^- or SO_4^{2-} is called an **anion**.

AVOGADRO NUMBER

The Avogadro number, usually denoted by *N*, is defined in a wide variety of ways; however, it is simply the conversion factor between the *gram* and the *atomic mass unit.* It has been determined by electrochemical and crystallographic techniques to have a value of 6.02×10^{23}. Thus, 1 gram = 6.02×10^{23} amu.

Formula

A **formula** *conveys information about the composition and structure of a compound.* An **empirical formula** states only *the* ***relative*** *numbers of atoms in a compound* and says nothing about its structure. By convention an empirical formula includes symbols of the elements making up a compound and subscripts affixed to the symbols to designate the relative numbers of atoms. Thus, the empirical formula for both ethylene and cyclohexane is CH_2, meaning that there is one carbon atom for every two hydrogen atoms in the compound.

A **molecular formula** states *the* **actual** *numbers of atoms in a molecule*, but says nothing about its structure. A molecular formula includes symbols of the elements that make up a molecule with subscripts affixed to the symbols to designate the actual numbers of atoms. The molecular formula of ethylene is C_2H_4, and the molecular formula of cyclohexane is C_6H_{12}, both simple multiples of their empirical formula, CH_2.

A **structural formula** gives *the actual numbers of atoms in a molecule, indicates which atoms are bonded to each other, and may indicate the three dimensional configuration of the molecule.* Thus, to indicate which atoms are bonded to each other in ethylene and cyclohexane, we write

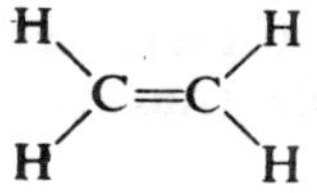

```
                                     H   H
                                      \ /
                                 H     C     H
                                  \  /   \  /
 H       H                          C       C
  \     /                         /  |      | \
   C = C          and            H   |      |   H
  /     \                        H   |      |   H
 H       H                         \ |      | /
                                    C       C
                                  /  \     /  \
                                 H     C     H
                                      / \
                                     H   H
```

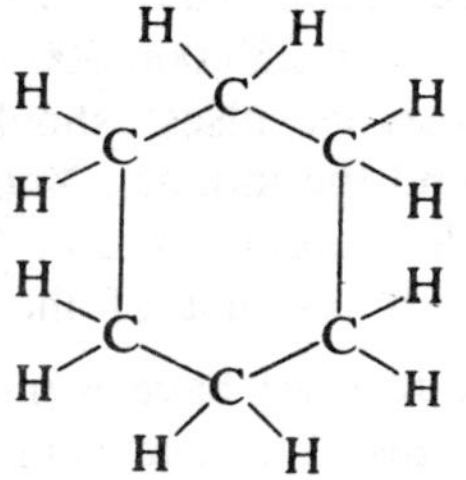

In cyclohexane there are six C atoms forming a six-membered ring. In both ethylene and cyclohexane each C atom has two H atoms bonded to it.

The Mole

Chemists have devised the concept of the mole to express large numbers of these particles.

One **mole** *of atoms, molecules, formulas, or ions is the Avogadro number of atoms, molecules, formulas, or ions,* respectively. One mole of titanium atoms is the Avogadro number of 6.02×10^{23} titanium atoms. One mole of sulfur dioxide molecules is the Avogadro number of sulfur dioxide molecules. One mole of H_2O_2 formulas is the Avogadro number of H_2O_2 formulas. One mole of nitrate ions is the Avogadro number of nitrate ions.

When using the mole concept, you must exercise great care in keeping track of what particle you are counting. You must specify the particle in words or by formula. For example, one mole of sulfur might refer to either the Avogadro number of sulfur atoms or the Avogadro number of S_8 molecules (a sulfur molecule consists of eight sulfur atoms arranged in the form of a crown-shaped ring). The former weighs 32.06 grams; the latter weighs 256.48 grams.

The mole is 6.02×10^{23} *particles of anything.* Usually the mole is used in referring to atoms or molecules or ions or electrons, *e.g.* 1 mole of copper atoms contains 6.02×10^{23} atoms, and 1 mole of water molecules contains 6.02×10^{23} molecules. The idea of a mole can be applied to anything although, of course, it is normally applied only to these small particles met with in science.

To give some idea of what an enormous number of particles are contained in a mole, try to imagine a mole of people *i.e.* 6.02×10^{23} people. If they stood packed tightly together over the whole surface of the earth (land and sea) there would be enough left over to cover another 250,000,000 earths (or if they all stood on one earth on everyone elses' shoulders the columns of people would stretch beyond the moon!). You might like to guess (or if you can, work out) the volume of a mole of footballs. (Answer: They have a volume 5 times that of the earth.)

In contrast we have a mole of hydrogen atoms (6.02×10^{23} of them) in just 1 g, and a mole of water molecules (6.02×10^{23} water molecules) in 18 g or a large tablespoonful (≡ 5 teaspoonfuls).

Put another way—there are as many molecules of water in a teaspoonful as there would be footballs in the volume of the earth.

This gives you some idea of just how small are the building blocks of chemistry—atoms and molecules—that make up all matter.

Molar Mass

For the mole to be a useful quantity that can be weighed in the laboratory, it must be related to mass. The **molar mass** *of any particle* (atom, molecule, formula, or ion) is *the sum of the average atomic masses of all atoms forming that particle.* (Many texts refer to this quantity as molecular weight.) The molar mass, frequently symbolized by M, can be computed if the formula of the particle is known or can be determined by a number of experimental methods if the formula of the particle is unknown.

The molar mass of an atom is its average atomic mass expressed in grams. It is the mass of one mole of such atoms. One mole of titanium atoms weighs 47.90 grams and consists of the Avogadro number of titanium atoms.

The molar mass of a molecule is the sum of the average atomic masses of all the atoms comprising the molecule. It is the mass of one mole of such molecules. One mole of CO_2 molecules is 12·0 grams of carbon plus 2 (15.9994 grams) of oxygen, or 44.0 grams, and consists of the Avogadro number of CO_2 molecules.

Many substances in the solid and liquid state are not constructed of discrete molecules, but rather the crystal or the liquid may be considered as the giant molecule. Table salt, sodium chloride, is an example of an ionic substance in which an infinite array of positively charged ions is associated with an infinite array of negatively charged ions. The ions are in such ratios that the total number of negative charges. No discrete molecules can be discerned; thus, molecular formula has no meaning in this case. The empirical formula, NaCl, shows what ions are present and states the ratio of these ions as simply as possible. The molar mass is the empirical-formula mass in calculations involving such substances.

The molar mass of a formula is the sum of the average atomic masses of all atoms comprising the formula. It is the mass of one mole of such formulas. One mole of $TiCl_2$ formulas weighs 47.90 grams of titanium plus 2(35.453 grams) of chlorine, or 118.81 grams, and consists of the Avogadro number of $TiCl_2$ formulas.

The molar mass of an ion is the sum of the average atomic masses of all atoms comprising the ion. It is the mass of the mole of such ions. One mole of nitrate ions, NO_3^-, weighs 14.0067 grams of nitrogen plus 3(15.9994 grams) of oxygen, or 62.0049 grams, and consists of the Avogadro number of nitrate ions. Note that the mass of one additional electron which gives rise to the —1 charge is negligible and is neglected.

To emphasize the points that have been developed during the last few pages the following Table 3.1 lists them all for some atoms and molecules.

Solved Examples

***Example* 3.2**

For a 250 gram sample of titanium metal, determine (a) the number

Table 3.1

Atom	*Symbol*	*Relative Atomic mass*	*Molar mass*	*Number of atoms in a mole*
Hydrogen	H	1	1 g mol^{-1}	6.02×10^{23}
Oxygen	O	16	16 g mol^{-1}	6.02×10^{23}
Carbon	C	12	12 g mol^{-1}	6.02×10^{23}
Sodium	Na	23	23 g mol^{-1}	6.02×10^{23}
Sulphur	S	32	32 g mol^{-1}	6.02×10^{23}

Molecule	*Formula*	*Relative molecular mass*	*Molar mass*	*Number of molecules in* a mole
Hydrogen	H_2	2	2 g mol^{-1}	6.02×10^{23}
Oxygen	O_2	32	32 g mol^{-1}	6.02×10^{23}
Water	H_2O	18	18 g mol^{-1}	6.02×10^{23}
Hydrogen chloride	HCl	36.5	36.5 g mol^{-1}	6.02×10^{23}
Carbon dioxide	CO_2	44	44 g mol^{-1}	6.02×10^{23}
Sodium chloride	NaCl	58.5	58.5 g mol^{-1}	6.02×10^{23}
Sulphur dioxide	SO_2	64	64 g mol^{-1}	6.02×10^{23}

of moles of titanium atoms present, (b) the number of titanium atoms present, and (c) the average mass of one titanium atom.

(a) The molar mass of a Ti atom is 47.90 g/mole. Determine the number of moles of Ti atoms by dividing the mass of Ti by the molar mass of a Ti atom.

$$\text{Number of moles of Ti atoms} = \frac{\text{mass of Ti}}{\text{molar mass of Ti}}$$

$$= \frac{250 \text{ g}}{47.90 \text{ g / mole}}$$

$$= 5.22 \text{ moles of Ti atoms}$$

(b) One mole of Ti atoms consists of the Avogadro number of Ti atoms. Determine the number of Ti atoms by multiplying the number of moles of ti atoms by the Avogadro number.

$$\text{Number of Ti atoms} = (\text{number of moles of Ti})\ (\text{Avogadro number})$$

$$= (5.22 \text{ moles})\ (6.023 \times 10^{23} \text{ atoms mole})$$

$$= 3.14 \times 10^{24} \text{ Ti atoms}$$

(c) One mole of Ti atoms weighs 47.90 grams and consists of the Avogadro number of Ti atoms. Determine the mass of one Ti atom by dividing the molar mass of a Ti atoms by the Avogadro number.

$$\text{Mass of a Ti atoms} = \frac{\text{molar mass of Ti}}{\text{Avogadro number}}$$

$$= \frac{47.90 \text{ g / mole}}{6.023 \times 10^{23} \text{ atoms / mole}}$$

$$= 7.953 \times 10^{-23} \text{ g/atom}$$

Alternatively, it was calculated above that 250 grams of Ti consists of 3.14×10^{24} Ti atoms. Therefore,

$$\frac{250 \text{ g}}{3.14 \times 10^{24} \text{ atoms}} = 7.96 \times 10^{-23} \text{ g / atom}$$

Note that dimensions attached to the various quantities are carried through the same mathematical operations as the numbers themselves. Also note that there are (after calculations in accordance with the rules in Section 1.3-c-2) four significant figures in the answer in (c) using the method requiring only constants from tables but only three significant figures in the method using experimental data.

***Example* 3.3**

For a 250 gram sample of sulfur dioxide, determine (a) the number of moles of SO_2 molecules present, (b) the number of moles of sulfur atoms and oxygen atoms present, (c) the number of SO_2 molecules present, (d) the number of sulfur atoms and oxygen atoms present, (e) the average mass of one SO_2 molecule.

(a) The molar mass of a SO_2 molecule is 64.06 g/mole. Determine the number of moles of SO_2 molecules by dividing the mass of SO_2 by the molar mass of a SO_2 molecules.

$$\text{Number of moles of } SO_2 \text{ molecules} = \frac{\text{mass of } SO_2}{\text{molar mass of } SO_2}$$

$$= \frac{250 \text{ g}}{64.06 \text{ g / mole}}$$

$$= 3.90 \text{ moles of } SO_2 \text{ molecules}$$

(b) The molecular formula, SO_2, indicates that there are one mole of S atoms and two moles of O atoms in each mole of SO_2 molecules.

Number of moles of S atoms = number of moles of SO_2 molecules
= 3.90 moles of S atoms

Number of moles of O atoms = (2)(number of moles SO_2 molecules)
= 7.80 moles of O atoms

(c) One mole of SO_2 consists of the Avogadro number of SO_2 molecules. Determine the number of SO_2 molecules by multiplying the number of moles SO_2 molecules by the Avogadro number.

Number of SO_2 molecules =(number if moles of SO_2) (Avogadro number)
= (3.90 moles) (6.023 × 10^{23} molecules/mole)
= 2.35 × 10^{24} SO_2 molecules

(d) The molecules formula, SO_2, indicates that there are one S atom and two O atoms in each SO_2 molecule.

Number of S atoms = number of SO_2 molecules
= 2.35 × 10^{24} S atoms

Number of O atoms = (2)(number of SO_2 molecules)
= 4.70 × 10^{24} O atoms

(e) One mole of SO_2 molecules weighs 64.06 grams and consists of the Avogadro number of SO_2 molecules. Determine the mass of one SO_2 molecule by dividing the molar mass of a SO_2 molecule by the Avogadro number.

$$\text{Mass of a } SO_2 \text{ molecule} = \frac{\text{molar mass of } SO_2}{\text{Avogadro number}}$$

$$= \frac{64.06 \text{ g / mole}}{6.023 \times 10^{23} \text{ molecules/ mole}}$$

$$= 1.064 \times 10^{-22} \text{ g/molecule}$$

Alternatively, it was calculated above that 250 grams of SO_2 consists of 2.35 × 10^{24} SO_2 molecules. Therefore:

$$\frac{250 \text{ g}}{2.35 \times 10^{24} \text{ molecules}} = 1.06 \times 10^{-22} \text{ g / molecule}$$

Percent Composition

The percent composition by mass of a compound is obtained by dividing the mass of each element present in the compound by the total mass of the compound, then converting to percent by multiplying by 100%. The masses of each element present must be determined experimentally by chemical analysis.

Example 3.4

If 2.44 grams of titanium combines with 3.60 grams of chlorine to form a compound, what is the percent composition of this compound?

The total mass of the compound equals 2.44 grams + 3.60 grams, or 6.04 grams.

$$\text{Percent Ti} = \left(\frac{\text{mass of Ti}}{\text{mass of compound}}\right)(100\%) = \left(\frac{2.44\ g}{6.04\ g}\right)(100\%) = 40.4\%\ \text{Ti}$$

$$\text{Percent Cl} = \left(\frac{\text{mass of Cl}}{\text{mass of compound}}\right)(100\%) = \left(\frac{3.60\ g}{6.04\ g}\right)(100\%) = 59.6\%\ \text{Cl}$$

If you know the percent composition of a compound, you can calculate the mass of any element present in a given mass of the compound by converting the percent to a decimal and multiplying the decimal by the given mass of the compound.

Example 3.5

Titanium forms two other compounds with chlorine in which the percents of titanium are 31.05% and 25.25%, respectively. What mass of titanium is present 6.04 grams of each of these compounds?

Change the percents to decimals by dividing by 100%. Thus, 31.05% = 0.3105 and 25.25%= 0.2525. Then multiply by the compound.

Mass of Ti = (6.04g)(0.3105) = 1.88 g of Ti

Mass of Ti = (6.04 g)(0.2525) = 1.53 g of Ti

Formulas

The empirical formula of a compound can be determined using molar masses of constituent atoms and analytical data for the compound, either in the form of masses of each element present or percent composition. Both methods involve calculation of the ratio of moles of constituent atoms as illustrated in the following three examples.

Formula Weights

Since atoms possess mass (weight), common sense tells us that molecules must also weigh something. The weight of a molecule is simply the sum of the atomic weights of the individual atoms making up the molecule. This weight is referred to as the *formula weight.* The formula weight of hydrogen chloride (HCl) is the sum of the atomic weights of hydrogen and chlorine:

1.008 amu + 35.453 amu = 36.461 amu

The formula weight of a carbon dioxide molecule is 44 amu. This is found by adding the weights of one carbon atoms and two oxygen atoms:

$$12.011 \text{ amu} + 2(15.999 \text{ amu}) = 44.00 \text{ amu}$$

Although ionic compounds do not occur as molecules, it is frequently useful to know the formula weights of these compounds. The procedure is the same as for molecular compounds: Sum the weights of the individual atoms in the formula.

***Example* 3.6.**

Calculate the formula weight of Calcium phosphate, $Ca_3(PO_4)_2$,

Weight of calcium	3×40.08 amu	=	120.24 amu
Weight of phosphorus	2×30.97 amu	=	61.94 amu
Weight of oxygen	8×15.999 amu	=	127.99 amu
Formula weight of $Ca_3(PO_4)_2$		=	310.17 amu

Ans. (a) 58.44; (b) 141.98; (c) 96; (d) 180.06; (e) 16488.93

***Example* 3.7**

Calculate the molar mass of Copper (II) sulphate crystals, $CuSO_4.5H_2O$:

Cu = 63·5; S = 32; O = 16; H=1

This compound contains water of crystallization and the formula indicates that five molecules of water the combined with each $CuSO_4$ unit.

$CuSO_4.5H_2O$	= Cu S O_4 $5H_2O$
molecular mass	= 63.5 + 32 + (4×16) + 5(2 +16)
	= 63.5 + 32 + 64 + 90 =249.5
Molar mass	= 249.5g mol^{-1}

CALCULATION OF PERCENTAGE COMPOSITION BY MASS FROM FORMULAE

If the preceding paragraphs have been understood, it is easy to see how the percentage composition of a compound may be calculated. Taking water as our example, we find that it has a molar mass of 18 g mol^{-1}. Of these 18g, 16g are accounted for by the oxygen content, and the remaining 2 g by the hydrogen.

i.e. $\frac{16}{18}$ of the total mass of water is oxygen.

i.e. $\frac{16}{18} \times 100$ per cent = 88.9 per cent.

And $\frac{2}{18}$ of the total mass of water is hydrogen.

i.e. $\frac{2}{18}\times100$ per cent = 11.1 per cent.

***Examples* 3.8**

Calculate the percentage composition by mass of sodium chloride Na=23; Cl=35.5

NaCI =Na CI

Molar mass $=23+35.5 = 58.5 gmol^{-1}$

Thus in every 58.5 g of sodium chloride there are 23 g of sodium and 35.5 g of chlorine.

$\therefore$ Fraction of Sodium in sodium chloride $= \frac{23}{58.5}$

and fraction of chlorine in sodium chloride $= \frac{35.5}{58.5}$

$\therefore$ *Percentage of sodium* $= \frac{23}{58.5}\times 100 = 39.3\%$

and *percentage of chlorine* $= \frac{35.5}{58.5}\times 100 = 60.70\%$

***Example* 3.9**

Calculate the percentage composition by mass of lead (II) nitrate. Pb= 207; N = 14; O = 16.

$Pb(NO_3)_2$ = Pb N_2 O_6

Molar mass $= 207 + (2\times 14) + (6\times 16)$

$=207 + 28 + 96 = 331\ g\ mol^{-1}$

Proceeding as in lase example;

Percentage of lead $\frac{207}{331}\times 100 = 62.5\%$

Percentage of nitrogen $\frac{28}{331}\times 100 = 8.5\%$

Percentage of oxygen $\frac{96}{331}\times 100 = 29.0\%$

***Example* 3.10**

Calculate the percentage of water of crystallization in washing soda crystals, sodium carbonate-10-water, $Na_2CO_3 . 10H_2O$

$Na_2CO_3 . 10H_2O$	= Na_2	C	O_3		$10H_2O$
Molar mass	= $(2 \times 23) + 12 + (3 \times 16) + 10(2 + 16)$				
	= 46	+ 12	48	+	180
	= 286 g mol^{-1}				

i.e. in every 286 g of washing soda there are 180 g of water.

$$\text{Percentage of water} = \frac{180}{286} \times 100 = 62.9\%$$

DETERMINATION OF FORMULAE

In the last sections the compositions by mass of compounds have been calculated from the formulae of compounds.

The reverse is obviously possible. If the masses of each element present in a certain mass of a substance is known (or if the percentage composition by mass of a substance is known), *the chemical formula can be calculated.* Chemists have a number of ways of determining experimentally the composition by mass of a compound and it is in this way *that chemical formulae are experimentally derived.*

For example, the percentage composition by mass of a certain oxide of sulphur is; oxygen, 50 per cent; sulphur, 50 per cent. Obviously the percentage of either of these elements depends on

(*a*) the atomic mass of that element,

(*b*) The number of atoms of that element in each molecule.

If the atomic mass of oxygen were equal to that of sulphur it is obvious that there would be one atom of oxygen present for each atom of sulphur. In actual fact, the atomic mass of oxygen is 16 and of sulphur 32, i.e., the atomic mass of oxygen is half the atomic mass of sulphur. For the oxide to contain 50 per cent by mass of each element there must, therefore, be two atoms of oxygen for each atom of sulphur. This means that the simplest formula for the oxide if sulphur is SO_2 but it is not possible , with the data given, to say that this is the actual formula. If the formula were S_2O_4, S_3O_6 (i.e.,S_nO_{2n}) it would still give the same percentage composition by mass.

Another oxide of sulphur contains 60 per cent oxygen and 40 per cent sulphur by mass. Knowing the atomic masses of oxygen and sulphur to be 16 and 32 respectively, the formula can again be calculated. By similar reasoning there are $\frac{60}{16}$ oxygen atoms for every $\frac{40}{32}$ sulphur atoms in a molecule of this oxide:

$$\frac{60}{16}:\frac{40}{32}=\frac{60}{1}:\frac{40}{2}=3:1$$

i.e. there are three oxygen atoms for each sulphur atom, giving the simplest formulas SO_3

In general, we can find that relative numbers of each type of atom in a molecule of compound, and hence its simplest formula, by dividing the percentage by mass, or the actual masses, of the various elements by their atomic masses. Only rarely are the figures so convenient as in the examples quoted above. The following problems should, however, make the method clear.

***Example* 3.11**

A compound contains 29.1 per cent sodium, 40.5 per cent sulphur and 30.4 percent oxygen. Calculate its formula (Na = 23; S = 32; O = 16)

The relative numbers of atoms of each element are obtained thus:

$$\text{Relative number of atoms} = \frac{\%}{\text{atomic mass}}$$

Percentage of sodium	= 29·1.	$\frac{29\cdot1}{23} = 1\cdot27$
Percentage of sulphur	= 40·5.	$\frac{40\cdot5}{32} = 1\cdot27$
Percentage of oxygen	= 30·4	$\frac{30\cdot4}{16} = 1\cdot90$

Therefore the molecules of the compound must contain whole numbers of sodium, sulphur and oxygen atoms in the ratio 1·27:1·27:1·90. The simplest whole number ratio agreeing with these figures is probably obvious to you. If not it can be found by dividing all the figures by the lowest one,

i.e. $1\cdot27 : 1\cdot27 : 1\cdot90 = \frac{1\cdot27}{1\cdot27}:\frac{1\cdot27}{1\cdot27}:\frac{1\cdot90}{1\cdot27} = 1:1:1\cdot5$

Therefore the simplest whole number ratio is 2:2:3. (N.B. The ratio is not absolutely exact because the figures given in the question are subject of experimental error.)

From these data we obtain $Na_2S_2O_3$ as the simplest formula for the compound. Such a formula is known as an *empirical formula.*

An empirical formula may be defined as the simplest formula which can be obtained for a compound using its composition by mass.

***Example* 3.12**

A compound containing water of crystallization has the following percentage composition: iron; 20·1%: sulphur, 11·5%; oxygen, 23.0%; water 45·%. Calculate the empirical formula of this compound. (fe=56, s=32, O=16, H=1.)

Here a molecule of water can be treated as a unit, in the same way as an atom. The molecular mass of water is 18.

		Relative number of atoms
Percentage of iron	= 20·1	$\frac{20\cdot1}{56} = 0{\cdot}36$
Percentage of sulphur	= 11·5	$\frac{11\cdot5}{32} = 0{\cdot}36$
Percentage of oxygen	= 23·0	$\frac{23\cdot0}{16} = 1{\cdot}44$
Percentage of water	= 45·3	$\frac{45\cdot3}{18} = 2{\cdot}52$

This indicates a formula $Fe_{0.36}S_{0.36}O_{1.44}.2.52\ H_2O$. As before, the simplest ratio of atoms becomes obvious by dividing the relative numbers of atoms obtained by the smallest number.

$$0{\cdot}36 : 0{\cdot}36 : 1{\cdot}44 : 2{\cdot}52 : = \frac{0\cdot36}{0\cdot36}:\frac{0\cdot36}{0\cdot36}:\frac{1\cdot44}{0\cdot36}:\frac{0\cdot67}{0\cdot67}:\frac{2\cdot52}{0\cdot36}$$

$$= 1 : 1 : 4 : 7$$

Therefore the simple formula for this compound is $FeSO_4.7H_2O$

***Example* 3.13**

20·00 g of a compound containing carbon, hydrogen, and oxygen are found to contain 8·00 g of carbon and 1·33 g of hydrogen. Calculate the empirical formula of this compound. (C = 12, H= 1, O=16.)

Since the compound contains only carbon, hydrogen, and oxygen, the mass of oxygen in 20·00 g of the compound

= 20·00 — (8·00 + 1·33)g

= 10·67 g

Although we are not dealing with percentages in this compound, the method of solution is the same.

		Relative number of atoms
Mass of carbon	= 8·00 g	$\frac{8 \cdot 00}{12} = 0{\cdot}67$
Mass of hydrogen	= 1·33 g	$\frac{1 \cdot 33}{1} = 1{\cdot}33$
Mass of oxygen	= 10·67 g	$\frac{10 \cdot 67}{16} = 0{\cdot}67$

$0{\cdot}67 : 1{\cdot}33 : 0{\cdot}67 = \frac{0 \cdot 67}{0 \cdot 67} : \frac{1 \cdot 33}{0 \cdot 67} : \frac{0 \cdot 67}{0 \cdot 67} = 1 : 2 : 1$

Therefore the empirical formula is CH_2O.

***Example* 3.14**

It has been found by experiment that 1·68 g of a copper oxide contains 1·34 g of copper, that 0·50 g of oxygen combine with 6·45 g of lead to give lead oxide, that 0·61 g of oxygen combine with 6·01 g of lead to give a second lead oxide and that 0·95 g of oxygen combine with 6·09 g of lead to give a third lead oxide. Obtain the formulae of the copper oxide and of the three lead oxides. (Cu = 63·5, Pb=207. O = 16.)

Copper oxide

1·34 g of copper combine with (1·68 — 1·34) g of oxygen

1·34 g of copper combine with 0·34 g of oxygen

		Relative number of atoms
Mass of copper	= 1·34 g	$\frac{1 \cdot 34}{63 \cdot 5} = 0{\cdot}0211$
Mass of oxygen	= 0·34 g	$\frac{0 \cdot 34}{16} = 0{\cdot}0212$

0·0211 : 0·0212 = 1 : 1

i.e. the empirical formula is CuO

First lead oxide

0·50 g of oxygen combine with 6·45 g of lead

		Relative number of atoms
Mass of lead	= 6·45 g	$\frac{6.45}{207} = 0.0312$
Mass of oxygen	= 0·50 g	$\frac{0.50}{16} = 0.0312$
0·0312 : 0·0312	= 1 : 1	

i.e. the empirical formula is PbO.

***Example* 3.15**

Determine the empirical formula of an alcohol that has a percent composition by mass of 60·0% carbon, 13·4% hydrogen, and 26·6% oxygen.

Assume 100 grams of the compound. The percent composition indicates the number of grams of each element present in 100 grams of the compound.

Mass of C = (100 g)(0.600) = 60.0 g of C

Mass of H = (100 g)(0.314) = 13.4 g of H

Mass of O = (100 g)(0.266) = 26.6 g of O

$$\text{Number of moles of C atoms} = \frac{\text{mass of C}}{\text{molar mass of C}}$$

$$= \frac{60.0\ g}{12.011\ g/\text{mole}}$$

= 5.00 moles of C atoms

$$\text{Number of moles of H atoms} = \frac{\text{mass of H}}{\text{molar mass of H}}$$

$$= \frac{13.4\ g}{1.0079\ g/\text{mole}}$$

= 13.3 moles of H atoms

$$\text{Number of moles of O atoms} = \frac{\text{mass of O}}{\text{molar mass of O}}$$

$$= \frac{26.6\ g}{15.9994\ g/\text{mole}}$$

= 1.66 moles of O atoms

$$C : H : O = \frac{5 \cdot 600}{1 \cdot 66} : \frac{13 \cdot 3}{1 \cdot 66} : \frac{1 \cdot 66}{1 \cdot 66} = 3.01 : 8.01 : 1.00$$

Therefore, the empirical formula is C_3H_8O, and the alcohol is 1-propanol

Frequently fractional ratios of moles of constituent atoms are obtained. In such cases multiply by a number that will reduce the ratio to small whole numbers. This process is illustrated in the following example.

***Example* 3.16**

Hematite, and iron-containing ore, is composed of 69.9% iron and 30.1% oxygen. What is its empirical formula?

Assume 100 grams of the compound. The percent composition indicates the number of grams of each element present in 100 grams of the compound.

$$\text{Mass of Fe} = (100\text{g})(0.699) = 69.9 \text{ g of Fe}$$

$$\text{Mass of O} = (100\text{g})(0.301) = 30.1 \text{ g of O}$$

$$\text{Number of moles of Fe atoms} = \frac{\text{mass of Fe}}{\text{molar mass of Fe}}$$

$$= \frac{69.9 \text{ g}}{55.847 \text{ g/mole}}$$

$$= 1.25 \text{ moles of Fe atoms}$$

$$\text{Number of moles of O atoms} = \frac{\text{mass of O}}{\text{molar mass of O}}$$

$$= \frac{30.1 \text{ g}}{15.9994 \text{ g/mole}}$$

$$= 1.88 \text{ moles of O atoms}$$

$$Fe : O = \frac{1 \cdot 25}{1 \cdot 25} : \frac{1 \cdot 88}{1 \cdot 25} = 1.00 : 1.50$$

Reduce the ratio of moles of constituent atoms, 1.00:1.50, to small whole numbers by multiplying by two. This gives a simplest ratio of 2.00:3.00, and the empirical formula for hematite is Fe_2O_3.

If the empirical formula of compound is known, the molar mass of that empirical formula and the percent composition of the compound can be calculated.

***Example* 3.17**

What are the molar mass and the percent composition of titanium (III) sulfate, $TI_2(SO_4)_3$?

Molar mass of a $Ti_2(SO_4)_3$ formula

$$= 2(47.90g) \text{ of Ti} + 3(32.06g) \text{ of S} + 12(15.9994g) \text{ of O}$$
$$= 383.97 \text{ g /mole}$$

$$\text{Percent of Ti} = \left[\frac{\text{mass of Ti}}{\text{molar mass of } Ti_2(SO_4)_3}\right](100\%) = \left[\frac{2(47.90 \text{ g})}{383.97 \text{ g}}\right](100\%)$$

$$= 24.95\% \text{ Ti}$$

$$\text{Percent S} = \left[\frac{\text{mass of S}}{\text{molar mass of } Ti_2(SO_4)_3}\right](100\%) = \left[\frac{3(32.06 \text{ g})}{383.97 \text{ g}}\right](100\%)$$

$$= 25.05\% \text{ S}$$

$$\text{Percent O} = \left[\frac{\text{mass of O}}{\text{molar mass of } Ti_2(SO_4)_3}\right](100\%) = \left[\frac{12(15.9994 \text{ g})}{383.97 \text{ g}}\right](100\%)$$

$$= 50.002\% \text{ O}$$

In the above examples molar masses of constituent atoms and analytical data in the form of masses of each element present or percent composition were required to determine the empirical formula. To determine the molecular formula, and experimentally determined molar mass of the molecule is also required.

***Example* 3.18**

Ascorbic acid (vitamin C), found in citrus fruits, tomatoes, and green vegetables, is composed of 40.9% carbon, 4.6% hydrogen, and 54.5% oxygen. Determine (a) the empirical formula of ascorbic acid and (b) the molecular formula if the experimentally determined molar mass of an ascorbic acid molecule is 176.13 g/mole.

(a) Assume 100 grams of the compound. The percent composition indicates the number of grams of each element present in 100 grams of the compound.

$$\text{Mass of C} = (100g)l(0.409) = 40.9 \text{ g of C}$$

$$\text{Mass of H} = (100 \text{ g})(0.046) = 4.6 \text{ g of H}$$

$$\text{Mass of O} = (100 \text{ g})(0.545) = 54.5 \text{ g of O}$$

$$\text{Number of miles of C atoms} = \frac{\text{mass of C}}{\text{molar mass of C}}$$

$$= \frac{40.9 \text{ g}}{12.011 \text{ g/mole}}$$

$$= 3.41 \text{ miles of C atoms}$$

$$\text{Number of moles of H atoms} = \frac{\text{mass of H}}{\text{molar mass of H}}$$

$$= \frac{4.6\ g}{1.0079\ g/\text{mole}}$$

= 4.6 moles of H atoms

$$\text{Number of moles of O atoms} = \frac{\text{mass of O}}{\text{molar mass of O}}$$

$$= \frac{54.5\ g}{15.9994\ g/\text{mole}}$$

= 3.41 moles of O atoms

$$C : H : O = \frac{3\cdot41}{3\cdot41} : \frac{4\cdot6}{3\cdot41} : \frac{3\cdot41}{3\cdot41} = 1.00 : 1.3 : 1.00$$

= 3.00 : 3.9 : 3.00

Note that it is reasonable to round off 3.9 to 4 when converting to whole numbers in this example because the fact that there are tow significant figures in 3.9 implies that the error is at least ±0.1. Therefore, the empirical formula of ascorbic acid is $C_3H_4O_3$.

Note additionally that it would have been unreasonable to round off 1.50 to 2 in Example 4.8 because three significant figures imply that the error is in the hundredths place.

(b) The molecular formula is some integral multiple of the empirical formula. Compute the molar mass of the empirical formula and determine what multiple the experimental molar mass of the molecule is of the molar mass of the empirical formula.

Molar mass of the empirical formula

= 3(12.011 g) of C + 4(1.0079 g) of H + 3(15.9994 g) of O

= 88.063 g/mole

$$\frac{\text{Molar mass of the molecule}}{\text{Molar mass of the empirical formula}} = \frac{176.13\ g/\text{mole}}{99.063\ g/\text{mole}}$$

= 2.0000

Therefore, two empirical formulas comprise the molecular formula, and the molecular formula of ascorbic acid is $C_6H_8O_6$.

CHEMICAL EQUATIONS

An equation is a convenient method for representing a chemical change by using formulae and symbols. we have already represented a number of chemical reactions by using "word equations", and the process of building up equations will be understood by converting a number of these into normal chemical equations.

A **chemical equation** *specifies the substances used up (reactants) and the substances formed (products) in a chemical reaction,* as established by experimentation. By convention, formulas for the reactants, written on the left side of the equation, and formulas for the products, written on the right side of the equation, are separated by a single arrow $\longrightarrow$, a double arrow $\rightleftharpoons$, or an equal sign =. Thus an *unbalanced* chemical equation describing the reaction between solid iron and copper sulphate solution to form solid copper ,Cu, and $FeSO_4$ solution, is written as follows.

$$Fe(s) = CuSO_4\ (aq) \longrightarrow Cu(s) + FeSO_4\ (aq)$$

The subscript notation (s) indicates the solid phase and (aq) the aqueous phase.

How to Construct Equations

(a) Mercury (II) oxide $\longrightarrow$ Mercury + Oxygen.

To construct a chemical equation for this reaction, the formula for each substance involved must be known:

Mercury (II) oxide. . . . . HgO
Mercury . . . . .Hg
Oxygen. O2

A provisional equation is then written:

$$HgO \longrightarrow Hg + O_2$$

This equation does not "balance", i.e. the numbers of atoms and molecules involved must be so adjusted that the total numbers of each type of atom on either side of the equation are equal. To give two atoms of oxygen, two molecules of mercury (II) oxide are required. These will also give two atoms of mercury. Therefore the equation becomes;

$$2HgO \longrightarrow 2Hg + O_2$$

i.e., two molecules of mercury (II) oxide two atoms of mercury and one molecule of oxygen. Note that the figure 2 in 2HgO applied to the whole molecule, but that when a small figure is used as a subscript, e.g. in $PbCl_2$ and $Pb(NO_3)_2$, it applies only to the atom or ion preceding it. You might be tempted to alter a formula as an easy method of making an equation balance. *This must never be done. A particular compound has a definite and unalterable formula.*

(b) Potassium chlorate → Potassium chloride + Oxygen.

Inserting the formulate for these substances we get:

$$KClO_3 \rightarrow KCl + O_2$$

To balance this equation there must be an even number of oxygen atoms on the left-hand side of the equation (why?). Therefore it is possible that two molecules of potassium chlorate will be required.

Trying this, we see that two molecules of potassium chloride and three of oxygen will be obtained, and the equation then balances.

$$2KClO_3 \rightarrow 2KCl + 3O_2$$

(c) Iron + Copper (II) sulphate → Copper + Iron (II) sulphate.

Inserting the formulae we get:

$$Fe + CuSO_4 \rightarrow Cu + FeSO_4.$$

The equation balances without further adjustment.

(d) Hydrogen + Oxygen → Water.

Inserting the formulae we get:

$$H_2 + O_2 \rightarrow H_2O.$$

Two molecules of water will be formed from one molecule of oxygen. This means that two molecules of hydrogen will be required Therefore the balanced equation is:

$$2H_2 + O_2 \rightarrow 2H_2O.$$

(e) Zinc + Hydrochloric acid → Zinc chloride + Hydrogen.

Inserting the formulae we get:

$$Zn = HCl \rightarrow ZnCl_2 + H_2$$

To give one molecule of zinc chloride, two atoms of chlorine, obtainable from two molecules of hydrochloric acid, are required. The equation then becomes:

$$Zn + 2HCl \rightarrow ZnCl_2 + H_2$$

This now balances.

The construction of equations becomes easy with practice, and to help with this your teacher will no doubt provide a number of examples for you to do. Do not try to learn correct equations in a parrot fashion. There are so many equations that have to be known that this method is never successful. A knowledge of the chemical reaction concerned, and of the way equation are built up, is all that is required.

State Symbols

Substances can exist as solids, liquids or gases and all three states can take part in chemical reactions.

Thus in the reaction:

$$2H_2 + O_2 \rightarrow 2H_2O$$

two gases (hydrogen and oxygen) are reacting to form a liquid (water). In the reaction:

$$2HgO \rightarrow 2Hg + O_2$$

a solid (mercury (II) oxide) is reacting to give two gases (mercury vapour and oxygen).

It is often desirable to show the state (solid, liquid or gas) of the substances represented by a chemical equation. This is easily done by using the symbols s, for solid, I, for liquid, and g, for gas, and putting these symbols (known as *state symbols*) in brackets after the formula.

Thus the above two equations become:

$$2H_2(g)+O_2(g) \rightarrow 2H_2O(1)$$

and

$$2HgO(s) \rightarrow 2Hg(g) + O_2(g).$$

Very often we study reaction in solution in water (aqueous solutions). For example:

$$Zn + 2HCl \rightarrow ZnCl_2 + H_2$$

is a reaction between solid zinc and an aqueous solution of hydrogen chloride (*i.e.* hydrochloric acid) to give an aqueous solution of zinc chloride and hydrogen gas. The hydrogen chloride here is not a solid, a liquid or a gas but it is in aqueous solution, and a state symbol for aqueous solution is needed. The symbol used is (aq).

So the equation is:

$$Zn(s) + 2HCl(aq) \rightarrow ZnCl_2(aq) + H_2(g)$$

It must be remembered that many substances can exist as solids, liquids or gases depending on their temperature. Hence water may be formed as liquid water, but very often it is formed as steam (gaseous water). Hence the state symbol used may change if the reaction is carried out at a different temperature.

The reaction

$$2H_2(g) + O_2(g) \rightarrow 2H_2\ O(l)$$

is usually carried out by burning hydrogen in oxygen, and there is no doubt that in the hot flame the water formed is a gas. The above equation is better written.

$$2H_2(g) + O_2(g) \rightarrow 2H_2O(g)$$

Another commonly met situation where the substance formed is not in its usual (i.e. at room temperature) state is when metals are extracted from their ores. Here high temperatures are often used and the metal is produced as a liquid. In such a situation the appropriate equation would be, for example:

$$Fe_2O_3(s) + 3CO(g) \rightarrow 2Fe(l) + 3CO_2(g).$$

Calculations Involving Equations

(a) Moles

Chemical equations are of immense value to the chemist, as they tell him combining ratios of different substances and the ratios of the products formed.

Consider the equation:

$$Fe + CuSO_4 \rightarrow Cu + FeSO_4$$

This equation says that *1 atom of iron* reacts with *1 molecule of copper (II) sulphate* to give *1 atom of copper* and *1 molecule of iron (II) sulphate.*

It also says that *1 mole of iron* reacts with *1 mole of copper (II) sulphate* to give *1 mole of copper* and *1 mole of iron (II) sulphate.*

There is 1 mole of each of the reactants and 1 mole of each of the products. *This 1 : 1 ratio does not always occur.*

For example, consider now the equation:

$$2H_2 + O_2 \rightarrow 2H_2).$$

This equation says that 2 molecules of hydrogen react with one molecule of oxygen to give 2 molecules of water, or that *2 moles of hydrogen* react with *one mole of oxygen* to give *2 moles of water.*

The equation always tells us how many *moles* of each *reactant* are needed and how many *moles* of each *product* are formed. Of course, to be correct the equation must be properly balanced. It is therefore essential for you to learn how to balance chemical equations as we have just described.

(b) Masses

As the equation tells us the number of moles of reactants and of products we can, by using atomic masses, calculate the masses of reactants and products.

Consider the equation:

$2HgO$	$\rightarrow 2Hg$	$+O_2$
2 molecules	2 atoms	1 molecule
2 moles	2 moles	1 mole

The molar mass of mercury (II) oxide is (201 + 16) g mol^{-1} = 217 g mol^{-1}.

The molar mass of mercury is 201 g mol^{-1}.

(2×217) g mercury (II) oxide → (2×201) g mercury + 32 g oxygen. i.e.., 434 g mercury (II) oxide → 402 g mercury + 32 oxygen.

***Example* 3.19**

Calculate the masses of each substance formed when 10 g of calcium carbonate react with excess of hydrochloric acid. (Ca = 40, C = 12, O = 16, CI = 35.5, H = 1.)

First a balanced equation must be produced and then the molecular masses added:

$$\begin{array}{ccccccccc} CaCO_3 & + & 2HCl & \rightarrow & CaCl_2 & + & H_2O & + & CO_2 \\ \underbrace{40+12+(3\times16)}_{100} & + & \underbrace{2(1+35\cdot5)}_{73} & \rightarrow & \underbrace{40+(2\times35\cdot5)}_{111} & + & \underbrace{(2\times1)+16}_{18} & + & \underbrace{12+(2\times16)}_{44} \end{array}$$

i.e., 1 mole (100g) of calcium carbonate reacts with 2 moles (73g) of hydrochloric acid to give 1 mole (111g) of calcium chloride, 1 mole (18g) of water and 1 mole (44g) of carbon dioxide.

Therefore, by simple proportion,

10 g calcium carbonate give $\frac{111}{100} \times 10 = 11.1$ g calcium chloride,

$\frac{18}{100} \times 10 = 1.8$ g water,

and $\frac{44}{100} \times 10 = 4.4$ g carbon dioxide.

Example 3.20

What mass of lead (II) nitrate must be heated to give 22.3 g of lead (II) oxide? (Pb = 207, N = 14, O = 16).

The equation is:

$$\underbrace{\underset{2[207+2\{14+(3\times16)\}]}{2Pb(NO_3)_2}}_{662} \rightarrow \underbrace{\underset{2(207+16)}{2PbO}}_{446} + \underbrace{\underset{4[14+(2\times16)]}{4NO_2}}_{184} + \underbrace{\underset{(2\times16)}{O_2}}_{132}$$

In this problem we are concerned only with the masses of lead (II) oxide and lead (II) nitrate. There was, therefore, no need to work out the molar masses of the other substances.

From the equation.

2 moles (446 g) of lead (II) oxide are obtained form 2 moles (662g) of lead (II) nitrate.

22.3 g of lead (II) oxide are obtained from $\frac{662}{446} \times 22{\cdot}3$ g of lead (II) nitrate

= <u>33.1 g of lead (II) nitrate</u>

Example 3.21

What is the maximum mass of silver chloride that can be precipitated from 8·5 of silver nitrate by the addition of hydrochloric acid? (Ag = 108, N = 14, O = 16, Cl = 35.5).

The equation is:

$$\underbrace{\underset{[108+14+(3\times16)]}{AgNO_3}}_{170} + HCl \rightarrow \underbrace{\underset{(108+35{\cdot}5)}{AgCl}}_{143{\cdot}5} + HNO_3$$

From the equation:

170 g of silver nitrate yield 143·5 g of silver chloride

8·5 g of silver nitrate yield $\frac{143{\cdot}5}{170} \times 8{\cdot}5$ g of silver choloride

= <u>7·17 g silver chloride</u>

***Example* 3.22**

Calcium carbonate which is insoluble in water, has a mass of 15 g. It is placed in a solution containing 12·6 g of nitric acid. What mass of crystal will remain after reaction has ceased? (Ca = 40, C= 12, O = 16, N = 14, H = 1).

The equation is :

$$\underbrace{\overset{CaCO_3}{[40+12+(3\times16)]}}_{100} + \underbrace{\overset{2HNO_3}{2[1+14+(3\times16)]}}_{126} \rightarrow Ca(NO_3)_2 + H_2O + CO_2$$

From the equation:

126 g of nitric acid react with 100 g of calcium carbonate

12·6 g of nitric acid react with $\frac{12\cdot6}{126}\times100$ g of calcium carbonate

= 10 g of calcium carbonate

Mass of calcium carbonate remaining = (15—10) g = 5 g

***Example* 3.23**

The enzyme mixture called zymase, present in bread yeast, catalyzes the fermentation of glucose, $C_6H_{12}O_6$, to ethyl alcohol, C_2H_5OH, and gaseous carbon dioxide, CO_2, as represented by the following balanced chemical equation.

$$C_6H_{12}O_6 \xrightarrow{\text{zymase}} 2C_2H_5OH + 2CO_{2(g)}$$

Determine how many molecules of CO_2 and what mass of C_2H_5OH are produced from 8.01×10^{20} $C_6H_{12}O_6$ molecules.

The balanced equation indicates that two CO_2 molecules are produced for each $C_6H_{12}O_6$ molecule consumed.

$$(8.01\times10^{20}\ \cancel{C_6H_{12}O_6\ \text{molecules}})\left(\frac{2CO_2\ \text{molecules}}{1\ \cancel{C_6H_{12}O_6\ \text{molecule}}}\right)$$

$=1.60\times10^{21}$ CO_2 molecules

The same number of C_2H_5OH molecules, 1.60×10^{21}, is produced. Determine the number of moles of C_2H_5OH molecules produced by dividing the number of C_2H_5OH molecules by the Avogadro number.

$$\text{Number of moles of } C_2H_5OH \text{ molecules} = \frac{1.60\times10^{21}\ \text{molecules}}{6.023\times10^{23}\ \text{molecules / mole}}$$

$= 2.66\times10^{-3}$ mole of C_2H_5OH molecules

Determine the mass of C_2H_5OH produced by multiplying the number of moles of C_2H_5OH molecules by the molar of a C_2H_5OH molecule.

$$\text{Mass of } C_2H_5OH = (2.66 \times 10^{-3} \text{mole})(46.069 \text{g/mole}) = 0.123 \text{ g of } C_2H_5OH$$

Example 3.24

Solid potassium nitrate, KNO_3, a common constituent in chemical fertilizers, decomposes on heating to solid potassium nitrate, KNO_2, and gaseous oxygen, O_2, (a) Write a balanced chemical equation for this reaction. (b) Determine how many moles of oxygen molecules and what mass of potassium nitrite formed from complete decomposition of 18.0 grams of potassium nitrate.

(a) Write and unbalanced chemical equation.

$$KNO_{3(s)} \rightarrow KNO_{2(s)} + O_{2(g)}$$

By inspection the equation is balanced if a coefficient of one-half is placed before O_2. However, since whole-number coefficients are desired, multiply coefficients through by two and obtain

$$2KNO_{3(s)} \rightarrow 2KNO_{2(s)} + O_{2(g)}$$

(b) Determine the number of moles of KNO_3 in 18.0 g since problems are most easily solved in terms of moles.

$$\text{Moles of } KNO_3 = \frac{\text{mass of } KNO_3}{\text{molar mass of } KNO_3} = \frac{18.0 \text{ g}}{101.105 \text{ g/mole}} = 0.178 \text{ mole of } KNO_3$$

The balanced equation indicates that one mole of O_2 forms for each two moles of KNO_3 consumed.

$$(0.178 \text{ mole of } KNO_3)\left(\frac{1 \text{ mole of } O_2}{2 \text{ moles of } KNO_3}\right) = 0.0890 \text{ mole of } O_2$$

The balanced equation also indicates that one mole of KNO_2 forms from one mole of KNO_3. Therefore, 0.178 mole of KNO_2 is produced. Determine the mass of KNO_2 produced by multiplying the number of moles of KNO_2 produced by the molar mass of KNO_2.

$$\text{Mass of } KNO_2 = (0.178 \text{ mole})(85.106 \text{ g/mole}) = 15.1 \text{ g of } KNO_2$$

Theoretical Yield, Actual Yield, and Percent Yield

The assumption that reactants are converted *completely* to desired products in a chemical reaction, frequently is not valid. There are several reasons for this. A state of equilibrium may develop such that significant amounts of reactant remain at the end of the reaction time. Side reactions may occur, to produce undesired products. An excess of one reactant may be used purposely in order to convert a second more expensive or difficulty prepared reactant quantitatively to products.

In the last step in the commercial preparation of aspirin excess acetic anhydride, $C_4H_6O_3$, along with sulphuric acid as a catalyst, is added to salicylic acid, $C_7H_6O_3$, to produce aspirin, $C_9H_8O_4$, and acetic acid, $C_2H_4O_2$.

$$C_7H_6O_3 + C_4H_6O_3 \rightarrow C_9H_8O_4 + C_2H_4O_2$$

The **theoretical yield** of aspirin in this reaction is *the maximum amount that can be produced based on the* ***limiting*** *reactant,* the one not in excess.

Example **3.25**

What is the theoretical yield of aspirin, $C_9H_8O_4$, if 5.00 ml of acetic anhydride, $C_4H_6O_3$ (density = 1.082 g/ ml), is stirred with 3.00 grams of salicylic acid, $C_7H_6O_3$?

Determine the number of moles of each of the reactants and which is the limiting reactant.

$$\text{Mass of } C_4H_6O_3 = (\text{volume})(\text{density}) = (5.00\,\cancel{\text{ml}})(1.082\text{ g}/\cancel{\text{ml}}) = 5.41\text{ g}$$

$$\text{Number of moles of } C_4H_6O_3 = \frac{\text{mass of } C_4H_6O_3}{\text{molar mass of } C_4H_6O_3} = \frac{5.41\ \cancel{g}}{102.090\ \cancel{g}/\text{mole}} = 0.0530 \text{ mole of } C_4H_6O_3$$

$$\text{Number of moles of } C_7H_6O_3 = \frac{\text{mass of } C_7H_6O_3}{\text{molar mass of } C_7H_6O_3} = \frac{3.00\ \cancel{g}}{138.123\ \cancel{g}/\text{mole}} = 0.0217 \text{ mole of } C_7H_6O_3$$

The balanced equation indicates that $C_7H_6O_3$ and $C_4H_6O_3$ react in a 1:1 mole ratio. Thus, $C_7H_6O_3$, with the smaller number of moles, is the limiting reactant, and the theoretical yield of aspirin that can be produced must be based on 0.0217 mole of $C_7H_6O_3$. The balanced equation also indicates that one mole of aspirin is produced for each mole of $C_7H_6O_3$ consumed. Therefore, 0.0217 mole of aspirin is produced.

$$\begin{aligned}\text{Mass of } C_9H_8O_4 &= (\text{number of moles of } C_9H_8O_4)(\text{molar mass of } C_9H_8O_4)\\ &= (0.0217\ \cancel{\text{mole}})(180.161\ \text{g}/\cancel{\text{mole}})\\ &= 3.91\ \text{g of aspirin}\end{aligned}$$

The **actual yield** of a product is *that recovered experimentally.* It is almost invariably less than the theoretical yield because of an unfavorable equilibrium, because of side reactions, or because losses may occur in separation and purification. Percent yield, defined as

$$\text{Percent yield} = \left(\frac{\textit{actual yield}}{\textit{theoretical yield}}\right)(100\%)$$

is used to express the degree of conversion of reactants to products.

***Example* 3.26**

Determine the percent yield of aspirin in Example 3.25 if 3.42 grams of aspirin is recovered.

$$\text{Percent yield} = \left(\frac{\text{actual yield}}{\text{theoretical yield}}\right)(100\%)$$

$$= \left(\frac{3.42\ \text{g}}{3.91\text{g}}\right)(100\%) = 87.5\%$$

4

Atomic Structure and the Periodic Table

DALTON'S ATOMIC THEORY

In 1808 John Dalton published his *New System of Chemical Philosophy*. In it he proposed an atomic theory based on the eighteenth-century quantitative laws discussed above and on numerous subsequent experiments. The basic postulates of his theory along with illustrative examples follow.

1. *Elements are composed of extremely small particles called atoms.* The element lead is made up of lead atoms.
2. *Atoms of a given element exhibit identical chemical properties; atoms of different elements exhibit different chemical properties.* All lead atoms exhibit identical chemical properties and all copper atoms exhibit identical chemical properties; however, lead atoms and copper atoms can be differentiated by their distinctly different chemical properties.
3. *No atom of one element is changed into an atom of another element in an ordinary chemical reaction.* When iron atoms are heated with sulfur atoms, only iron atoms and sulfur atoms are present in the iron (II) sulfide (FeS) product that is formed.
4. *Compound substances are formed when an integral number of atoms of one element combine with an integral number*

of atoms of one or more other elements. The relative numbers of atoms of different elements will be fixed in a given pure compound. In the compound iron pyrites there is always are atom of copper for every atom of iron and every two atoms of sulfur.

A test of any theory is how well it explains observed facts or scientific laws derived from these facts. We can understand the two quantitative law discussed previously on the basis of Dalton's atomic theory.

1. *Law of conservation of mass.* According to Dalton's postulate, atoms are conserved in an ordinary chemical reaction; thus, there can be no change in mass.
2. *Law of constant composition.* According to Dalton's fourth postulate, the relative numbers of atoms of different elements are constant. The relative masses or percents by mass must also be constant.

EVIDENCE OF SUBATOMIC PARTICLES

Electrolysis studies by Humphrey Davy and Michael Faraday, two British chemists, in the early part of the nineteenth century had indicated that matter was electrical in nature. However, only in the latter part of that century did gas discharge tube experiments provide concrete evidence for the existence of charged particles.

During the last half of the 1800's, several significant discoveries were made that showed that the atom was not the ultimate particle but itself was composed of subatomic particles. The initial discovery along these lines came from investigations using evacuated electrical discharge tubes of a design shows in Fig 4.1

William Crookes and others had demonstrated that a beam of rays traveled between the electrodes of a tube when the tube was evacuated. Furthermore, the beam could be *deflected* by a negative electric charge. Because like charges repel each other, this implied that the beam carried a negative charge.

Using data collected by others and from his own experiments, J. J. Thomson (1897) concluded that the beam consisted of tiny particles each bearing a unit of negative charge. The particles were alter named electrons. The mass of an electron was determined to be very small—about $\frac{1}{1837}$ that of the lightest atom (hydrogen).

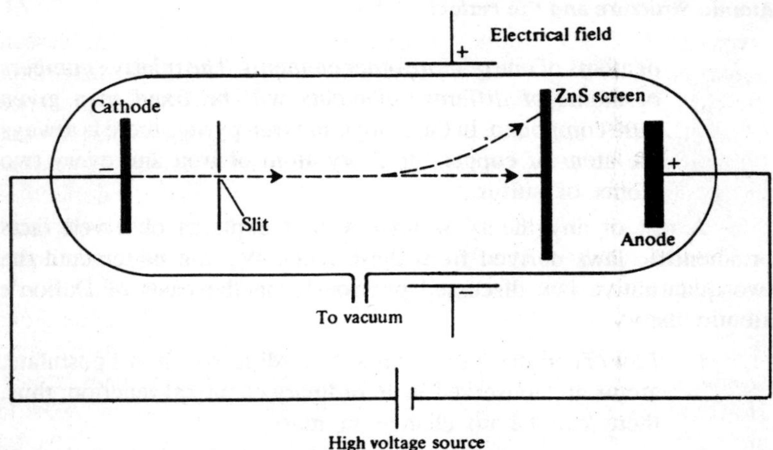

Figure 4.1 : Gas Discharge Tube for Studying Electrons

In somewhat more sophisticated experiments, Joseph Thomson, a British physicist, in 1897 and Robert Millikan, and American nuclear physicist, in 1909 determined the charge/mass ratio for an electron and the charge of an electron, respectively. From these results it was determined that the mass of an electron is 9.110×10^{-28} gram. It is assigned a relative charge of —1.

PROTONS

Gas discharge tubes of slightly different construction (Fig 4.2) led to the discovery of protons. When a high voltage was passed

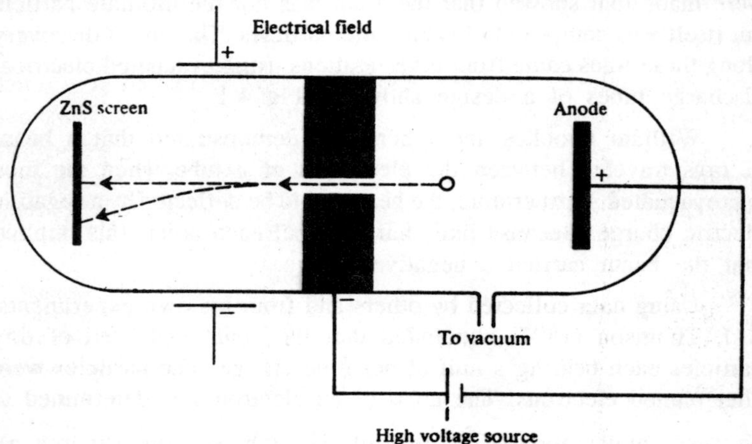

Figure 4.2 : Gas Discharge Tube for Studying Protons

across such a tube, flashes were noted on the zinc sulfide screen placed on the *opposite* side of the cathode from the anode. Furthermore, the rays were deflected toward the *negative* plate in the presence of an electrical field and the degree of deflection *depended* on what residual gas was present in the tube. When hydrogen gas was present in the tube, there was significant deflection, whereas the deflection was markedly less with oxygen gas or nitrogen gas. Such behaviour can be understood as follows. Electrons, emanating from the cathode, collided with residual gas molecules on the way to the anode. Some of these electrons had high enough energy to eject electrons from the gas molecules, thereby forming positive ions. Any positive ions would be accelerated toward the cathode since unlike charges attract. However, some positive ions would pass through the perforated cathode and impinge on the zinc sulfide screen. The positive ions were deflected on their way to the screen if an electrical field were present, the deflection being greater for positive ions of smaller mass.

Experiments similar to Thomson's experiment, requiring the use of both electrical and magnetic fields, could be used to determine the charge/mass ratio for the positive ions. Further, if they were formed by ejection of an electron from a neutral species, the positive ions could be assumed to have a charge of the *same* magnitude as, but *opposite* in sign to, that of the electron, and the mass of these positive ions could be determined. Protons (H^+ ions), which resulted if hydrogen were the residual gas into the tube, were found to be very much heavier than the electron (about 1836 times, or 1.673×10^{-24} gram).

NEUTRONS

Although the existence of neutrons was postulated very much earlier, it was not until 1932 that James Chadwick, a British physicist, was able to find direct evidence for such particles. The appropriate apparatus is shown in Figure 4.3. When Li, Be, or B sheets were bombarded with dipositively charged alpha particles (helium atoms

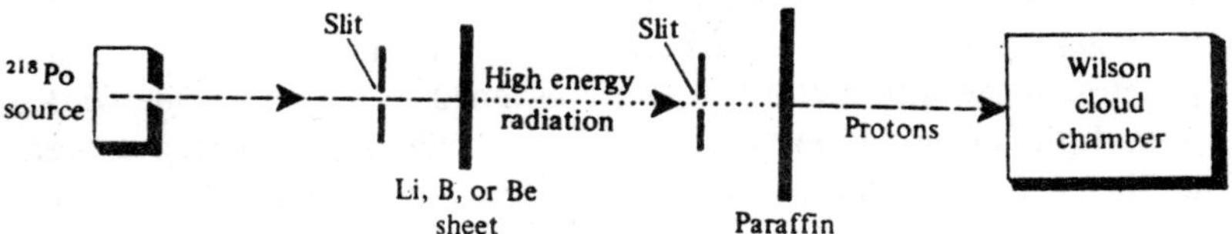

Figure 4.3 : Apparatus for Studying Neutrons.

from which the electrons have been removed), high energy particles were emitted that cold penetrate a 1 inch lead sheet and lose only half of their intensity. The path of these particles was unaffected by an electrical field, and thus the particles were uncharged. These particles were originally assumed by several workers to be a form of gamma radiation; however, they were much too energetic. It was observed that these particles ejected protons from a block of paraffin wax placed in their path. Protons were indicated by the fact that the ejected particles were deflected toward the negative plate of an electrical field and left a chain of water droplets in a Wilson cloud chamber. Chadwick suggested the name "neutron" and postulated that the mass of a neutron was similar to the mass of a proton. If this were the case, a neutron could eject a proton from the paraffin by direct collision. Francis Aston, a British scientist, later determined the mass of a neutron (1.675×10^{-24} gram) by means of a mass spectrometer.

The Nuclear Atom

In 1898, following his measurement of the charge/mass ratio for electrons, Thomson proposed a structure for an atom. According to his model the atom was a sphere of positive electricity containing negative electrons much as seeds are embedded within the pulp of a grape. Since protons and other cations had been found to be much more massive than electrons, Thomson postulated that most of the mass of an atom was associated with the positive electricity. Positively charged ions could presumably be formed by stripping away one or more electrons, thereby leaving most of the mass of the atom and a resultant positive charge.

An alpha-particle-scattering experiment, performed in 1911 by Ernest Rutherford, a British physicist, disproved the Thomson model and placed the concept of a nuclear atom on a firm basis. The important features of Rutherford's experiment are diagramed in Fig 4.4. A beam of alpha particles resulting from the radioactive decay of polonium was directed onto a thin metal foil by collimation through a pinhole in a lead plate. Most of the very energetic alpha particles easily penetrated the thin metal foil and continued to a zinc sulfide screen with little or no deflection. However, some alpha particles were deflected at quite large angles and a few were even *reflected* back from the foil. Rutherford was astounded by these results—"it was almost as incredible as if you fired a 15 inch shell at a piece of tissue paper and it came back and hit you".

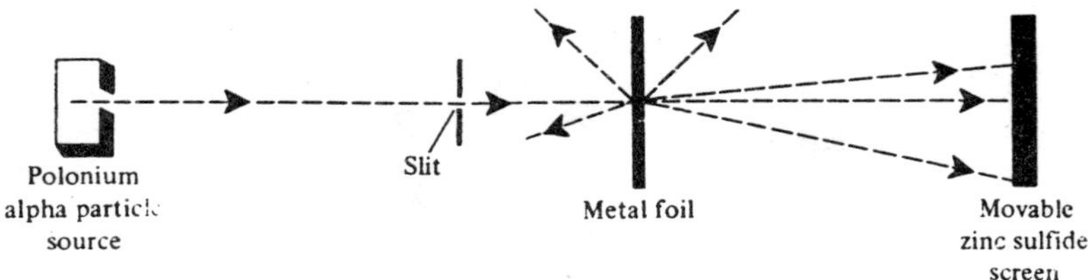

Figure 4.4 : Rutherford's alpha-particle-scattering experiment.

Rutherford recognized that an atom according to the Thomson model, with its mass and positive charge spread uniformly, would not provide significant repulsion of positively charged alpha particles and that, contrary to his experimental results, all alpha particles should pass through the metal foil with little deviation from their original path. However, Rutherford further recognized that if an atom's *positive electricity and mass* were concentrated in a very small region, most of the alpha particles would pass through with no deflection, but a few would directly encounter a concentration of positive charge (immovable because of its high mass) and would be repelled strongly and deflected considerably from their original path. These ideas are illustrated in Fig 4.5. Rutherford thus postulated that an atom has a very small nucleus in which positive charge and mass are concentrated and a relatively empty extranuclear region containing electrons. Quantitative alpha-particle-scattering studies indicated that the nucleus had a diameter of about 10^{-13} cm, whereas the diameter of an atom was close to 10^{-8} cm.

Atomic Number

Moseley, a student in Rutherford's laboratory, further clarified the concept of a nuclear atom and, at the same time, suggested a rational periodic law that is the basis of our modern periodic table. Using vacuum tubes like those used previously in gas discharge tube experiments, moseley bombarded replaceable anodes (positive electrodes) with electrons and determined the frequencies (number of pulses passing a given point per second) of x-rays emitted from the target anodes. He found a linear relationship between the square root of the frequency (for the highest-frequency x-ray of a series emitted for each element) and the order in which the element appeared in Mendeleev's periodic table. His result are plotted in Fig 4.6. Two significant observations emerge from his plot. First, gaps appeared

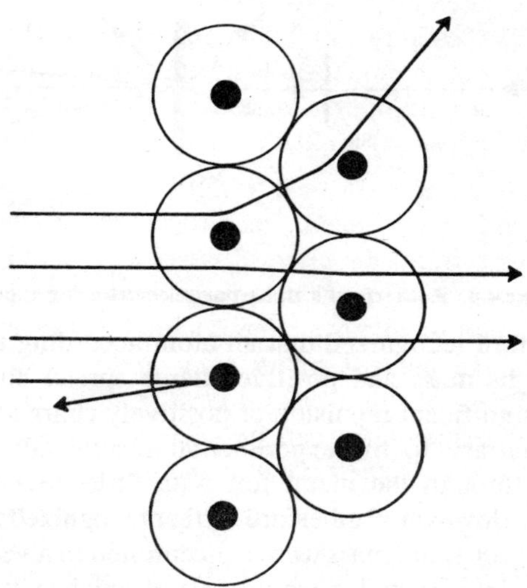

Figure 4.5 : Deflection of alpha particles by a nulcear atom.

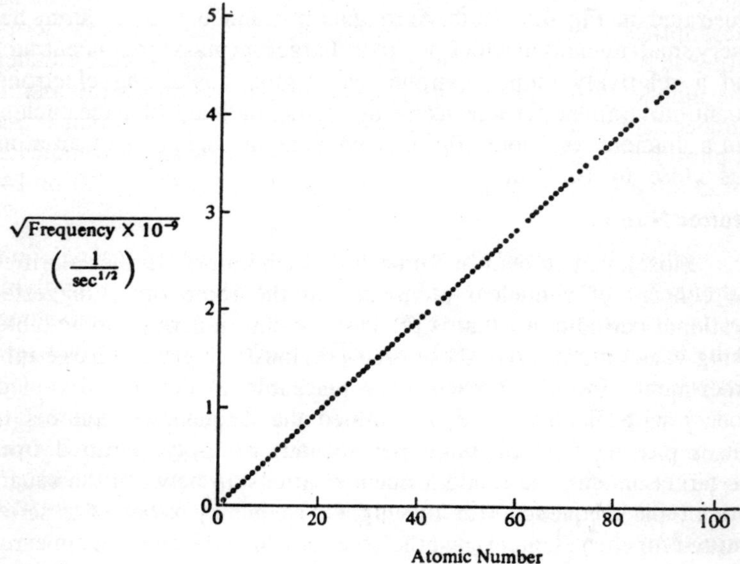

Figure 4.6 : Moseley's plot of x-ray frequencies versus atomic number.

corresponding to elements still undiscovered, but predicated earlier by Mendeleev. Second just as in Mendeleev's periodic table, Moseley found that iron, cobalt, and nickel and antimony, tellurium, and iodine (also chlorine, argon, and potassium) appeared in those orders on his graph in opposition to the order of increasing atomic mass.

On the basis of these observations, Moseley postulated, "there is in the atom a fundamental quantity which increases by regular steps as we pass from one atom to the next. This quantity can only be the charge on the central positive nucleus." Moseley called this quantity the atomic number. The **atomic number** is *the number of protons in the nucleus of an atom* and is also *the number of extranuclear electrons in a neutral atom.* On the basis of Moseley's work, a new periodic law can be stated—*properties of atoms are periodic functions of their atomic numbers.*

Isotopes

Isotopes are *atoms that have the same atomic number but different masses.* Most of the mass of an atom is in its nucleus which is composed of protons and neutrons. *Isotopes of a given atom have the same number of protons but different numbers of neutrons.* Isotopes are often designated using an *atomic number* (number of protons) subscript and a *mass number* (number of protons and neutrons) superscript, followed by the chemical symbol for the atom of that atomic number. The number of neutrons equals the mass number minus the atomic number. For example, $^{31}_{15}P$ indicates an atom with 15 protons, that is, phosphorus, and (31 — 15) or 16 neutrons. Likewise, $^{238}_{92}U$ indicates an atom with 92 protons, that is, uranium, and (238 — 92) or 146 neutrons.

All the isotopes of an element undergo similar chemical reactions although the rates of these chemical reaction (especially those involving bonds of hydrogen to another element) may differ. Only the isotopes of hydrogen have been given specific names — $^{1}_{1}H$, protium; $^{2}_{1}H$, deuterium; and $^{3}_{1}H$, tritium.

Isotopes are responsible for the fact many element have non integral atomic masses. For example, copper has two naturally occurring isotopes. Copper's atomic mass of 63.546 amu results from the average of $^{63}_{29}Cu$ and $^{65}_{29}Cu$ weighted appropriately for their natural abundances of 69.09% and 30.91%, respectively.

Isotopes are also responsible for the fact that atomic mass is not directly related to atomic number, for the fact that chlorine, argon, and potassium; iron, cobalt, and nickel; and antimony, tellurium, and iodine appear in the periodic table in order of increasing atomic number, but not in the order of increasing atomic mass. An overabundance of higher mass isotope or a lower mass isotope of an element (a function of nuclear stability) can occasionally cause an average atomic mass that is higher or lower than expected.

Atomic Spectra

The relationship of electron configurations in the extranuclear region of an atom and properties of the atom were discovered largely through interpretation of atomic spectra. A simple spectroscope, having the components shown in Fig 4.7, can be used for illustration. A *continuous spectrum* or a rainbow of colors is obtained on the film in Fig 4.7a if a beam from a white-light source such as the sun is dispersed by passage through a glass prism. Lower-energy or lower-frequency components (red) are bent the least and appear at the lower right edge of the film in Fig 4.7a; higher-energy or higher-frequency components (blue) are bent the most and appear at the upper left edge of the film in Fig 4.7a. If, on the other hand, the source is colored light emitted by a heated gaseous sample, a series of narrow lines of color (called a *line spectrum*) rather than a continuous blend of colors is obtained on the film. Thus, hydrogen atoms give four lines in the visible region of the spectrum as shown in Fig 4.7b. Furthermore, different elements can be shown to have their own characteristic pattern of lines. Interpretation of such line spectra resulted in models for the extranuclear region of the atom. Building such a model was no easy task. It could be likened to your trying to construct a harp when you had never seen one and your only knowledge of such an instrument was that you had heard tones emitted from it than its strings were plucked by a three-year-old child living in the apartment above yours.

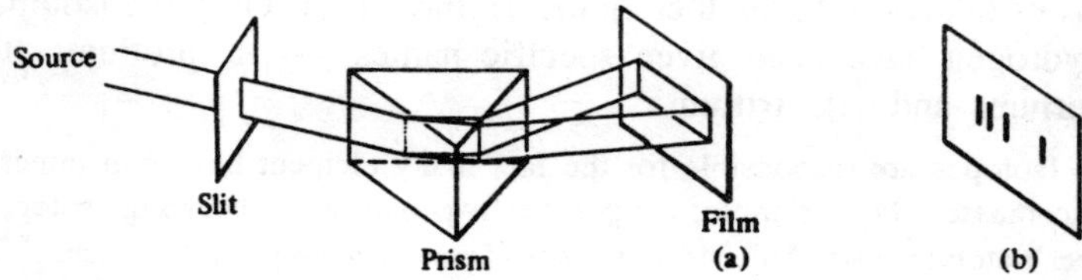

Figure 4.7 : Components of a simple spectroscope

In 1913 Niels Bohr, a Danish physicist, developed the first reasonable model for the electronic structure in the extranuclear region of the hydrogen atom. He explained four basic atomic spectral observations as follows.

1. *Heated gaseous atoms emitted a spectrum.* The gaseous atoms absorbed heat energy and were raised to higher energy states. Upon their return to lower energy states light energy was emitted. It was postulated that electrons in atoms were responsible for the spectra, that electrons were excited to higher energy states and then emitted light upon returning to lower energy states. A continuous spectrum (Fig 4.7a) would be expected by this mechanism assuming a continuous range of energy states existed.
2. *Line spectra were observed* (Fig 4.7b). Bohr postulated that electrons were present only in certain allowed (quantized) energy levels and that no electrons could have energies lying between energy levels. These energy levels represented the energies of electrons that were pictured a moving in circular paths called orbits around the nucleus. The frequencies of the emitted spectral lines were a measure of the spacing of these energy levels.
3. *Line spectra were characteristic for each element.* The number of protons (equal to the total positive charge in the nucleus and to the number of electrons in a neutral atom) differs for atoms of unlike elements. The attractive forces between the nucleus and electrons would then vary and could be expected to alter the spacings of the energy levels for atoms of different elements.
4. *Energy levels were closer together at higher energies.* If energy levels were equally spaced, only one spectral line would be observed rather than a series of spectral lines. Bohr developed a mathematical relationship that accounted for energy level spacings in the hydrogen atom.

QUANTUM MECHANICAL MODEL OF THE ATOM

Niels Bohr proposed that the energy of an electron in an atom is quantized; *i.e.* an electron has a definite, specific amount of energy. Depending on their energy, electrons in atoms occupy regions of energy called energy levels (shells). The lowest energy level is the *K* shell. Electrons of higher energy occupy higher energy levels such as *L, M, N,* etc. The change of an electron's energy is also quantized. An electron can gain or lose only a specific amount of energy in order to reach another energy level. Calculations of electron transitions based upon these postulates agreed with experimental data for the hydrogen atom. The mathematical relationships developed by Bohr to explain electron transitions generated the concept that in a hydrogen atom the electron circled the nucleus at certain fixed distances, much like a miniature solar system.

During the 1920's, experimental data indicated that energy levels are subdivided into sublevels. In searching for a new set of mathematical expressions that would permit calculating the energies of electron transitions between sublevels, physicists deduced that an electron's position and energy could not be *simultaneously* determined. The idea is based upon considering the electron as a particle. Alternatively, Louis de Broglie proposed that electrons exhibit wave behavior and studies using this approach would be fruitful. Using these concepts, physicists derived equations related to wave behavior (wave equations) that permitted calculation of electron transition energy in agreement with experimentally observed values. This new wave theory of atomic structure is called *quantum mechanics.**

Orbitals, Electrons, and Quantum Numbers

Central to this theory is the concept that the exact path along which an electron moves cannot be determined. Instead, for an electron having a particular energy, the wave equation allow calculation of the *probability* of the electron's location in relation to the nucleus. Considered over a period of time, this probability gives a close approximation of an electron's behaviour. The probability is related to a pattern called an *electron cloud.* The volumed and shape of the electron cloud about a nucleus where an electron having a *particular allowed energy* is most likely to be found is dependent on an electron's energy.

An electron possessing a given amount of energy occupies a given orbital in an atom. Because there are many allowed energies an electron can have, an orbital is present for each allowed energy. The energy of a electron in an atom is described by a set of four integers that are solutions to wave equations. These four values are

* The principal originators of quantum mechanics were Louis de Broglie, Werner Heisenberg, Erwin Schrodinger, and Paul Dirac, each of whom received a Nobel Prize for his contributions to this field.

called *quantum numbers* and are symbolized by *n, l,* m_l, and m_s collectively, these values relate to the orbital in which an electron is located.

The various energy levels are designated by numbers 1, 2,.........in order of increasing energy. Each of these numbers is called the *principal quantum number n* for that energy level. Each energy level actually contains sublevels. The number of sublevels in an energy level is equal to the principal quantum number of that level.

l Value	Sublevels
0	s
1	p
2	d
3	*f*

Sublevels differ in energy and other characteristics; this will be discussed shortly. Four types of sublevels—*s, p, d,* and *f*—are sufficient in describing all the elements currently known. Other sublevels in the fifth and higher energy levels are theoretically possible. Sublevels are designated by a second quantum number, *ℓ*. The general shape of the electron cloud for the electron is determined by *ℓ*. The following assignments are made:

Energy Level's Principal Quantum Number, *n*	Number of Sublevels Within the Energy Level
1	1
2	2
3	3
.	.
.	.
.	.

The relationship between *n* and *l* is *l* = O, 1, 2, 3,. . . . (n — 1). An electron occupying the *s* orbital of the first energy level is designated is 1s electron.

The energy of the lower-lying sublevels is shown in Fig. 4.8.

Energy and Electron distribution

The number of orbitals each type of sublevel contains varies depending on the sublevel. An *s* sublevel, regardless of its energy

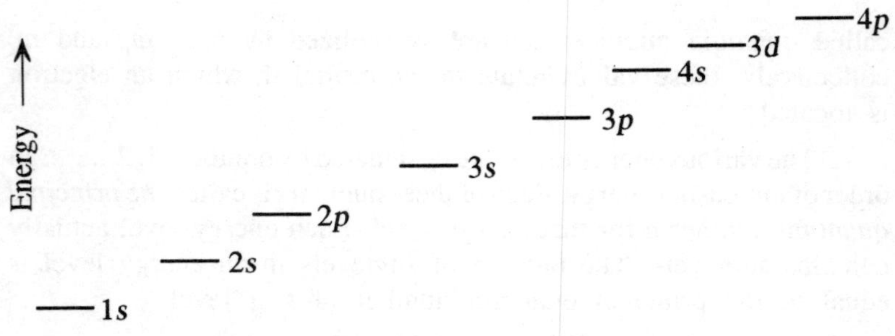

Figure 4.8 : Relative energies of lower-lying sublevels

level, contains only one orbital; *p* sublevels have three orbitals; *d* sublevels have five orbitals, and *f* sublevels have seven orbitals. The total number of orbitals in a given energy level equals n^2.

Energy Level's Principal Quantum Number, *h*	Number of Sublevels in Energy Level	Types of Sublevels $l = 0, 1, 2, ...(n—1)$	Total Number of Orbitals, n^2, in Energy Levels
1	1	$l = 0$; s	$1 = 1^2$
2	2	$l = 0$; *s*....................1 $l = 1$; *p*....................3	$4 = 2^2$
3	3	$l = 0$; *s*....................1 $l = 1$; *p*....................3 $l = 2$; *d*....................5	$9 = 3^2$
4	4	$l = 0$; *s*....................1 $l = 1$; *p*....................3 $l = 2$; *d*....................5 $l = 2$; *f*....................5	$16 = 4^2$

Each orbital can be occupied by a *maximum* of two electrons. Since n^2 gives the total number of orbitals in a given energy level, $2n^2$ equals the *maximum* number of electrons that a given energy level can accommodate.

$$\frac{2 \text{ electrons}}{\text{orbital}} \times \frac{n^2 \text{ orbitals}}{\text{energy level}} = \frac{2n^2 \text{ electrons}}{\text{energy level}}$$

This relationship is summarized in Tables 4.1 and 4.2

Table 4.1
Electron Distribution in the First Four Energy Levels

n	Type of Sublevle	Number of Orbitals in Sublevel	Maximum Number of Electrons in Sublevel	Maximum Number of Electrons in Energy Level
1	1*s*	1	2	2
2	2*s*	1	2	8
	2*p*	3	6	
3	3*s*	1	2	18
	3*p*	3	6	
	3*d*	5	10	
4	4*s*	1	2	32
	4*p*	3	6	
	4*d*	5	10	
	4*f*	7	14	

Table 4.2
Energy levels in atoms

n, principal quantum number of level	1	2	3	4	5	6	7
Maximum number of electrons (theoretical)	2	8	18	32	50	72	98
Number actually found in nature	2	8	18	32	32	10	2

The electron distribution of an atom is such that electrons occupy orbitals of the lowest energy sublevel available. If this sublevel is filled (already occupied by the maximum number of electrons allowed), additional electrons then occupy sublevels of higher energy. Fig 4.8 allows us to predict the order of filling of subshells for atoms of an element. This order is summarized in Table 4.3 for the first 20 elements. Note that the energy of the 4*s* sublevel is slightly lower than that of the 3*d*. Therefore, the 4*s* is filled before the 3*d* in potassium and calcium. The order of filling for a particular atom is called the atom's electron configuration. The superscript number refers to the number of electrons present in the sublevel.

Table 4.3
Electron configuration of elements 1 to 20

Element	*Atomic Number*	*Electron Configuration*	
Hydrogen	1	$1s^1$	
Helium	2	$1s^2$	
Lithium	3	$1s^22s^1$	(recall only two elctrons may occupy an s subshell)
Beryllium	4	$1s^22s^2$	
Boron	5	$1s^22s^22p^1$	
Carbon	6	$1s^22s^22p^2$	
Nitrogen	7	$1s^22s^22p^3$	
Oxygen	8	$1s^22s^22p^4$	
Fluorine	9	$1s^22s^22p^5$	
Neon	10	$1s^22s^22p^6$	(p sublevels can hold a maximum of six electrons)
Sodium	11	$1s^22s^22p^63s^1$	
Magnesium	12	$1s^22s^22p^23s^2$	
Aluminium	13	$1s^22s^22P^63s^23P^1$	
Silicon	14	$1s^22s^22P^63s^23p^2$	
Phosphorus	15	$1s^22s^22p^63s^23p^3$	
Sulfur	16	$1s^22s^22p^63s^23p^4$	
Chlorine	17	$1s^22s^22p^63s^23p^5$	
Argon	18	$1s^22s^22p^63s^23p^6$	
Potassium	19	$1s^22s^22p^63s^23p^64s^1$	
Calcium	20	$1s^22s^22p^63s^23p^64s^2$	

The *l* quantum number determines the general shape of the electron cloud. The third quantum number, m_l, relates to the orientation of the electron cloud in space. m_l and *l* are related: m_l = *l*, *l*—1, *l*—2,........, O, —1, —2,........, —*l*.

l	0	1	2	3
Sublevel	*s*	*p*	*d*	*f*
Number of Orbitals	1	3	5	7
Orbital m_l Assignment	1	1 0 −1	2 1 0 −1 −2	3 2 1 0 −1 −2 −3

The Pauli Exclusion Principle and Hund's Rule

The fourth quantum number, m_s, is related to the spin of an electron. An electron can be imagined to be spinning on its axis. Two directions of spin are possible—clockwise and counterclockwise. To occupy and orbital simultaneously, two electrons must have *opposite* spins—one clockwise and the other counterclockwise. The m_s values +½ and –½ respectively. We shall use the notation of an upward arrow, ↑, to denote +½; a downward arrow ↓, denotes –½.

Taken collectively, a set of the four quantum numbers uniquely defines an electron in any atom. In an atom, each electron has its own set of quantum numbers; i.e., no two electrons in an atom can have the *same set* of quantum numbers. This is called the *Pauli exclusion principle* in honor of the physicist Wolfgang Pauli who first proposed it. Table 4.4 summarizes the quantum number relationships. Several examples are given below.

Table 4.4

Allowed values of quantum number sets for electrons—first. three energy levels of atoms

n	l	m_l	m_s
1	0	0	+½ or –½
2	0	0	+½ or –½
2	1	1	+½ or –½
2	1	0	+½ or –½
2	1	–1	+½ or –½
3	0	0	+½ or –½
3	1	1	+½ or –½
3	1	0	+½ or –½
3	1	–1	+½ or –½
3	2	2	+½ or –½
3	2	1	+½ or –½
3	2	0	+½ or –½
3	2	–1	+½ or –½
3	2	–2	+½ or –½

***Example* 4.1**

In its lowest energy condition, the electron in a hydrogen atom is in the orbital of an s sublevel of the first shell, *i.e.* $1s^1$. The principal quantum number n equals l and because this is an s sublevel orbital. l equals zero and m_l is zero;

$$1s \underset{m_l = 0}{\uparrow} \quad -\frac{1}{2}$$

m_s is +½, Therefore the set 1, 0, +½ is assigned to this electron.

Example **4.2**

A helium atom has an electron configuration of $1s^2$. Each electron has a unique set of quantum numbers: The first is 1, 0, 0, +½; the other is 1,0,0 –½;

$$+\frac{1}{2} \qquad\qquad -\frac{1}{2}$$

$$1s \ \underline{\uparrow\downarrow} \quad m_l = 0$$

Example **4.3**

The *electron configuration for the lithium atom is* $1s^2 2s^1$. The sets of quantum numbers would be 1, 0, 0, +½; 1, 0, 0, –½; 2, 0, 0, +½.

$$2s \ \underline{\uparrow} \quad m_l = 0$$

$$1s \ \underline{\uparrow\downarrow} \quad m_l = 0$$

Example **4.4**

A boron atom's electron configuration is $1s^2 2s^2 2p^1$. For the electron in the *p* sublevel, the correct interpretation for its set of quantum numbers is $n = 2$; *p* sublevels have an l value of 1: there are three orbitals in a *p* sublevel—the first has an m_l value of 1: the first electron in an orbital is assigned an m_s value of –½. Therefore, the $2p^1$ electron's set of quantum numbers is 2, 1, 1, +½.

$$2p \ \underset{1}{\underline{\uparrow}} \ \underset{0}{\underline{\quad}} \ \underset{-1}{\underline{\quad}} = m_l$$

$$2s \ \underset{0}{\underline{\uparrow\downarrow}} = m_l$$

$$1s \ \underset{0}{\underline{\uparrow\downarrow}} = m_l$$

One further restriction is placed on electron occupancy of an orbital. In an uncombined atom, orbitals of the same sublevel are of equal energy, *e.g.*, the three orbitals of a *p* sublevel. *Each* orbital will acquire an electron before any of these orbitals acquires a second electron. This generalization is commonly called *Hund's rule*. In accordance with the Pauli exclusion principle, the second electron must have a spin opposite that of the first electron. The following examples illustrate Hund's rule and the buildup of electron occupancy of available orbitals.

Example **4.5**

Nitrogen atoms have $1s^2 2s^2 2p^3$ configuration.

$2p$ ↑ ↑ ↑
1 0 −1 = m_l

$2s$ ↑↓
$m_l = 0$

$1s$ ↑↓
$m_l = 0$

Note that each of the *p* orbitals is occupied by an electron.

Before periodic classifications, chemistry was a jumble of unrelated facts. During the 18th century the ancient art of *alchemy* had developed to the point where some alchemists (who we can call early scientists) were no longer only concerned with converting other elements into gold but were making careful observations of how different substances behaved. Then in the 19th century, following John Dalton, atomic masses of the elements were being determined.

It was from attempts to relate the chemical and physical properties of the elements and their compounds to their atomic masses that the periodic table was born. Dmitri Mendeleev, a Russian chemist, whose work was first published in 1869 and who could be considered the father of our modern periodic table.

A number of scientists had tried to develop systematic relationships among the known chemical elements by plotting various physical properties of the elements as a function of atomic mass Unaware of the efforts of most of these earlier workers, Mendeleev developed the modern periodic table by arranging the elements in order of increasing atomic mass and also in columns of elements with similar chemical characteristics. His insight in interpreting even the rather inaccurate data available to his was phenomenal. His ideas concerning the relationship of properties of group of elements an the logical patterns into which these elements fall have provided the basis for chemical progress ever since.

Four aspects of primary importance stand out in Mendeleev's work. First, he stated a periodic law—the properties of elements are periodic functions of their atomic masses. Second, he recognizes (as is reflected in our present periodic table) that hydrogen was somehow unique. Third, he suggested that chemical properties were more important in determining periodicity than were atomic masses. Thus, as shown in the classical short Mendeleev table Fig 4.9, he realized that Sb, Te, and I should fall in that order under As, Se, Br, even though order of atomic masses suggested Sb, I, and Te. This was because the chemical properties of Te were more like those of Se than those of Br, and the chemical properties of I were more like those of Br than those of Se. Likewise, Fe, Co, and Ni should be placed in that order, even though Co had a greater atomic mass than Ni, because the chemical properties of Co were more like those of Fe than were the chemical properties of Ni. Fourth, he made predictions on the basis of his table. Using the *unoccupied* positions in his table as a basis, not only did he predict the existence of at least a dozen undiscovered elements, but he also predicted properties of some of these elements. He realized that the properties of an undiscovered element would be intermediate between those of the four elements surrounding it in his table. With the discovery of gallium in 1875, scandium in 1879, and germanium in 1886, the accuracy of Mendeleev's predictions was quickly verified. A comparison of predicted and experimentally measured properties of germanium, shown in Table 4.5, illustrates Mendeleev's remarkable insight into relationships among properties of elements, particularly considering the fact that only three (Si, As, and Sn) out of the four elements surrounding Ge were known at the time.

Table 4.5
Predicted and Observed Properties of Germanium

Property	*Predicted*	*Observed*
Atomic mass (amu)	72	72.59
Atomic volume (cm^3)	13	13.6
Density (g/cm^3)	5.5	5.35 (20°)
Specific heat	0.073	0.076
Color	dirty gray	grayish white
Product of heating in air	white GeO_2	white GeO_2
GeO_2 properties	refractory solid	refractory solid
Molar volume (cm^3)	22	22.16
Density (g/cm^3)	4.7	4.228 (25°)
$GeCl_4$ properties	colorless liquid	colorless liquid
Molar volume (cm^3)	113	113.35
Density (g/cm^3)	1.9	1.844 (30°)
Boiling point (°C)	100	84

Period	Series	Group 1 R_2O	Group 2 RO	Group 3 R_2O_2	Group 4 RH_4 RO_2	Group 5 RH_2 R_2O_5	Group 6 RH_2 RO_3	Group 7 RH R_2O_7	Group 8 – RO_4
1	1	H=1							
2	2	Li=7	Be=9.4	B=11	C=12	N=14	O=16	F=19	
3	3	Na=23	Mg=24	Al=27.3	Si=28	P=31	S=32	Cl=35.5	
4	4	K=39	Ca=40	–=44	Ti=48	V=51	Cr=52	Mn=55	Fe=56 Co=59 Ni=59 Cu=63
	5	(Cu=63)	Zn=65	–=68	–=72	As=75	Se=78	Br=80	
5	6	Rb=85	Sr=87	?Yt=88	Zr=90	Nb=94	Mo=96	–=100	Ru=104 Rh=104 Pd=106 Ag=108
	7	(Ag=108)	Cd=112	In=113	Sn=118	Sb=122	Te=125	I=127	
6	8	Cs=133	Ba=137	?Di=138	?Ce=140	–	–	–	– – – – – –
	9	(—)	–	–	–	–	–	–	
	10	–	–	?Er=178	?La=180	Ta=182	W=184	–	Os=195 Ir=197 Pt=198 Au=199
	11	(Au=199)	Hg=200	Ti=204	Pb=207	Bi=208			
7	12	–	–	–	Th=231	–	U=240	–	– – – – – –

Fig. 4.9 Mendeleev's Periodic Table of 1871

Mendeleev's predictions convinced even the most skeptical of the value of his periodic classification. It only remained for henry Moseley, a British physicist, to find the explanation for the several atomic mass irregularities in Mendeleev's table and to develop a totally consistent periodic law. In order to appreciate Moseley's work more clearly, a number of experiments that indicated that atoms are divisible and electrical in nature should be examined.

Electronic Structure

We have seen earlier that the different elements have different numbers of electrons and that these are arranged in energy shells. The outermost shells are of highest energy, i.e. they are the most reactive. *It is these outermost electrons that take part in chemical reactions and determine the chemical properties of the elements.* It is, therefore, the *electronic structure of the atom* that is the fundamental feature on which the periodic table is based. O course,, Mendeleev did not know this, the electron not being discovered until 1897, 28 years after he had proposed his periodic table.

A periodic table of the form used these days is illustrated in Fig. 4.10. In this we have given the *names* and *symbols* of the elements together with their *atomic numbers* and *relative atomic masses.*

The Alkali Metals

Within a group the elements are similar.

Let us consider the elements of group 1, lithium, sodium, potassium, rubidium and caesium. They are all very reactive metals with a valency of one. Their electronic structures are:

	Atomic number	Number of Electrons in Shell (n)					
		1	2	3	4	5	6
Lithium	3	2	1				
Sodium	11	2	8	1			
Potassium	19	2	8	8	1		
Rubidium	37	2	8	18	8	1	
Caesium	55	2	8	18	18	8	1

Thus an atom of each of these elements has one electron in the outer shell, *i.e.* this electron is of higher energy, and therefore more easily removed, than the others. If it is removed the remaining species is a positively charged ion *e.g.*:

$$Na \rightarrow Na^{+} + e,$$

and the electronic structure now has 8 electrons in its highest energy shell. This, as we have seen, is a particularly stable number of electrons. The size of the atom increases with increase in the atomic number, hence the outer electron moves progressively further away from the nucleus on passing for lithium to caesium. Also the increasing number of inner electrons increasingly screen this outer electron. Because of these two factors it becomes progressively easier to remove the outer electron on moving from lithium to caesium, i.e. *the reactivity of the group ℓ elements increases on going down the group.* Therefore potassium reacts more vigorously with water than does sodium, and caesium reacts with explosive violence.

The hydroxides of the metals in group 1 all dissolve in water to give alkaline solutions, hence their name *alkali metals* (see section (*f*) for a definition of a metal).

The Halogens

Fluorine, chlorine, bromine an iodine are, like the alkali metals, reactive elements with a valency of one. This, too, is expected from their electronic structures.

	Atomic number	Number of Electrons in Shell (n) 1	2	3	4	5
Fluorine	9	2	7			
Chlorine	17	2	8	7		
Bromine	35	2	8	18	7	
Iodine	53	2	8	18	18	7

An atom of each of these elements has seven electrons in its outer shell. They are just one short of the stable octet, and they react by gaining an electron, e.g.: $Cl + e \rightarrow Cl^{-}$, to give a negatively charged ion.

Like the alkali metals, and the elements in all other groups, the size of the atom increases as the group is descended. The ability to attract electrons is greatest in the small fluorine atom and least with iodine, hence *the reactivity of the group 7 elements decreases on going down the group.*

All of these elements readily form salts with the elements in the groups (1, 2 and 3). Hence their name halogens (salt formers).

The Noble Gases

Helium, neon, argon, krypton and xenon are five elements that were not known to Mendeleev. They are very unreactive and form very few compounds. This lack of reactivity and the fact that they are gases means that they are found principally in the atmosphere. (Some of these elements are also found with natural gas in certain parts of the world). Their discovery did not take place until air could be verified during the early part of this century, although helium had been observed in the spectrum of the sun some years before.

Their lack of reactivity is readily understood in terms of their electronic structures as they all have a completely full outer octet.

Hydrogen

The first element in the periodic table, hydrogen, does not have properties that enable it to be placed satisfactorily in any group. This is due to two unique features. (i) The highest energy shell of hydrogen atom can hold only two electrons, in contrast to all the others (except helium) that can hold 8 or more (ii) When hydrogen loses its electron the ion formed, H^+ , is a bare nucleus. This is very very small in contrast with a positive ion of any other element which must have some electrons still surrounding the nucleus.

Metals and Non-metals

(i) We are now able to give a definition of metals and non-metals. *Metals* are those elements in the early groups of the periodic table.

Their atoms *tend to lose electrons* to form positive ions.

Non-metals are those elements in the later groups of the periodic table. Their atoms *tend to gain electrons* to form negative ions, rather than to lose electrons.

Metals and *non-metals* have other characteristics, such as metals being good conductors of electricity and non-metals being poor conductors of electricity.

(ii) Metallic characteristics increase as a group is descended. We have already commented that the ease of forming a positive ion (a metallic characteristic) increases from lithium to caesium, and the ease of forming a negative ion (a nonmetallic characteristic)

decreases from fluorine to iodine. If a group near the middle of the periodic table is studied this increase in metallic character on descending group is even more striking. Consider group 4.

Carbon — definitely a non-metal

Silicon — also a non-metal, but looks somewhat metallic.

Germanium — difficult to decide on inspection. It looks like a metal, but its structure is the same as carbon and silicon (it has a diamond structure)

Tin — definitely a metal

Lead — definitely a metal.

(iii) As a period is crossed there is a gradual change in the proprieties of the elements from metallic to non-metallic, e.g. in period 3 (Na, Mg, Al, Si, P, S, Cl Ar).

Sodium — a very reactive metal with a valency of 1 (*e.g.* NaH and Na_2O).

Magnesium — a reactive metal with a valency of 2(*e.g.* MgO).

Aluminium — a less reactive metal with a valency of 3 (*e.g.* Al_2O_3).

Silicon — an unreactive non-metal with a valency of 4 (e.g. SiO_2, SiH_4).

Phosphorus — a reactive non-metal with valencies of 3 (*e.g.* PH_3) and of 5 (*e.g.*P_2O_5).

Sulphur — a reactive non-metal with valencies of 2(*e.g.* H_2S) and 6 (*e.g.* SO_3).

Chlorine — a very reactive non-metal with a usual valency of 1(*e.g.* HCl) but sometimes 7 (*e.g.* Cl_2O7).

Argon — a noble gas with no known compounds.

Notice how the valency alters as the period is crossed, and how the valencies are connected with the number of electrons in the outer shell.

The Transition Elements

The elements, as you can see from the periodic table Fig 4.10 do not occur until period 4 and they are all *metals.* They show a close resemblance to one another. Generally they are hard, lustrous and have high melting points. Most of then have several valencies, and their ions in solution are coloured (in contrast to ions of all the other elements that are colourless). Many of these metals are useful as catalysts.

Figure 4.10 : The Periodic Modern Table

KEY: Relative atomic mass → 6355; Atomic number → 29Cu ← Symbol; Cooper ← Name

1.008 1H Hydrogen

Group → Period ↓	1	2	←				Transition Metals					→	3	4	5	6	7	0
																		4.003 2He Helium
2	6.939 3Li Lithium	9.012 4Be Beryllium											10.811 5B Boron	12.01 6C Carbon	14.00 7N Nitrogen	16.00 8O Oxygen	19.00 9F Fluorine	20.18 10Ne Neon
3	22.99 11Na Sodium	24.31 12Mg Magnesium											26.98 13Al Aluminium	28.09 14Si Silicon	30.97 15P Phosphorus	32.06 16S Sulphur	35.45 17Cl Chlorine	39.95 18Ar Argon
4	39.10 19K Potassium	40.08 20Ca Calcium	44.96 21Sc Scandium	47.90 22Ti Titanium	50.94 23V Vanadium	52.00 24Cr Chronium	54.94 25Mn Manganese	55.85 36Fe Iron	58.93 27Co Cobalt	58.71 28Ni Nickel	63.55 29Cu Copper	65.37 30Zn Zinc	69.72 31Ga Gallium	72.59 37Ge Germanium	74.92 33As Arsenur	76.96 34Se Ealenium	79.90 35Br Bonrine	83.80 36Kr Kypton
5	85.47 37Rb Rubidium	87.62 38Sr Stronium	88.91 39Y Yttrium	91.22 40Zr Zirconium	92.91 41Nb Niobium	95.94 42Mo Molybde-num	98.91 43Tc Technetium	101.1 44Ru Ruthenium	102.9 45Rh Rhodium	106.4 46Pd Palladium	107.9 47Ag Silver	112.4 48Cd Cadmium	114.8 49In Indium	118.7 50Sn Tin	121.8 51Sb Antimony	127.6 52Te Tellurium	126.9 53I Iodine	1313 54Xe Xenon
6	132.9 55Cs Caesium	137.3 56Ba Barium	(see below) 57.71	178.5 71Hf Hafnium	180.9 73Ta Tantalum	183.9 75Re Rhenium	186.2 75Re Rhenium	190.2 26Os Osnium	192.2 77Ir Iridium	195.1 18Pt Platinum	197.0 79Au Gold	200.6 80Hg Mercury	204.4 81Ti Thallium	207.2 82Pb Lead	209.0 83Bi Bismuth	12101 84Po Polonium	12101 35At Astatine	12221 86Rn Radon
7	(223) 87Fr Francium	226.0 88Ra Radium	(see below) 89.103															

Lanthanides →	138.9 57La Lanthanum	140.1 58Ce Cerium	140.9 59Pr Praseody-mium	144.2 60Nd Neodymium	(147) 61Pm Promethium	150.4 62Sm Samarium	152.0 63Eu Europium	157.3 64Gd Gadolinium	158.9 65Tb Teribium	162.5 66Dy Dysprosium	164.9 67Ho Holmium	167.3 68Er Erbium	168.9 68Tm Thulium	173.0 70Yb Yiterbium	175.0 71Lu Lutetium
Actinides →	(227) 89Ac Actinum	232.0 90Th Thorium	231.0 91Pa Protactinium	238,0 92U Uranium	237.0 93Np Neptunium	(242) 94Pu Plutonium	(243) 95Am Americium	(247) 96Cm Curium	(247) 97Bk Berkelium	(251) 98Cf Californium	(254) 99Es Einsteinium	(253) 100Fm Fermium	(256) 101Md Mendele-vium	(254) 102No Nobelium	(257) 103Lw Lawrencium

Atomic Masses, Atomic Numbers and the Mendeleev Classification

Mendeleev, as we have already explained, did not put all the elements in the position which seemed correct according to the atomic masses, but he put them where their chemical properties best fitted. He believed that such *atomic masses* had been inaccurately determined and would, in time, be corrected. In most cases this happened, but there are still four pairs of elements whose positions would need reversing to place them in order of their atomic masses. It should be clear to you by now that *atomic mass* is not the fundamental property on which the periodic table is based. The *fundamental property* is the *electronic structure*, or the *charge on the nucleus*. This is given by the *atomic number*. It was fortunate for Mendeleev that the order of atomic masses and atomic numbers are almost identical. Otherwise his classification would not have been possible.

Physical Properties of the Elements in Periods 1-3

At the same time that Mendeleev was working on his periodic table, the German chemist, Julius Lothar Meyer, also noticed the way in which certain physical properties of the elements varied with their atomic masses. He plotted curves of the values of various properties in the same way that Mendeleev had done with both chemical and physical properties.

Fig. 4.11–4.14 are a modern version of Lothar Meyer's curve. In them it can be seen that elements with similar chemical properties e.g. lithium and sodium, occupy similar positions on the curves. Notice that: *(i)* The melting points and boiling points rise as the period is crossed until the element in group IV is reached. The melting points and boiling points of the elements in the other groups are very low. *(ii)* The densities rise to group III and IV. After group IV they are of low value. *(iii)* The molar volumes fall as group III is approached. Then they rise again as the period is crossed.

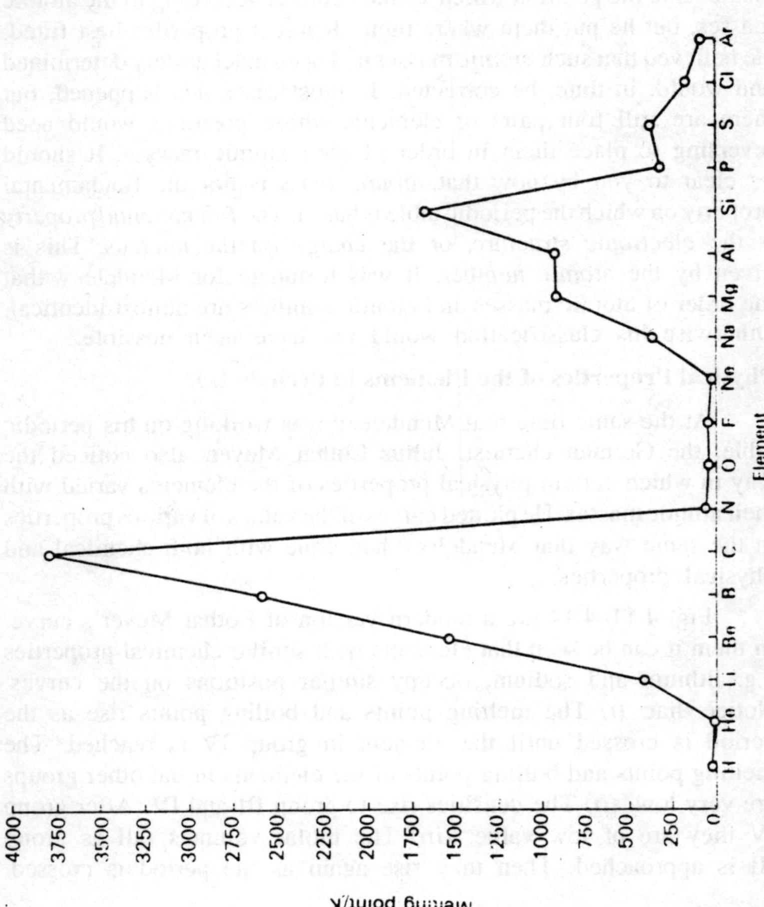

Fig 4.11 The melting points (K) of elements with atomic numbers 1-20.

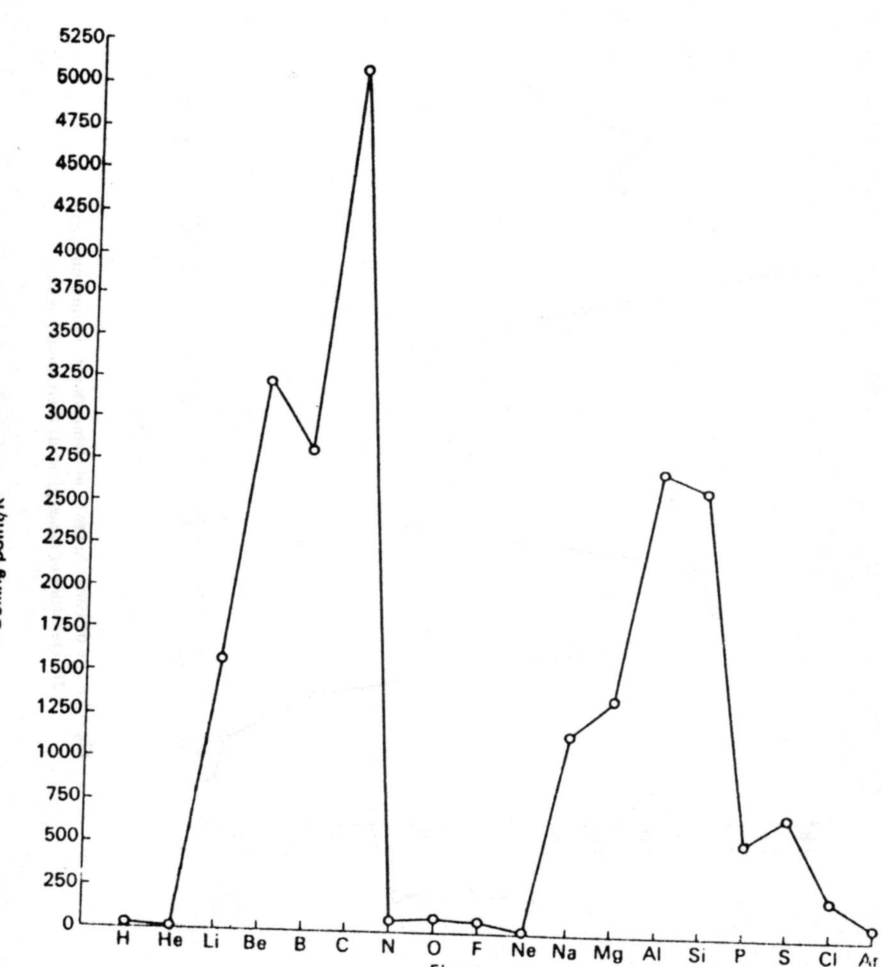

Fig 4.12 The boiling points (K) of elements with atomic numbers 1-20.

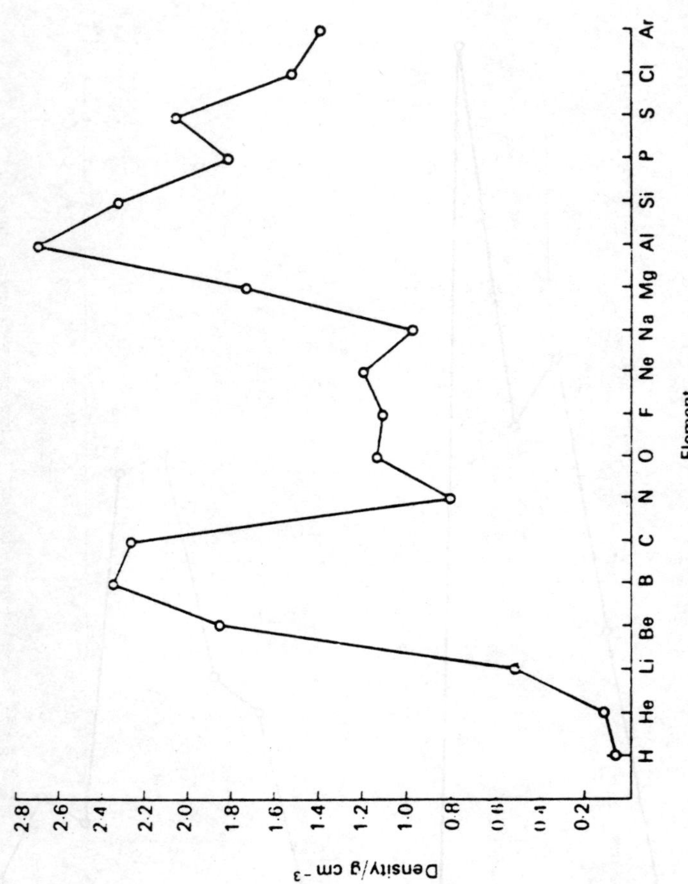

Fig 4.13 The densities (g cm $^{-3}$ of elements with atomic numbers 1-20 (Densities of gases given at their boiling points)

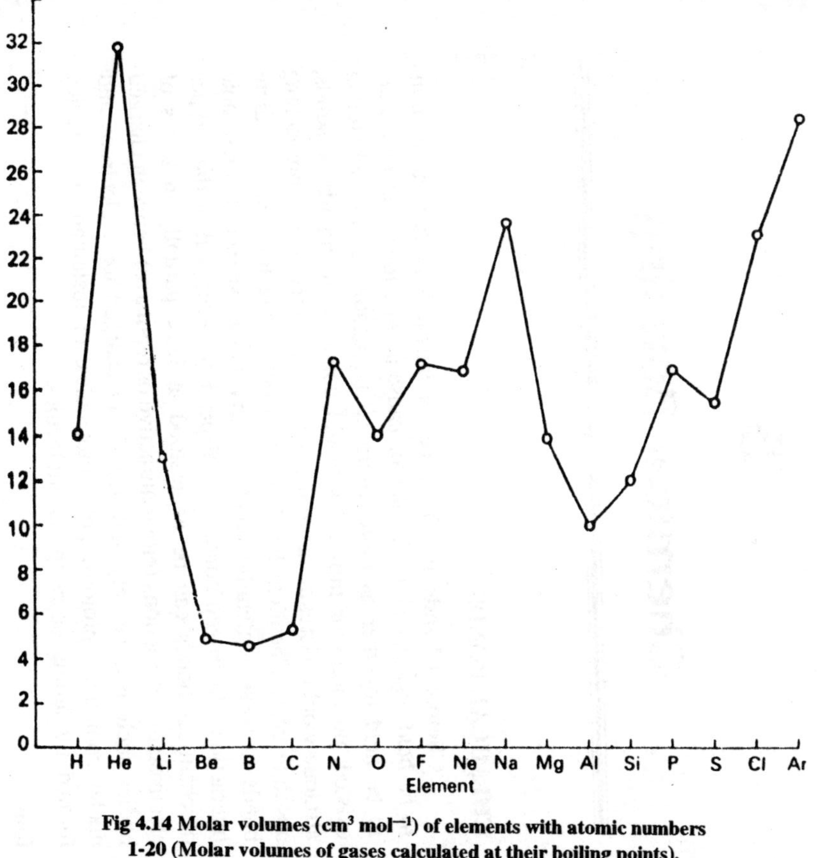

Fig 4.14 Molar volumes ($cm^3\ mol^{-1}$) of elements with atomic numbers 1-20 (Molar volumes of gases calculated at their boiling points).

5

Chemical Bonding

CHEMICAL BONDS

Chemical bonds are *the forces or interactions that cause atoms to be held together as molecules or cause atoms, ions, and molecules to be held together as more complex aggregates.* If no such forces existed, no atoms or molecules would condense to liquids or solids; all atoms would exist as monatomic gases, even at temperatures approaching absolute zero. Chemical bonds range from strong ionic and covalent bonds to weaker metallic bonds to very much weaker dipole-dipole interactions, hydrogen bonds, and dispersion forces. All of these types of chemical bonds can be understood at least partially in terms of electrostatic interactions (either attractive or repulsive) between charged bodies such as positively and negatively charged ions, electrons and nuclei containing protons, polar molecules, or instantaneous dipoles formed by atoms or nonpolar molecules.

Ions

An ion is *an atom or distinguishable group of atoms bearing one or more positive or negative charges due to the loss or gain of one or more electrons.* Common table salt, sodium chloride, is composed of Na^+ and Cl^- ions. These ions are formed from single atoms. Ammonium nitrate consists of ammonium ions (NH_4^+) and nitrate ions (NO_3^-). these ions are formed from distinguishable groups of atoms. *A positively charged ion* such as Na^+ or NH_4^+ is called a **cation**. A *negatively charged ion* such as Cl^- or NO_3^- is called an **anion**.

The formation of a cation requires a *loss* or removal of one or more electrons from a neutral atom. The **ionization energy** is *the energy required to remove an electron from a gas phase atom or ion.* It is measure of the hold that an atom or cation has on a given electron. This hold result from the electrostatic attraction between the positive nuclear charge and the negative charge of the electron. The first ionization energy for Mg is the energy required for the following process.

$$Mg_{(g)} \longrightarrow Mg^{+}_{(g)} + e^{-}$$

This step involves removal of an electron from a neutral gas phase Mg atom to form a Mg^{+} cation. A second step, involving removal of an electron from a positively charged gas phase cation, requires energy equivalent to the second ionization energy, and can be represented as

$$Mg^{+}_{(g)} \longrightarrow Mg^{2+}_{(g)} + e^{-}$$

With few exceptions, first ionization energies increase across periods and decrease down groups of the periodic table. In addition, second, third, and higher ionization energies are successively greater than first ionization energies. Thus, the first three ionization energies for magnesium are 176, 347 and 1848, kcal/mole, respectively. Note that it is much more difficult to remove electrons from an inner shell—as indicated by the high third, ionization energy—than it is to remove electrons from the outer shell—as indicated by the lower first and second ionization atoms only lose outer-shell electrons.

The formation of an anion requires a *gain* of one or more electrons by a neutral atom. **Electron affinity** is *the energy released when an electron is added to a gas phase atom or ion.* It is a measure of the hold that an atom or anion has on an *additional* electron. This hold results from the electrostatic attraction between the positive nuclear charge and the negative charge of the electron. The first and second electron affinities for oxygen are the energies released in the following steps.

$$O_{(g)} + e^{-} \longrightarrow O^{-}_{(g)}$$

$$O^{-}_{(g)} + e^{-} \longrightarrow O^{2-}_{(g)}$$

Being dependent upon the same factors, electron affinities show approximately the same trends as ionization energies—they increase across periods and decrease down groups. However, they are not easily measured experimentally, and accurate values for only a few elements are known. Second, third, and higher electron affinities are successively much less favorable than first electron affinities, and atoms gain at most two or three electrons in ordinary chemical reactions.

Many commonly occurring ions have noble gas or filled-shell electron configurations. Atoms frequently gain or lose sufficient electrons to attain noble gas configurations.

Ionic Bonds

The simplest kind of electrostatic attraction, that between two oppositely charged ions, gives rise to an ionic bond. An **ionic bond** results from *an electrostatic attraction between a positively charged ion* (cation) *and a negatively charged ion* (anion). The formation of an ionic bond involves three steps:

1. The loss of one or more electrons by one element to form a cation (requires energy).
2. The gain of one or more electrons by another element to form an anion (releases energy).
3. The attraction of the cation and anion formed (releases energy).

An ionic bond is favored when an atom with low ionization energy and an atom with high electron affinity combine. Sodium chloride, formed by the reaction of Na atoms and Cl atoms, is an example of an ionic compound. Little energy is required to remove an electron form a gaseous Na atom to form a Na^+ cation in step 1 since Na has a very low ionization energy for the process

$$Na_{(g)} \longrightarrow Na^+_{(g)} + e^-$$

Significant energy is released when an electron is added to gaseous Cl at atom to form a Cl^- anion in step 2 since Cl has a high electron affinity for the process.

$$Cl_{(g)} + e^- \longrightarrow Cl^-_{(g)}$$

Considerable energy is also released when the two ions are brought together in step 3 and their opposite charges attract each other to form an ionic bond.

$$Na^+_{(g)} + Cl^-_{(g)} \longrightarrow Na^+Cl^-_{(g)}$$

The net result in the formation of a fully ionic bond is complete electron transfer from the atom with low ionization energy to the atom with high electron affinity.

Whether atoms gain or lose one, two or more electrons in the formation of ionic compounds depends upon the relative energies released for each case in the three steps. The compound that has the greatest release of energy during its formation will be the favored one and the most stable one. Thus, energy calculations can predict that MgF_2 is formed, but that Mg_2F. MgF, and MgF_3 are not. Frequently, as in this case, the stable salt formed in one whose ions have noble gas or filled-shell electron configuration.

Ionic compounds are favored between elements on the far left of the periodic table, having low ionization energies, and elements on the far right of the periodic table, having high electron affinities. Examples include potassium chloride (KCI), cesium bromide (CsBr), and magnesium oxide (MgO). Relatively high electron affinities also result from the formation of some of the common oxyanions such as hydroxide (OH^-), nitrate (NO_3^-) carbonate (CO_3^{2-}), sulfate (SO_4^{2-}), and phosphate (PO_4^{3-}). Therefore, these oxyanions form ionic compounds with cations derived from elements on the far left of the periodic table. Examples include rubidium nitrate ($RbNO_3$) calcium carbonate ($CaCo_3$), and barium sulfate ($BaSO_4$). Formation of some transition metal cations with low charge requires relatively *low* ionization energies. Therefore, these cations form ionic compounds with anions derived from elements on the far right of the periodic table. Example include manganese (II) chloride ($MnCl_2$) and iron (II) fluoride (FeF_2).

Nondirectional Nature of Ionic Bonds

Electrostatic interactions are not specific for one direction; they occur in all directions. Therefore, ionic bonds are non directional. A given cation may be attracted simultaneously by a number of anions and vice versa. Thus, sodium chloride has a crustal lattice (Fig 5.1) consisting of Na^+ ions each surrounded by six Cl^- ions, and Cl^- ions, each surrounded by six Na^+ ions. Sodium chloride, like all ionic compounds, consists of an infinite array of positively and negatively charged ions in such ratios that the total number of positive charges is just balanced by the same total number of negative charges. No isolated NaCl until can be identified within this infinite array. However, we can still consider NaCl as the simplest formula describing sodium chloride and can define a bond length for NaCl as the centre-to-centre distance between adjacent Na^+ and Cl^- ions.

Since electrostatic forces of attraction between oppositely charged ions are strong in all directions, ionic compounds are usually quite

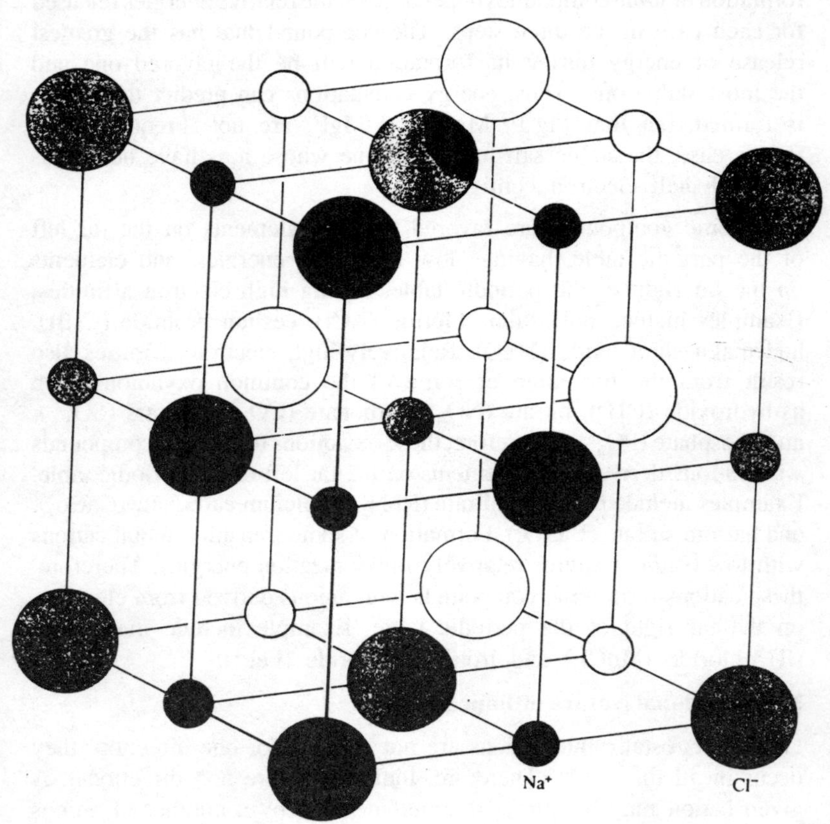

Fig. 5.1 : Crystal lattice of sodium chloride. Darkly shaded ions are close to you and lightly shaded ions far from you. Na^+ and Cl^- ions essentially touch other but have been reduced in size for clarity.

hard, fairly dense, high melting, solids. However, many ionic compounds dissolve readily in water, a property responsible for their importance in biological processes. Salts of ions such as Na^+, K^+, Cl^-, ingested daily by each of us, dissolve readily in and are easily transported by our blood and other body fluids.

Covalent Bonds

A **covalent bond** arises *from the attraction between two atoms as a result of sharing one or more outer shell electrons by the two atoms.* A single covalent bond results when two atoms are held together by the sharing of a pair of electrons with each atom normally contributing one electron to the pair.

The hydrogen molecule is the simplest case involving a covalent bond.

Lewis' Theory of Covalent Bonding

How can the attraction between atoms in molecules be explained without the formation of ions?

Shortly after Bohr proposed his model of the atom, G.N.Lewis provided the first suitable explanation of covalent bonding. He noticed that most covalent compounds contain and even number of electrons. For example, water (H_2O) contains 10 electrons, 1 from each hydrogen plus 8 from the oxygen. Similarly, carbon dioxide (CO_2) contains 22 electrons, 8 from each oxygen atom plus 6 from carbon.

This observation suggested to Lewis that covalent bonding is a result of a tendency of electrons to pair during the bonding process. From this reasoning, he concluded *a covalent bond is the result of the sharing of a pair of electrons between two bonded atoms.* Initially he envisioned that each of the combining atoms contributed one electron to the shared pair.

The simplest covalent compound is hydrogen gas (H_2). Recalling the dot diagrams, we may picture the formation of an H_2 molecule from hydrogen in the following manner:

$$H\cdot + H\cdot \longrightarrow H:H$$

In this way, lewis has given us a basic understanding of the distinction between covalent and ionic bonding. Both types of bonding are of an electrical nature. The distinction between bond types is in the way the electrical forces operate. *Ionic bonding is the result of attraction between ions of unlike charge; covalent bonding is the result of attraction between positively charged nuclei and negatively charged pairs of electrons.*

Multiple Covalent Bonds

Multiple covalent bonds occur when more than one electron pair is shared between two atoms. When *two electron pairs are shared between two atoms*, a **double bond** is formed. It is stronger than a single bond, but not twice as strong, because the two electron pairs repel each other and cannot become fully concentrated between the two atoms. Its bond length is shorter than of a single bond. When ***three** electron pairs are shared between two atoms*, a **triple bond** is formed. It is stronger than a single bond and a double bond, but not three times as strong as a single bond. Its bond length is shorter than that for a double bond. A quadruple bond, involving the sharing of four electron pairs, is known for $Re_2Cl_8^{2-}$. *Saturated* molecules possess only single bonds; *unsaturated* molecules possess one or more double or triple bonds.

Nitrogen gas consists of nonpolar, diatomic, covalent molecules having triple bonds. Each N atom has I$s^2 2s^2 2p^3$ electron configuration. Each N atom shares three $2p$ electrons with another N atom to form a triple bond.

$$:\underset{\cdot}{\dot{N}}\cdot + \cdot\underset{\cdot\cdot}{\dot{N}}: \longrightarrow :N:::N: \quad \text{or} \quad :N{\equiv}N:$$

It is a nonpolar bond because both atoms are identical and have the same attraction for the three electron pairs.

Multiple bonds are formed commonly by carbon, nitrogen, and oxygen atoms and occasionally by boron, silicon, phosphorus, sulfur, and other atoms. Multiple bonds arise whenever there are too few outer shell electrons to provide each atom with a stable noble gas electron configuration using a completely single-bonded structure. Electronic structures can be used to predict whether or not multiple bonds exist in simple molecules.

Writing Electronic Structures

Electronic structures commonly referred to as Lewis electron dot structures are essential for determining whether or not multiple bonding exists in a molecule or complex ion, and more important, for determining the three dimensional spatial configuration of a molecule or complex ion. For many molecules for which you do not know the bonding picture and structure, several possible electronic structures can be written. It is relatively easy to decide which of the several electronic structures is most valid. At this stage we are not concerned with the specific geometry of several atoms around another atom.

Electronic structures in which there are two electrons around hydrogen, eight electrons around second period elements, and frequently (but not always) *eight electrons around representative elements in the third through seventh periods, are preferred.* Shared electrons forming covalent bonds are counted as being around both of the atoms that share them when applying this rule. This rule results from the fact that atoms tend to share sufficient electrons to attain the very stable electron configuration of the next noble gas. They do this by forming a number of covalent bonds equal to the number of *additional* electrons required. A hydrogen atom, H ·, shares *one* additional electron to achieve the electron configuration of He: and forms *one* covalent bond. A fluorine atom, :F̤̈· (and usually atoms of other elements of group VII), shares *one* additional electron and forms *one* covalent bond; an oxygen atom, :Ö̤· (and usually atoms of other elements of group VI), shares *two* additional electrons and forms *two* covalent bonds; a nitrogen atom, :Ṅ· (and usually atoms of other elements of group V), shares *three* additional electrons and forms three covalent bonds; and a carbon atom, :Ċ· (and usually atoms of other elements of group IV), shares *four* additional electrons and forms *four* covalent bonds—all to achieve the :N̤̈e: electron configuration.

This rule also leads to the formation of the greatest possible number of covalent bonds—*one* for hydrogen and group VII elements, *two* for group VII elements, *three* for group V elements, and *four* for group IV elements. The maximum number of covalent bonds provides the greatest possible bond energy and results in the most stable molecule or the one with the lowest energy. The maximum number of covalent bonds can be achieved by any combination of single and/or multiple bonds.

Hydrogen will combine with all the nonmetals (except boron) to form compounds that follow the rules. Consider the hydrogen fluoride molecule HF. The dot diagrams of hydrogen and fluorine are H· and .F̤̈:. If the hydrogen and fluorine atoms each contribute an unpaired electron to forming a shared electron pair, both atoms appear to achieve a noble gas electron arrangement.

H· + .F̤̈: ⟶ H :F̤̈:

This way of representing the HF molecule is known as a *Lewis dot structure.*

Oxygen has only six outer electrons. To attain the desired eight electrons, it must form two covalent bonds. Since hydrogen forms only one bond, two hydrogen atoms are necessary to satisfy each oxygen atom.

$$H\cdot + \cdot\ddot{\underset{..}{O}}\cdot + \cdot H \longrightarrow H:\ddot{\underset{..}{O}}:H$$

Carbon, with only four outer electrons, needs to form four bonds with hydrogen to acquire a share in eight outer electrons.

$$\begin{array}{ccc} & H & \\ & \cdot & \\ H\cdot & \cdot\dot{\underset{\cdot}{C}}\cdot & \cdot H \\ & \cdot & \\ & H & \end{array} \longrightarrow \begin{array}{c} H \\ H:\ddot{\underset{..}{C}}:H \\ H \end{array}$$

Rather than always using dots, chemists have simplified the procedure by *using a line to represent a pair of electrons.* Using this procedure, the molecules HF, H_2O, and CH_4 are represented

$$H{-}\overline{\underline{F}}| \qquad H{-}\overline{\underline{O}}{-}H \qquad \begin{array}{c} H \\ | \\ H{-}C{-}H \\ | \\ H \end{array}$$

Bonding between nonmetals doesn't always involve hydrogen. For example, carbon will combine with fluorine atoms.

$$\begin{array}{ccc} & :\ddot{F}: & \\ & \cdot & \\ :\ddot{\underset{..}{F}}\cdot & \cdot\dot{\underset{\cdot}{C}}\cdot & \cdot\ddot{\underset{..}{F}}: \\ & \cdot & \\ & :\underset{..}{F}: & \end{array} \longrightarrow \begin{array}{c} :\ddot{F}: \\ :\ddot{\underset{..}{F}}:\ddot{\underset{..}{C}}:\ddot{\underset{..}{F}}: \\ :\underset{..}{F}: \end{array} \text{ or } \begin{array}{c} |\overline{F}| \\ | \\ |\overline{\underline{F}}{-}C{-}\overline{\underline{F}}| \\ | \\ |\underline{F}| \end{array}$$

Similarly, sulfur will bond with chlorine to form SCl_2.

$$:\ddot{\underset{..}{Cl}}\cdot \quad \cdot\ddot{\underset{..}{S}}\cdot \quad \cdot\ddot{\underset{..}{Cl}}: \longrightarrow :\ddot{\underset{..}{Cl}}:\ddot{\underset{..}{S}}:\ddot{\underset{..}{Cl}}: \text{ or } |\overline{\underline{Cl}}{-}\overline{\underline{S}}{-}\overline{\underline{Cl}}|$$

With the exception of hydrogen and boron, we can determine the minimum number of covalent bonds a nonmetal will form by subtracting the number of outermost electrons from eight.

***Example* 5.1**

Determine the correct electronic structures for the following molecules: (a) CH_3l, (b) CO_2, and (c) C_2H_6O.

Although electrons are indistinguishable, it is helpful to designate outer shell electrons of different atoms with different symbols until you have determined the correct electronic structure.

(a) $:\dot{C}\cdot$ share 4 additional electrons and forms 4 covalent bonds.

H ∘ shares 1 addition electron and forms 1 covalent bond.

ˣˣ I ˣ (with 7 × electrons) share 1 additional electron and forms 1 covalent bond.

Usually the atom capable of forming the most covalent bonds is the central atom. With the exception of hydrogen, which is always a terminal atom, the central atom is also the least electronegative atom. Try a structure in which 3 H atoms and 1 I atom are bonded to C to provide C with 4 covalent bonds.

```
    H
    o·  xx
H ∘ C × I x
          x
    H
```

Note that each H atom has 2 electrons around it and forms 1 bond, the C atom has 8 electrons around it and forms 4 bonds, and the I atom has 8 electrons around it and forms 1 bond, according to the previous rules. This is the correct electronic structure for CH_3l.

The following structure is an incorrect one.

```
    H
    o·  xx
H ∘ C x I ∘ H
    ··  xx
```

It obeys the general rule of having 2 electrons around H and 8 around C and I and it has 4 covalent bonds like the previous structure. However, it has one too few bonds for .C and one too many bonds for I.

(b) $:\dot{C}\cdot$ shares 4 additional electrons and forms 4 bonds.

ˣˣ O ˣ (with 6 × electrons) shares 2 additional electrons and forms 2 covalent bonds.

The C atom can form more covalent bonds, is the less electronegative atom, and is probably the central atom. Therefore, try a structure with the C atom bonded to 2 O atoms and with single bonds. Always start with a single-bonded structure first unless you know that double bonds exist.

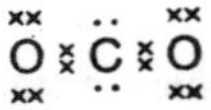

xx .. xx
O ẋ C ẋ O
xx .. xx

This is a poor structure because the O atoms do not have an octet of electrons and the C atom forms only 2 bonds and the O atoms only one bond (none of the atoms form the maximum number of bonds of which they are capable).

The following structure suffers from the same two difficulties except that it is the C atom rather than the O atoms that lacks and octet of electrons.

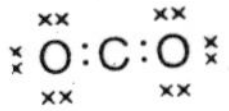

xx xx
ẋ O : C : O ẋ
xx xx

The preceding difficulties can be alleviated by the use of multiple bonding. The following structures are only marginally better than previous ones because

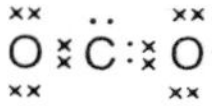

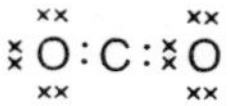

xx .. xx   xx xx
O ẋ C : ẋ O   ẋ O : C : ẋ O
xx xx   xx xx

they still suffer partially from the two problems already mentioned. However, introduction of a second C=O double bond solves both problems.

Note that each atom has 8 electrons around it and that each atom has formed its maximum number of bonds. This is the correct electronic structure for CO_2.

Perhaps you can tell simply by looking at previous structures that the following one is incorrect even though all atoms have an octet of electrons and 4

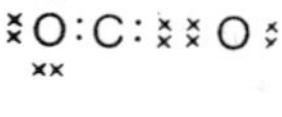

xx
ẋ O : C : ẋ ẋ O ẋ
xx

covalent bonds are formed, it is incorrect because one O atom forms 1 bond whereas the other forms 3 bonds; both normally form 2 bonds.

Note that no previous knowledge other than the molecular formula was required to obtain the correct electronic structure for CO_2.

(c) : Ċ · shares 4 additional electrons and forms 4 covalent bonds.

H∘ shares 1 additional electron and forms 1 covalent bond.

xx
ẋ O x shares 2 additional electrons and forms 2 covalent bonds.
x

A longer chain of atoms is required to write a correct electronic structure for C_2H_6O. Use the C atom and the O atom as central atoms in the chain since they are capable of forming more covalent bonds. An H atom can only be a terminal atom in an electronic structure because it can only form one bond. The chains are possible, C—C—O or C—O—C.

```
  .  .  xx            .  xx  .
 .C:C:O x            .C x O x C.
  .  .  xx            .  xx  .
```

After writing the chains (including outer shell electrons), simply place the H atoms in terminal positions wherever there remain unpaired electrons.

```
   H H                 H       H
   .o .o  xx           .o  xx  .o
 H o C:C x O o H     H o C x O x C o H
   o. o.  xx           o.  xx  o.
   H H                 H       H
```

Note that in both structures there are 2 electrons around each H atom and electrons around each C and O atom. Each H atom forms 1 bond, each O atom forms 2 bonds, and each C atom forms 4 bonds. Both structures are correct for C_2H_6O. These are two isomers (see chapter 11) , ethyl alcohol and dimethyl ether.

Exception to the Octet Rule

Atoms with Less Than Eight Electrons

An atom with fewer than four outer shell electrons may have less than eight electrons around it after sharing its outer shell electrons to form covalent bonds. The boron atom is an example. Boron trichloride (BCl_3) is a colorless gas with a boiling point slightly below room temperature. It funes from reaction with moisture in air and is a useful starting material for the preparation of many interesting boron compounds. It can be represented as forming from one B atom and three Cl atoms. The BCl_3 molecule is stable even though B has only six electrons

```
                                 xx
                               x Cl x
             xx             xx   x.   xx
:B· + 3 x Cl x  ——→     x Cl x B x Cl x
             xx             xx        xx
```

associated with it. Other boron compounds including BF_3. BBr_3, and $B(CH_3)_3$ have similar electronic structures.

An atom like boron in the above stable molecules readily achieves an octet of electrons when the molecules containing boron react further with other molecules having an unshared pair of electrons. Thus, boron trichloride reacts with ammonia (NH_3) to form a compound in which the nitrogen donates both electrons that forms the shared-pair covalent bond joining the two originally stable molecules.

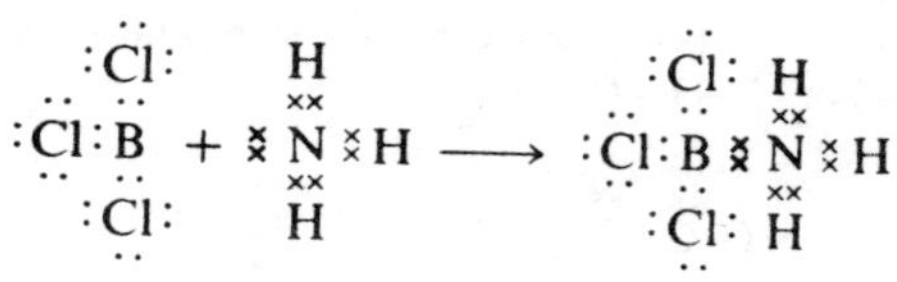

$$\begin{array}{ccc} & & \\ \ddot{:Cl:} & & H \\ :\ddot{Cl}:B & + & \overset{\times\times}{\underset{\times\times}{\times N \times}}H \\ :\ddot{Cl}: & & H \end{array} \longrightarrow \begin{array}{ccc} \ddot{:Cl:} & & H \\ :\ddot{Cl}:B & \overset{\times}{\times} & \overset{\times\times}{\underset{\times\times}{N}}\times H \\ :\ddot{Cl}: & & H \end{array}$$

Similar shared-pair covalent bonds to which one atom donates both electrons are formally constituted between transition metal atoms or cations that are capable of accepting electron pairs and anions or molecule that are capable of donating electron pairs. Metal complexes, which contain a central metal atom or cation covalently bound to a cluster of two, four, or six anions or molecules, are commonly formed. For example, in the presence of ammonia, silver (I) forms a colorless, soluble, complex ion, $[Ag(NH_3)_2]^+$, that is a part of Tollens' reagent for the oxidation of aldehydes (Section 18.5-c-2).

$$Ag^+ + 2\ :NH_3 \rightarrow [H_3N:Ag:MH_3]^+$$

Likewise, in basic solution in the presence of citrate ion, copper (II) forms a blue, soluble, complex ion, $[Cu(citrate)_2]^{6-}$, that is a part of Benedict's solution for the oxidation of aldehydes and to test for the presence of reducing sugars.

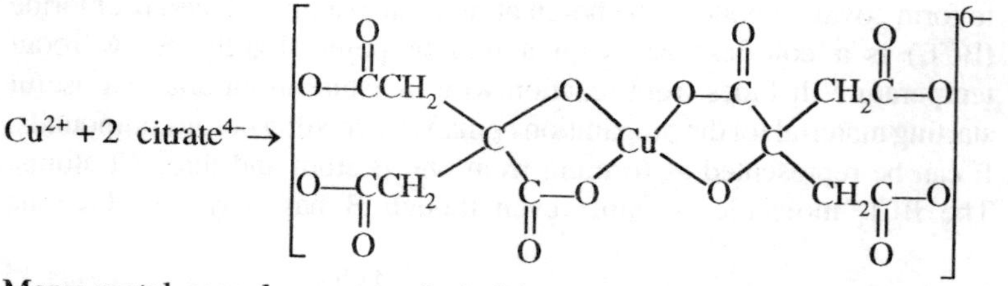

$Cu^{2+} + 2\ citrate^{4-} \rightarrow$ [structure: two citrate ligands, each with O—C(=O)CH$_2$ groups, CH$_2$C(=O)—O groups, central C atoms bonded via O and C(=O)—O to Cu] $^{6-}$

Many metal complexes are of vital importance in biological systems.

A number of molecules posses unpaired electrons, and one or more atoms in these molecules will not have eight electrons, associated with it. Electronic structures are inadequate for determining the nature of bonding and the shape of many such molecule. For example, the best electronic structure for nitric oxide (a colorless gas formed when dilute nitric acid, HNO_3, is used as an oxidizing agent) is $:\dot{N}::\ddot{O}$, but NO can be shown experimentally to have a bond energy and bond distance appropriate for 2½ covalent bonds rather than just a double bond. Likewise, the best electronic structure for the common oxygen molecule is $\ddot{\underset{..}{O}}::\ddot{\underset{..}{O}}$, but O_2 is experimentally to have two *unpaired* electrons. Thus, we represent O_2 as $:\dot{\underset{..}{O}}:\dot{\underset{..}{O}}:$, although this structure disobeys the generalizations discussed previously (eight electrons around O and O forming two bonds) and does not predict observed double bond character.

Atoms with More Than Eight Electrons

Representative elements in the third through seventh period may have more than eight electrons associated with them, and transition elements almost always have more than eight electrons associated with them in typical molecules. Phosphorus pentafluoride (PF_5, a colorless as), thionyl chloride ($SOCl_2$, a colorless liquid), xenon tetrafluoride (XeF_4, a white solid), and chromium hexacarbonyl [$Cr(CO)_6$, a white solid and a useful starting material for preparing compounds with carbon bonded to Cr] illustrate this point. Third through seventh period elements need not always restrict themselves to an octet of electrons because *d* orbtals can be utilized to hold additional electrons.

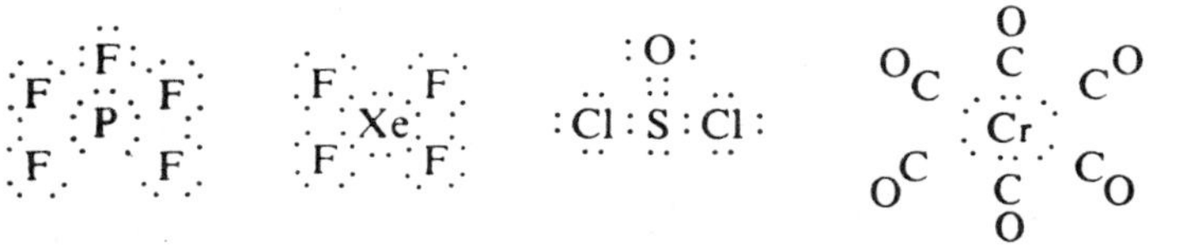

Plus 3 more pairs of *d* e⁻'s around Cr

The bonding in all compounds is not so simple as those example we have been studying above. In addition to the carbon hydrogen compound CH_4, there also exists the compound with molecular formula C_2H_6. Let us see how we can write lewis structure for this compound. A little thought reveals that the two carbon atoms must be bonded to each other. The reason for this is that *hydrogen can form only one bond.* If hydrogen formed more than one bond, it would have more electrons around it than the noble gas helium. Thus, the bonding arrangement in C_2H_6 must be

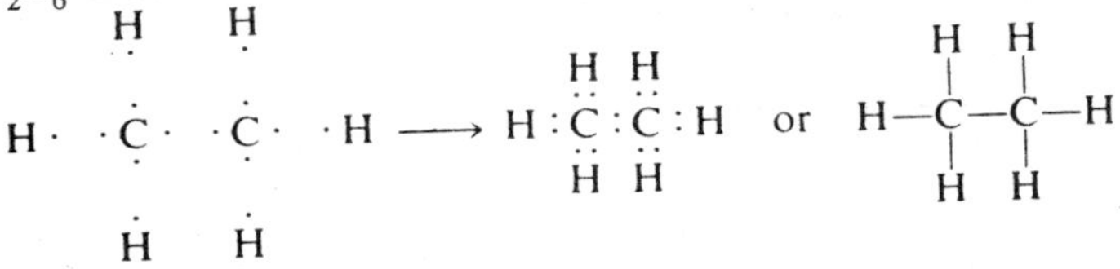

$$H\cdot \quad \cdot\underset{\dot{H}}{\overset{H}{\dot{\underset{\cdot}{C}}}}\cdot \quad \cdot\underset{\dot{H}}{\overset{H}{\dot{\underset{\cdot}{C}}}}\cdot \quad \cdot H \longrightarrow H:\underset{H}{\overset{H}{\underset{..}{\overset{..}{C}}}}:\underset{H}{\overset{H}{\underset{..}{\overset{..}{C}}}}:H \quad \text{or} \quad H-\underset{H}{\overset{H}{\underset{|}{\overset{|}{C}}}}-\underset{H}{\overset{H}{\underset{|}{\overset{|}{C}}}}-H$$

In addition to C_2H_6, the compound C_2H_4 can also be formed by combination of carbon and hydrogen. We know that hydrogen will form only one bond and that carbon atoms want to form four bonds. How then can we write a Lewis structure for C_2H_4? The answer lies in allowing two bonds to form between the two carbon atoms. If one bond is a shared pair of electrons, then two bonds must be two shared pairs of electrons or four electrons. The Lewis dot structure of C_2H_4 must be

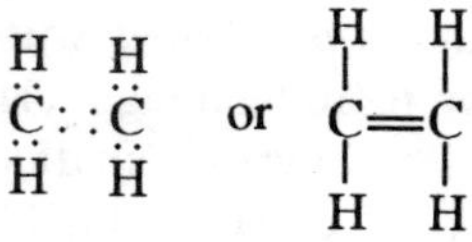

```
H  H           H   H
..  ..         |   |
C::C     or    C===C
..  ..         |   |
H  H           H   H
```

This is the only way in which two carbon atoms and four hydrogen atoms can combine to satisfy all the rules of writing Lewis dot structures. We can visualize this in the following way.

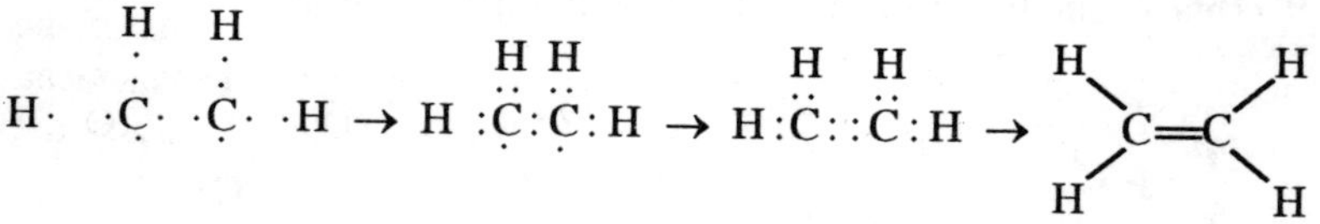

```
   H   H
   .   .                HH            H  H           H        H
H· ·C· ·C· ·H  →  H :C:C:H  →  H:C::C:H  →        \      /
   .   .                               ..  ..             C==C
                                                          /      \
                                                         H        H
```

A two-electron bond is known as a *single* bond. A four-electron bond is called a *double* bond*. Other molecules containing double bonds are carbon dioxide (CO_2) and formaldehyde (CH_2O). Their structures using the line representation for pairs of electrons are

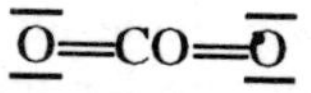

```
                          H
                           \
O=C=O        and           C=O
                           /
                          H
```

Carbon dioxide **formaldehyde**

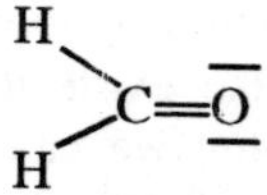

Sometimes a double bond isn't sufficient to satisfy all the rules for writing Lewis structures and we need to consider triple bonds. A triple bond *is three shared electron pairs.* Acetylene with the molecular formula C_2H_2 is such a case. The Lewis structure of acetylene is

$$H—C\equiv C—H$$

* A double bond is two pair of electrons shared between two atoms. Electrons shared by atoms are called bonding electrons. Outer shell electrons that are not shared are called nonbonding electrons.

It is significant to note that in general only carbon, nitrogen, oxygen, and sometimes sulphur and phosphorus form multiple bonds.

Coordinate covalent bonds

So far, we have only considered a contribution of one electron by each of the atoms sharing the electron pair. It has been have found a number of molecules that cannot be explained in this manner. Carbon monoxide is such a molecule. If we only allow each atom to contribute a single electron toward the formation of each bond, we find it impossible to obey all the requirements for writing Lewis structures. For example, two attempts are illustrated below:

1. $\cdot\dot{\underset{\cdot}{C}}\cdot + \cdot\ddot{\underset{\cdot}{O}}: \longrightarrow |C{=}\overline{\underline{O}}$

2. $\cdot\dot{\underset{\cdot}{C}}\cdot + \cdot\ddot{\underset{\cdot}{O}}: \longrightarrow C{\equiv}O|$

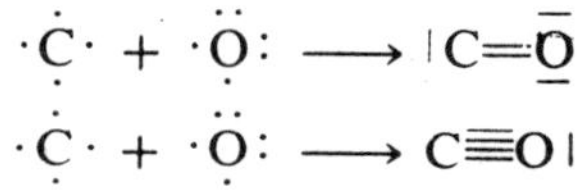

Both attempts are unsatisfactory; in reaction 1 the carbon is surrounded by only 6 electrons; in reaction 2 the oxygen is surrounded by 10 electrons. If we envision one atom as donating both electrons of the shared pair, however, we can quickly arrive at a suitable electron arrangement. In order to visualize this more readily, the electrons on carbon are represented by dots and the electrons on oxygen, by crosses.

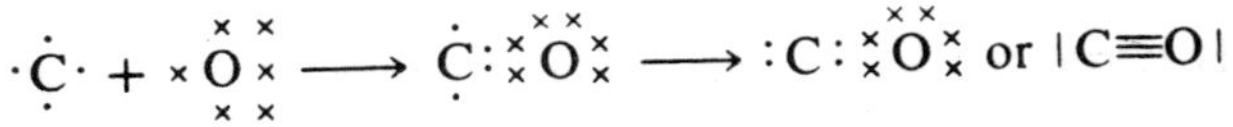

$$\cdot\dot{\underset{\cdot}{C}}\cdot + \times\overset{\times\times}{\underset{\times\times}{O}}\times \longrightarrow \dot{\underset{\cdot}{C}}{:}\,{}^{\times}_{\times}\overset{\times\times}{O}{}^{\times}_{\times} \longrightarrow :C:{}^{\times}_{\times}\overset{\times\times}{O}{}^{\times}_{\times} \text{ or } |C{\equiv}O|$$

A bond formed by the contribution of two electrons from a single atom is called a *coordinate covalent bond.*

This view of bonding should be accepted with a great deal of caution. All electrons are identical. There is no way we can determine which atom the electrons came from or how the bond is actually formed. We have merely constructed a convenient way of keeping track of electrons and developed some minimal understanding of why certain molecules exist. A covalent bond is the sharing of an electron pair between two atoms. Whether we envision the formation of the bond as resulting from the contribution of one electron by each atom or both electrons by a single atom, the result is the same: shared electron pairs.

Let us see how we can use the idea of a coordinate covalent bond to explain an important biological process—the transport of oxygen from the lungs to the cells of animals, including man, Hemoglobin in the red blood cells is responsible for the movement of oxygen molecules in these organisms. We are in a position to investigate in more detail how this occurs.

Each hemoglobin molecule contains four subunits called *heme*. The heme unit itself is rather complex with an Fe^{2+} ion at its center. It is this iron ion that actually does the work of carrying the oxygen molecule. By the use of some sophisticated techniques, chemists have found heme to have the following structure:

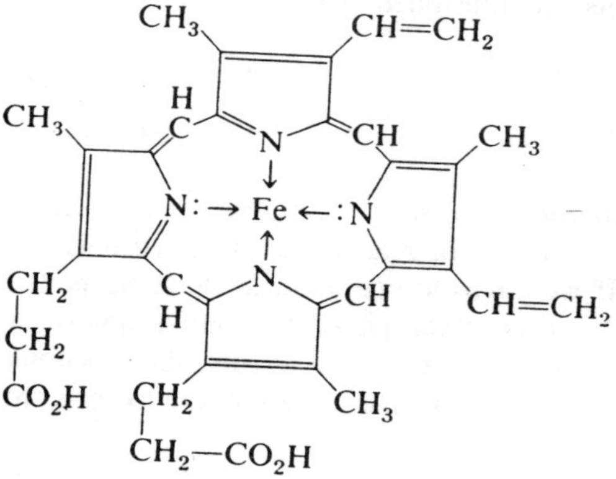

CH_3 $CH{=}CH_2$

H

CH_3 C N CH CH_3

N: → Fe ← :N

N

CH_2 C CH $CH{=}CH_2$

H

CH_2

CO_2H CH_2 CH_3

$CH_2{-}CO_2H$

The iron ion is bonded to four nitrogen atoms contained in a complex molecule known as *protoporphyrin IX*. The iron ion is capable of accepting a pair of electrons from an oxygen molecule to form a coordinate covalent bond. The exact arrangement of the oxygen molecule around the iron is yet unknown. One arrangement that has been suggested is

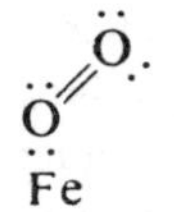

O

O

Fe

In the lungs, molecular oxygen bonds to the iron ions of the heme units in hemoglobin to produce oxyhemoglobin. The oxyhemoglobin is carried to other regions of the body where it releases the oxygen molecule for use by the cells. The manner in which the body regulates the bonding and release of oxygen will be discussed later.

Unfortunately, the iron in hemoglobin will also form a coordinate covalent bond with certain other smaller molecules if they are present in the lungs. Carbon monoxide is one such molecule and, even worse, the bond formed between hemoglobin and carbon monoxide is more stable than the bond between hemoglobin and oxygen. *If hemoglobin is bonded to CO, it cannot bond to O_2.* This means that hemoglobin will preferentially bond with carbon monoxide rather than oxygen if both are present in the lungs. So great is this preference, that when only one out of every thousand gas molecules in this lungs is carbon monoxide, a sufficient amount of hemoglobin is bonded to carbon monoxide to paralysis or death because of a lack of oxygen movement to the cells.

Experiment evidence indicates that this structure is incorrect, however. The oxygen molecule actually has two unpaired electrons. The following structure shows two unpaired electrons:

$$|\overline{\underset{\cdot}{O}}-\overline{\underset{\cdot}{O}}|$$

Unfortunately, this structures is also unsatisfactory. Besides not obeying the rules for writing Lewis structure, it does not agree with other experimental evidence indicating that the oxygen molecule actually has a double bond.

This example serves to point out the inadequacies of the Bohr model of the atom and of Lewis structures. While these ideas are useful, they are not infallible. There is nothing wrong with the oxygen molecule; what is wrong is the inability of the theories we have been using to explain the nature of the oxygen molecule. Lest you think that scientists are incapable of explaining the oxygen molecule, let us note that a different theory of atomic and molecular structure can explain the nature of oxygen. Quantum mechanics yields a satisfactory answer to the question of electron arrangement in oxygen molecules. It is a much more difficult theory to understand and an explanation is beyond the scope of this text.

POLAR COVALENT BONDS

A nonpolar covalent bond always arises between two *identical* atoms or between two *different* atoms having identical electronegativities. **Electronegativity** is *a measure of the tendency of an atom in a molecule to attract shared electrons to itself.* It indicates the relative tendency of an atom to attain a negative condition when forming a bond. It is a measure of the hold that an atom has on *shared* electrons. This hold arises from electrostatic attraction between the shared electrons and the protons in the nucleus of an atom. Electronegativities are dependent upon the same factors as ionisation energies and increase across periods and decrease down groups (Table 5.1).

Table 5.1 : Electronegativity Values of the Elements (according to Pauling)

H 2.1																	He —
Li 1.0	Be 1.5											B 2.0	C 2.5	N 3.0	O 3.5	F 4.0	Ne —
Na 0.9	Mg 1.2											Al 1.5	Si 1.8	P 2.1	S 2.5	Cl 3.0	Ar —
K 0.8	Ca 1.0	Sc 1.3	Ti 1.5	V 1.6	Cr 1.6	Mn 1.5	Fe 1.8	Co 1.8	Ni 1.8	Cu 1.8	Zn 1.6	Ga 1.6	Ge 1.8	As 2.0	Se 2.4	Br 2.8	Kr —
Rb 0.8	Sr 1.0	Y 1.2	Zr 1.4	Cb 1.6	Mo 1.8	Tc 1.9	Ru 2.2	Rh 2.2	Pd 2.2	Ag 1.9	Cd 1.7	In 1.7	Sn 1.8	Sb 1.9	Te 2.1	I 2.5	Xe —
Cs 0.7	Ba 0.9	57-71 1.1-1.2	Hf 1.3	Ta 1.5	W 1.7	Re 1.9	Os 2.2	Ir 2.2	Pt 2.2	Au 2.4	Hg 1.9	Tl 1.8	Pb 1.8	Bi 1.9	Po 2.0	At 2.2	Rn —
Fr 0.7	Ra 0.9																

When two atoms with the same electronegativity bond together, both have equal ability to attract electrons and an equal sharing of electrons results. Bonded atoms having the same electronegativity form nonpolar covalent bonds. If the bonded atoms have unequal electronegativities, the electrons are unequally shared. The bonding pair of electrons will be closer to the atom with the greater electronegativity. This unequal sharing creates a negative end of the bond and a positive end of the bond. Remember this is *not* a complete transfer of an electron; it is merely a movement of shared electrons toward the more electronegative element.

Consider the molecule HCl. The electronegativities of hydrogen and chlorine are 2.1 and 3.0, respectively. Thus, the bonding pair of electrons are, on the average, closer to the chlorine atom than to the hydrogen atom. The result is a *polar covalent bond.* The direction of the electron shift is indicated by an arrow. The head of the arrow paints towards the more negative end of the bond: $\overset{\delta+ \rightarrow \delta-}{\text{H—Cl}}$. The lowercase Greek latter delta (δ) is used to represent a partial (incomplete) unit of charge.

Electronegativity difference gives a rough idea of the bond type:

Difference in electronegativity between bonded atoms	*Bond type*
0.0	Nonpolar covalent
Greater than 0.0 to 1.7	Polar covalent
Greater than 1.7	Ionic

Dipole Moments

The presence of polar bonds within a molecule may give rise to a dipolar molecule, depending on the molecular shape. A dipolar molecule is one that will take a preferred orientation in the presence of an electrical field. *Dipole moments are caused by an unequal sharing of electrons by Atoms in molecules.*

Consider the H—Cl molecule. The hydrogen-chlorine bond is polar, giving the chlorine a partial negative charge and the hydrogen a partial positive charge. In the presence of an electrical field, the hydrogen chloride molecule would be expected to orient itself such that the chlorine is pointed toward the positive pole of the field and the hydrogen toward the negative pole.

	δ^- δ^+	δ^- δ^+	δ^- δ^+	
+	Cl—H	Cl—H	Cl—H	−
+	Cl—H	Cl—H	Cl—H	−
+	Cl—H	Cl—H	Cl—H	−
+	Cl—H	Cl—H	Cl—H	−
+				−
+				−
+				−

A molecule that will orient itself in an electrical field is said to possess a *permanent dipole moment* and is called a *polar molecule.*

Shape is an important factor in determining whether a molecule has a permanent dipole moment. Beryllium fluoride, for example, has *polar bonds* but because the molecules is linear, it does *not* possess a permanent dipole moment.

$$\overset{\delta^-}{F}\overleftarrow{\;—\;}\overset{\delta^+}{Be}\overrightarrow{\;—\;}\overset{\delta^-}{F}$$

Both ends of the molecule are negatively charged and the molecule can find no preferential orientation in an electric field.

In contrast to BeF_2, water molecules are V-shaped and as such find a preferred alignment in the field (as shown on p. 120).

We can say, then, that polar bonds are a necessary, but not sufficient, condition to give a molecule a permanent dipole moment.

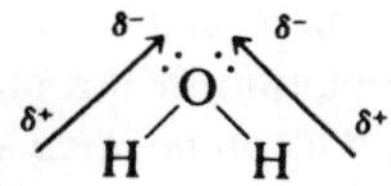

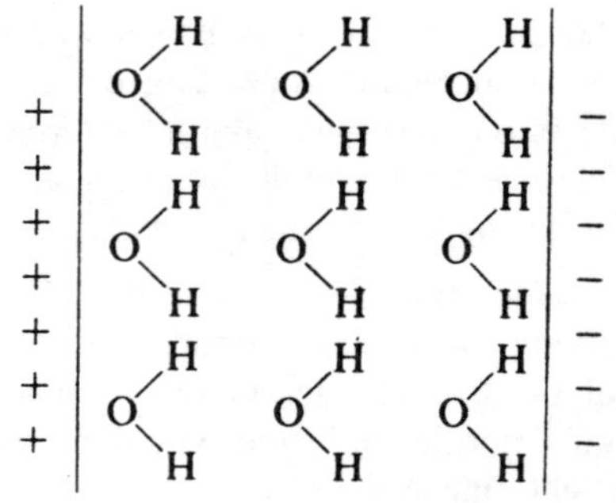

Polar Molecules

A **polar molecule** is *one in which there is a net separation of positive and negative charge*, as in $^{\delta+}H—Cl^{\delta-}$. A polar diatomic molecule arises from two covalently bound atoms having different electronegativities, whereas a nonpolar diatomic molecule results from two covalently bound atoms that are identical or have identical electronegativities.

For a molecule composed of more than two atoms, it is more difficult to predict whether it is polar or nonpolar. You consider both the polarity of individual bonds and the orientation of these bonds in space with respect to each other. You must examine the shape of the molecule and determine how the dipoles of individual polar bonds either add to or cancel each other.

Symmetrical molecules have a structure exhibiting a regular repeating pattern of bonds and are nonpolar even though individual bonds within the molecules are polar. For example, in carbon dioxide the bond dipoles are equal in magnitude of charge separation and opposite in orientation (indicated by arrow). Thus, the bond dipoles cancel each other, leading to a nonpolar molecule.

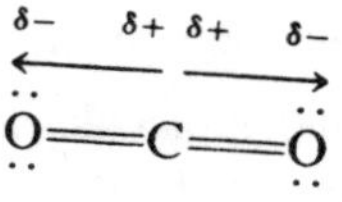

In symmetrical molecules there is no net charge separation for each molecule taken as a whole, and the molecules are nonpolar. For example the two bond dipoles in $BeCl_2$ cancel each other; the three bond dipoles in BF_3 cancel each other; the four bond dipoles in CH_4 cancel each other; the five bond dipoles in PF_5 cancel each other; and the six bond dipoles in SF_6 cancel each other.

Unlike CH_4, NH_3 and H_2O are not symmetrical, and have net dipoles.

Intermolecular Dipole Forces and Hydrogen Bonding

Oxygen and hydrogen chloride are both diatomic molecules of similar molecular weight. The boiling and melting points of these substances should be similar. The forces holding the molecules together in the liquid and solid states are weak. The data in Table 5.3 reveal that the melting and boiling points of these compounds are considerably different.

Table 5.3 : Normal Melting and Boiling Points of HCl and O_2

Substance	Molecular Weight	MP (°C)	BP (°C)
O_2	32	—218	—183
HCl	36.5	—112	— 84

Consideration of the bonding within these molecules allows us to given an explanation for the differences in their melting and boiling points. The oxygen-oxygen bond in O_2 is nonpolar; whereas the hydrogen-chlorine bond in hydrogen chloride is polar. In the liquid and solid states of these substances the molecules are relatively close to one another. We know ions arrange themselves in the solid so as to place oppositely charged ions as close together and like charged ions as far apart as possible. Thus, it is reasonable to expect polar molecules to aliegn them-selves so as to place the positive end of one molecule as close as possible to the negative end of a nearby molecule, as shown in Fig. 5.2.

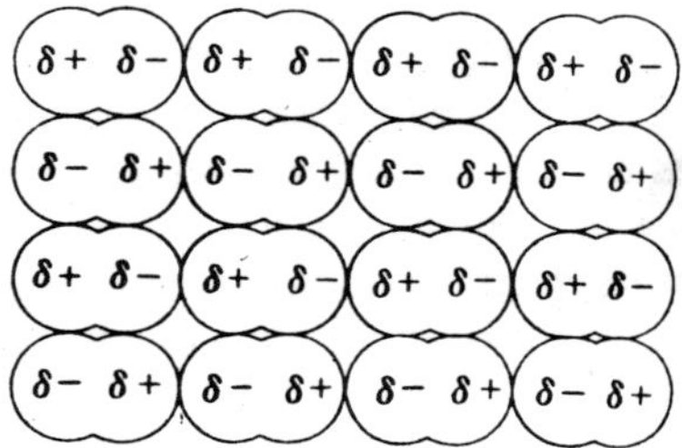

Figure 5.2 : Arrangement of Polar Molecule Solid.

This attraction between adjacent molecules means that more energy is required to separate them. Consequently, the melting and boiling points should be higher for polar molecules than for nonpolar molecules of approximately the same size and weight. The melting and boiling points of ammonia, water, and hydrogen fluoride are considerably higher than would be expected even from consideration of their polar character. The relatively high boiling points of these three substances implies the presence of an abnormally strong intermolecular force. X-ray examination of the arrangement of water molecules in an ice crystal suggests the presence of this intermolecular force. The observed arrangement indicates attraction between the hydrogen atoms of one molecule and the oxygen atom of the neighboring water molecule. Experimentation has shown this attraction is stronger than normal dipole attractions. This is termed a *hydrogen bond.* Hydrogen bonds occur between a hydrogen atom of one molecule and a very electronegative atom of another molecule. Hydrogen bonds are frequently represented by a dotted line to indicate that attraction is weaker than true covalent bond but stronger than normal dipolar attractions.

Hydrogen bonding seems to be a special case of dipolar attraction in which the hydrogen atom of one molecule is attracted toward a nonbonding pair of electrons on an atom of an adjacent molecule. It results from the very polar character of bonds between hydrogen and nitrogen, oxygen, or fluorine.

In addition to explaining the relatively high melting and boiling points of NH_3, H_2O, and HF, the extreme polarity of the hydrogen oxygen bond in water explains the solubility of ionic substances in water. We shall also see in later chapters that hydrogen bonding plays an extremely important role in chemistry of life process. Hydrogen bonding is, for example, responsible for the unique properties of proteins.

Resonance

The phenomenon in which no single classical electronic structure adequately describes experimentally observed bonding parameters (bond energy and bond distance) is called **resonance**. Ozone, O_3 (a gas with a sharp, penetrating odor that is formed in significant amounts in the atmosphere during electrical storms), requires the use of resonance structures. Structure I is a valid electronic structure according to guidelines given in this section.

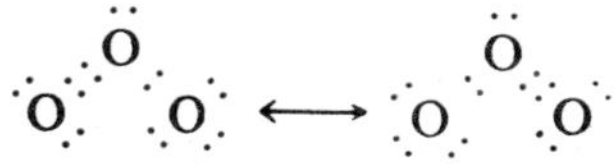

$$\ddot{O}: \ddot{O} = \ddot{O} \longleftrightarrow \ddot{O} = \ddot{O} : \ddot{O}$$

I II

However, structure I incorrectly implies one single O—O bond (longer) and one double O═O (shorter). Experimentally the two bonds are found to be identical and equivalent to about 1½ bonds. The actual electron distribution is *intermediate* between these depicted by two contributing resonance structures I and II and is never like that depicted by either contributing structure.

Resonance is a misnomer because it implies oscillation form one structure to another or oscillation of an electron pair from one bond to another. *This is not true.* A mule, the offspring of a male donkey and a mare, can be considered a hybride of a donkey and a horse. However, it is not a horse at one instant and a donkey at another; it is always a mule. Likewise, a molecule or complex ion exhibiting resonance has *one* real structure, which is never any of the rather unsophisticated electronic structures used to describe it.

***Example* 5.2**

The nitrate ion, NO_3^-. exhibits resonance. Write the contributing electronic structures for NO_3^-. What information about bonding do these structures convey?

$\dot{N}$: shares 3 additional electrons

$\overset{\times\times}{\underset{\times\times}{\times O \times}}$ shares 2 additional electrons.

The charge, —1, provides one additional electron.

Add to N with its 5 outer shell electrons the one additional electron from the charge. Then let N share an electron pair with each of 3 O atoms. The resulting structure leaves N with only 6 electrons around it whereas 8 would provide a more

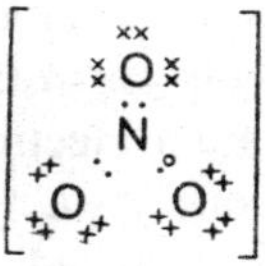

$$\left[\begin{array}{c} \overset{\times\times}{\times O \times} \\ : \\ N \\ O \quad\quad O \end{array} \right]$$

stable structure. This problem can be alleviated by introducing one double bond. This can be accomplished in any of the three bonds producing three equivalent resonance structures. These three resonance structures taken together imply that all three N—O bond energies and bond distances are identical and correspond to about $1\frac{1}{3}$ bonds.

Shapes of Simple Molecules

Even simple molecules show a great variety of three-dimensional geometries. Fig 5.3 shows a few examples. Molecules containing two atoms, such as I_2 and HBr, are necessarily linear (two points determine a straight line). Molecules containing three atoms are necessarily planar (three points determine a plane); however, they may be linear like CO_2 or non linear (bent) like H_2S. Molecules containing four atoms may be planar or nonplanar. The planar molecules may be linear like $(CN)_2$, non linear like HN_3, triangular like BBr_3, or T-shaped like BrF_3. The nonplanar molecules may be pyramidal like PH_3 or non linear like H_2O_2. As the number of atoms in a molecule increases to five and higher, the variety of configurations and their complexity increase.

Dispersion Forces (van der Waals Forces)

None of the bonding interactions discussed thus far account for bonding *between* nonpolar molecules like H_2 or atoms like Ar and thus the existence of these species in the liquid or solid phase. One further type of interaction called dispersion forces or van der Waals forces provides the explanation.

Dispersion forces *arise from attraction between temporary or instantaneous dipoles.* On the average an Ar atom has a spherical distribution of electron density and no separation of charge Fig 5.4 . However, at any given instant, the electron cloud may be more concentrated at one side of the atom producing an instantaneous dipole as shown in Fig 5.4b. Electrons of one atom electrons of adjacent atoms and cause a correlation of their motions. As a result, the first instantaneous dipole induces a similar dipole in an adjacent molecule. These two dipoles oriented in the same direction Fig 5.4 produce an attractive force.

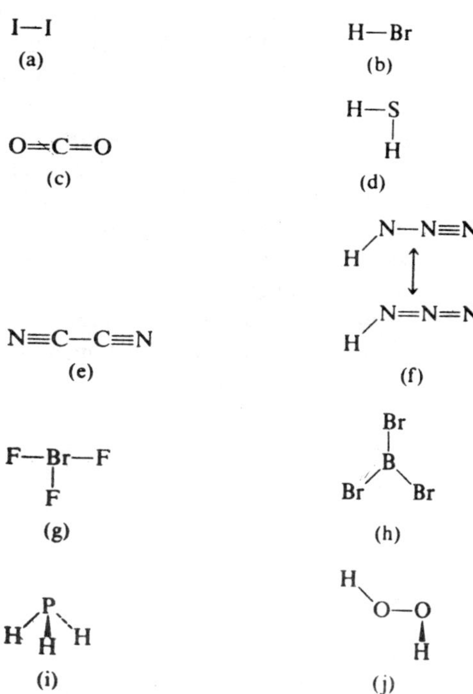

Fig 5.3 Geometries of few simple molecules. (a) Iodine (b) hydrogen bromide; (c) carbon dioxide; (d) hydrogen sulfide; (e) cyanogen; (f) hydrazoic acid; (g) bromine trifluoride; (h) boron tribromide; (i) phosphine; (j) hydrogen peroxide. All atoms and bonds are in the plane of the paper except those shown with a dashed line, which are back of the paper (away from the reader), and those shown with a darkened wedge, which are out of the paper (toward the reader).

Figure 5.4 : Model of dispersion forces. (a) Spherical distribution of electron density of an isolated atom; (b) instantaneous dipole produced by distortion of the electron cloud; (c) attraction of two instantaneous dipoles.

Dispersion forces are very weak compared to ionic or covalent interactions and depend upon the ease with which the electron cloud can be distorted or polarized. As a first approximation, dispersion forces increase with increasing numbers of electrons in the atom or molecule. Consequently, properties depending primarily upon interatomic or intermolecular dispersion forces show a similar trend. Thus, the melting points and boiling points of CH_4, SiH_4, GeH_4, and SnH_4 increase almost linearly with the increasing number of electrons in these molecules (10, 18, 36, and 54, respectively). Since the number of protons and neutrons increases as the number of electrons in a series of molecules increases, the masses in that series must also increase. Thus, as a second approximation, one can say that dispersion forces increase with increasing mass in a series of atoms or molecules.

6

Gases, Liquids and Solids

GASES

Gases are matter and have **mass**. However, owing to their very low densities and difficulties in handing, gases are not commonly measured quantitatively in terms of mass. They must be contained in sealed vessels, which are often difficult to weigh on ordinary chemical balances, in amounts that are so small as to be near the sensitivity limits of many chemical balances. Instead, quantities of gases are described more conveniently in terms of three interrelated variables—volume, temperature, and pressure

Volume

The **volume** *of a substance is the space occupied by that substance.* A gas occupies the *total* volume of its container, unlike a liquid, which occupies a fixed volume regardless of its container.

Volume is usually sufficient to specify the amount of a solid or liquid because the volumes of solids or liquids change very little with changes of temperature and pressure. Such is not the case for gases; temperature and pressure must also be specified. Thus, 1 liter of oxygen at a certain temperature and pressure has a different mass and contains a different number of moles from 1 liter of oxygen at the same temperature but a different pressure or at the same pressure but a different temperature.

Temperature

Temperature is a measure of the intensity of heat. Temperature determines the direction heat flow. Thus, heat flows spontaneously from a region of higher temperature to a region of lower temperature.

Pressure

Pressure is *the exertion of a force on one body by another.* The pressure of a gas is the force that the gas exerts on the walls of its container. Pressure is a property that determines the direction in which mass flows. When you pump air into a deflated bicycle tire, the tire expands because the pressure on the inside of the tire is greater than the pressure on the outside. When you open the valve and allow air to escape from the tire, the air moves from a region of higher pressure to a region of lower pressure.

The pressure at any given point in a gas or a liquid is the same in all directions. Thus, atmospheric pressure exerted by air on a mountain climber is the same in the directions. Likewise, at any given depth in a quiet pond, the pressure exerted by water on an underwater swimmer is the same in all directions. For the mountain climber or the swimmer, there is no net push in any direction. However, in the presence of a strong wind in the air or a strong current in the water, the mountain climber and the swimmer experience a push from the direction of high pressure to the direction of low pressure.

Atmospheric pressure and water pressure both vary with depth. As the mountain climber ascends to higher altitudes, the atmospheric pressure decreases and his body is compressed by a lesser column of air above it. At high enough altitudes the pressure of oxygen gas becomes too small to support respiration. As the swimmer dives deeper, the water pressure increases and his body is compressed by a greater column of water above it acted on by the force of gravity. At great enough depths the increased pressure on his eardrums creates a sensation of pain.

A *barometer* is used to measure atmospheric pressure.

Pressure changes with altitude and temperature; therefore, a standard reference point for measuring gas pressures has been defined. **One atmosphere** is *the pressure that supports a 760 mm column of mercury at O°*. Since pressures are commonly measured using mercury barometers and manometers, pressures are frequently expressed in millimeters of mercury (1 atm = 760mm Hg).

A mm Hg is frequently referred to as a *torr* (named after Evangelista Torricelli, the inventor of the barometer). The SI unit of measuring pressure is Pascal (Pa). Conversion factors between these various units of pressure are :

1 atm=760 mm Hg=760 torr=101·3 kilopascal (1kilo pascal=1000 pascal).

THE GAS LAWS

Boyle's Law

By considering gases to be composed of small particles in constant motion, it is possible to explain the variation of the volume of a gas with pressure.

Suppose that a certain mass of gas enclosed in a container of volume V exerts a pressure P. This pressure is caused by the particles which make up the gas hitting the walls of the container. Provided the temperature of the gas does not alter, the number of collisions with the wall in a particular length of time remains constant.

Now suppose that the same mass of gas is put into a container only half as big, *i.e.* of volume $\frac{V}{2}$. The number of particles in a particular volume will then be twice as great. This means that collisions with the walls will be twice as frequent and the pressure doubled, i.e. $2P$.

In other words, if the volume of a fixed mass of gas is halved, its pressure is doubled. This was first realized by Robert Boyle in 1662, and is summarized in Boyle's Law:

At constant temperature the volume of a given sample of a gas is inversely proportional to the pressure exerted on the gas.

i.e. Pressure × Volume = Constant

The inverse relationship is illustrated in Fig. 6.1

If V_1 and V_2 are the different volumes of a fixed mass of gas at pressures P_1 and P_2, then

$$P_1, V_1 = P_2V_2$$

Example **6.1**

A sample of oxygen gas occupies a volume of 400 mL at a pressrue of 720mm. What will be its volume at a pressrue of 760 mm, assuming the temperature remains constant ?

The pressure is increasing. The volume of a gas varies inversely with the pressrue of the gas; therefore, the volume decreases as the pressure increases. From the expression $P_1V_1 = P_2V_2$ we have

$$P_1 = 720 \quad V_1 = 400\text{mL}$$
$$P_2 = 760 \quad V_2 = ?$$

Thus :

V_2, volume of sample of oxygen at 760 mm=(400 ml) $\left(\frac{720 \text{ mm}}{760 \text{ mm}}\right)$=379mL

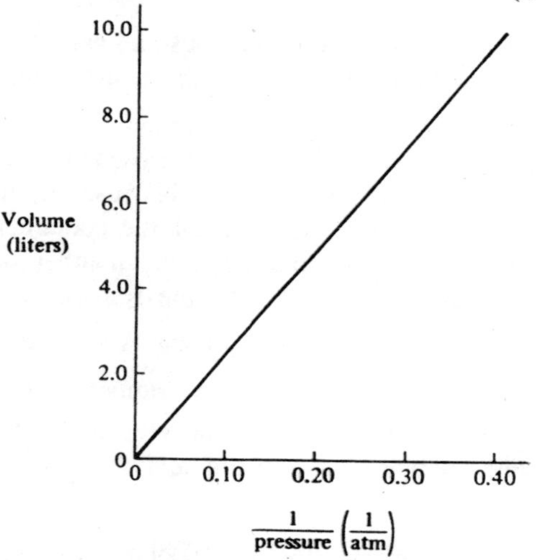

Fig 6.1 Variation of volume of one mole of an ideal gas at 25°C as a function of reciprocal pressure. Temperature and number of moles remain constant.

If the original volume and pressure of a given sample of gas are known, the pressure that the gas will exert when its volume has changed (at constant temperature) can be calculated.

Charles's Law

To explain the expansion of gases when they are heated, consider a fixed mass of gas. When the temperature of the gas is raised more energy is supplied to the particles, and they therefore move more rapidly. This means that particles will hit the walls of a container more frequently and, unless the volume is increased, this will result in an increase in pressure.

Charles observed that gases maintained at constant pressrue expanded equally for equal rises in temperature. His law states:

A fixed mass of gas, at constant pressure, expands $\frac{1}{273}$ *of its volume at O°C, for each degree Centigrade rise in the temperature,* i.e. a gas with a volume of 273 mL at O°C would expand by 1 mL for each degree rise in temperature and would contract 1mL for each degree fall in temperature (Fig 6.2).

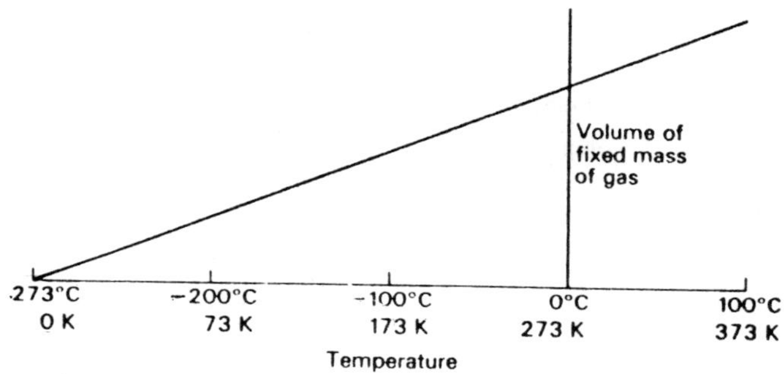

Fig. 6.2 : Graph illustrating Charles' Law.

Theoretically gases would have a zero volume* at —273°C but in practice they all liquefy above this temperature. This temperature of –273 °C, at which a gas that did not liquefy would have no volume, is the lowest temperature possible. It is therefore known as the *absolute zero. The absolute scale of temperature* is obtained merely by adding 273° to the Centigrade temperatures i.e.

$$x\ ^\circ c = (x + 273)\ k$$

Thus —273°C becomes 0 K, 0 °c becomes 273 K, etc. (K is the symbol for the *Kelvin*, the temperature unit on the absolute scale.)

Charles' law may therefore be stated more conveniently for our purpose:

The volume of a fixed mass of gas, at constant pressrue, is directly proportional to its absolute temperature.

$$i.e.\ \frac{\text{Volume}}{\text{Temperature in K}} = \text{Constant}$$

If V_1 and V_2 are the different volumes of a fixed mass of gas at absolute temperatures T_1 and T_2, then:

$$\frac{V_1}{T_1} = \frac{V_2}{T_2}$$

***Example* 6.2**

A sample of oxygen gas occupies a volume of 2.10 litres at 25°C. What volume will this sample occupy at 150°C assuming the pressure remains constant?

* This concept of zero volume of a gas at —273 °C is unrealistic.

All quantitative calculations involving temperature changes of gases require the use of absolute temperatures.

T_1 = Initial temperature : 25 °C + 273° = 298 K

T_2 = Final temperature: 150 °C + 273° = 423 K

V_1 = Initial volume = (2.10) litres

V_2 = Final volume = ?

$$V_2 = \text{Volume of sample of oxygen at } 150^\circ\text{C} = V_1\left(\frac{T_2}{T_1}\right)$$

$$= (2.10 \text{ litres})\left(\frac{423 \text{ K}}{298 \text{ K}}\right) = 2.98$$

***Example* 6.3**

At what temperature will the initial volume of oxygen in Example 5.3 occupy 0.750 litre assuming the pressure remains constant ?

T_1 = initial temperature = 2.98 K

V_1 = initial volume = 2.10 litres

V_2 = final volume = 0.75 litre

$V_1/T_1 = V_2/T_2$

T_2 = Temperature of sample of oxygen occupying 0.750 litre $T_1(V_2/V_1)$

$$= (298 \text{ K})\left(\frac{0.750 \text{ litre}}{2.10 \text{ litre}}\right)$$

$$= 106 \text{ K}$$

The general Gas Equation

By combining the laws of Boyle and Charles, an expression relating the temperature, pressure and volume of a fixed mass of gas is obtained:

$$\frac{P_1V_1}{T_1} = \frac{P_2V_2}{T_2}$$

This is very important equation. It means, for example, that if the volume of a given mass of gas under one set of conditions is known its volume under a different set of conditions may be calculated. *Always remember that the temperature used must be on the Absolute scale.*

Avogadro's Law

In 1811 Amadeo Avogadro, and Italian physicist, *proposed that equal volumes of all gases at the same temperature and pressure contain equal numbers of molecules.* Stated alternatively, the volume of a gas at constant temperature and pressrue is directly proportional to the number of molecules of the gas. Since a mole is related to the number of molecules only by a constant Avogadro member, it can also be stated that *at constant temperature and pressure the volume of a gas is directly proportional to the number of moles of the gas.* This direct relationship is illustrated schematically in Fig. 6.3. When the initial number of moles is doubled, the initial volume is doubled. When the initial number of moles is reduced to half, the initial volume also decreases to half the original value. This direct relationship is illustrated graphically in Fig. 6.4. Volume plotted against number of moles yields a straight line.

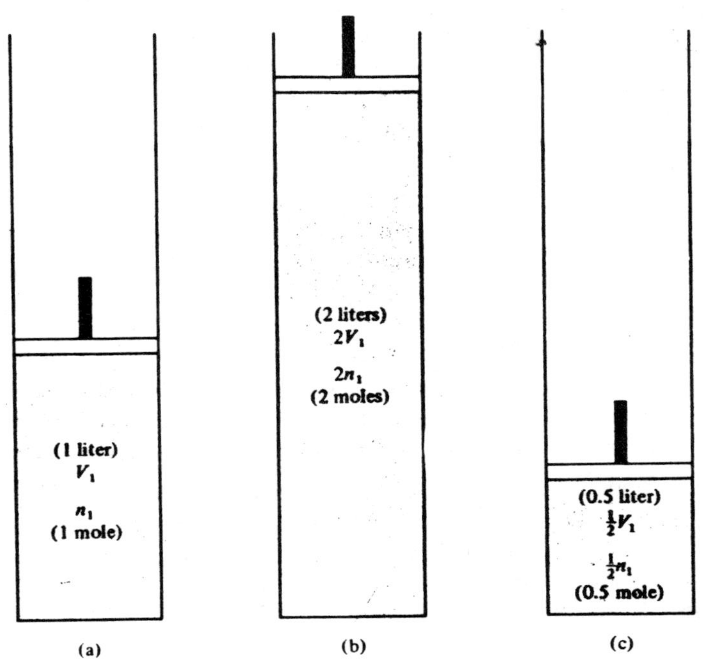

Fig. 6.3 : The variation of the volume of a gas at constant temperature and pressure in a cylinder fitted with a movable piston. V = volume; n = number of moles. (a) Initial conditions; (b) V_1 doubles when n_1 is doubled; (c) V_1 halves when n_1 is halved.

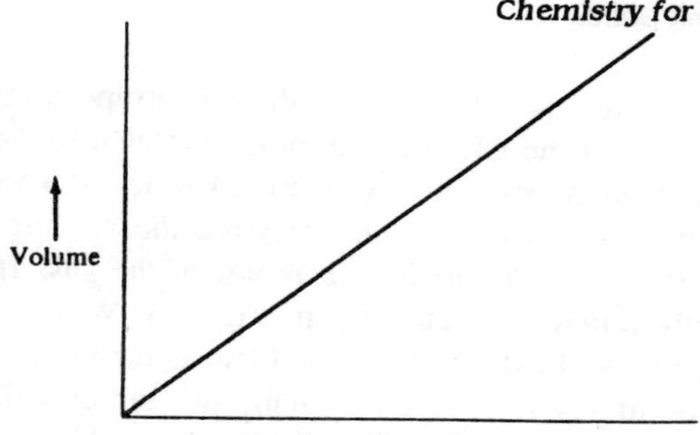

Figure 6.4 : The variation of volume of a gas with number of moles of the gas, temperature and pressure remaining constant.

Dalton's Law of Partial Pressures

In 1801 John Dalton summarized the behaviour of two or more *nonreacting* gases in the same container—*the total pressure exerted by a mixture of gases is equal to the sum of the partial pressures of the various gases.* Partial pressure is the pressure a gas would exert if it were the only gas in the container. For air in an inflated balloon Dalton's law of partial pressures can be written as

$$P_{total} = P_{N_2} + P_{O_2} + P_{CO_2} + P_{H_2O} + P_{Ar} + \ldots\ldots$$

where the subscripts denote the various gases occupying the same volume of the balloon and at the same temperature.

The Diffusion of Gases

If a girl using perfume enters a room the smell is rapidly detected all over the room. Less pleasantly, if some hydrogen sulphide is prepared in a laboratory its smell is immediately noticed by everyone present.

In both cases the *spontaneous spreading of a gas throughout any volume accessible to it* is called **diffusion**. The rapid random motion of gas molecules causes them to move from a region of relatively high concentration or partial pressure to a region of lower concentration of partial pressure.

Graham's law of diffusion

These experiments indicate that the rate of diffusion of a gas depends on its density, i.e. gases of low density diffuse more rapidly than gases of high density. Thomas Graham investigated this relationship and formulated his *Law of Diffusion.*

At a constant temperature and pressure the rate of diffusion of a gas is inversely proportional to the square root of its density.

Using symbol, this becomes :

$$R \alpha \frac{1}{\sqrt{d}} \text{ or } R = \frac{k}{\sqrt{d}}$$

(where R = rate of diffusion in cm^3 s^{-1} or other suitable units, and d = density).

If the rates of diffusion of two gases, A and B, are compared, it follows that:

$$\frac{R_A}{R_B} = \frac{\sqrt{d_B}}{\sqrt{d_A}}$$

Alternatively, the times taken for equal volumes of two different gases to diffuse, under similar conditions, may be compared. Then:

$$\frac{T_A}{T_B} = \frac{\sqrt{d_A}}{\sqrt{d_B}}$$

Molecular masses of gases are directly proportional to their densities Therefore :

$$\frac{R_A}{R_B} = \frac{\sqrt{M_B}}{\sqrt{M_B}}$$

i.e. Grahm's Law affords a means for determining molecular masses.

Many biochemical processes are dependent upon the passage of gas molecules across semipermeable membranes. It has been shown experimentally that for a given gas at constant temperature the rate of diffusion is proportional to the partial pressure gradient across the membrane; that is, the greater the partial pressure difference, the faster will be the net passage of molecules across the membrane. A given gas diffuses in response to its own partial pressure gradient independent of what other gases are present, and continues to diffuse until its partial pressures are equal on both sides of the membrane.

Kinetic Molecular Theory of Gases

When dust particles in a room are illuminated from the side by sunlight from a window, it is observed that the particles do not settle to the floor but rather undergo irregular zigzag motion (called *Brownian motion*) in space. Furthermore, the motion of the particles becomes more rapid as the temperature of the room increases. Apparently the minute dust particles are bombarded constantly by molecules comprising air. Such observations provide evidence that the molecules comprising air are constantly in motion and that the velocities of these molecules are related to temperature.

On the basis of the behavior of gases James Maxwell, a Scottish physicist, and Ludwig Boltzmann, an Austrian physicist, proposed in 1859 the kinetic molecular theory. The basic assumption of this model are

1. *Gases consist of tiny, discrete molecules.*
2. *The average distance between gas molecules is so large compared to the size of the molecules themselves that the volume of a gas consists mostly of empty space.* Calculations show that in air, for example, more than 99.9% of the total volume is empty space. Thus, air can be compressed significantly upon the exertion of pressures much greater than atmospheric pressrue.
3. *There are essentially no attractive forces between gas molecules so that they are completely independent of each other.* Gases expand spontaneously to fill any volume accessible to them so that there can be no significant binding forces between one molecule and another.
4. *Gas molecules are in perpetual, rapid, random, straight - line motion, colliding frequently with each other and with the walls of the container.* Gas molecules bombard minute particles giving rise to Brownian motion and bombard the walls of a container resulting in the exertion of pressure. Gas molecules expand into larger volumes and diffuse from one container to another.
5. *Collisions between molecules and between molecules and container walls are elastic; thus, there is no loss of kinetic energy on collision.* Gases do not change temperature spontaneously on standing, indicating that gas molecules do not lose energy by collisions. In the course of a collision, one molecule may transfer energy to another molecule, but the average energy of the pair of molecules remain constant.
6. *The average kinetic energy of all as molecules is directly proportional to the absolute temperature.* Temperature is a measure of the intensity of heat and rises when additional heat is added. Heat is a form of energy which imparts increased

velocity and higher kinetic energy to the gas molecules. Molecules will exhibit a range of kinetic energies at any temperature because collisions continually change the velocity and kinetic energy of any given molecule.

Gases whose behavior is consistent with the postulates of the kinetic molecular theory are said to be **ideal gases.**

Boyle's Law

The pressrue exerted by a gas depends only on the number of molecular impact on a unit wall area per second. When the volume is reduced, the molecules collide more frequently with a unit wall area, and the pressure increases. The converse is true when the volume is increased.

Charles' Law

When the temperature is increased, the molecules move with greater velocities, collide harder and more frequently with the container walls, and would produce a greater pressure. To compensate and maintain the pressure constant, the gas expands the volume and maintains the number of molecular impacts on a unit wall area per second at a constant value. The converse is true when the temperature is decreased.

Avogadro's Law

When the number of moles (and thus molecules) is increased, the number of molecular impact on a unit wall areas increases, and the pressure would increase. To compensate and maintain the pressrue constant, the gas expands the volume and maintains the number of molecular impacts on a unit wall area per second at a constant value. The converse is true when the number of moles is decreased.

Dalton's Law of Partial Pressures

If attractive forces between molecules are negligible, each molecule of a gas mixture hits the container walls the same number of times per second and with the same force as it would if no other gas molecules were present. Thus, the partial pressure of a given gas is unaffected by the presence of other gases.

Diffusion

The average kinetic energy for a collection of molecules is fixed for a given temperature and is expressed quantitatively by

$$\text{Average kinetic energy } (KE) = \tfrac{1}{2}M\upsilon^2$$

where M is the molar mass and υ is the average velocity of the molecules. The rate of diffusion of gas molecules is related to their average velocity. Solving from above,

$$\upsilon = \sqrt{\frac{2KE}{M}}$$

Thus, velocity and rate of diffusion are inversely proportional to the square root of the molar mass of the molecules.

Stoichiometric Calculations Involving Gases

Standard Temperature and Pressure

Volumes of gases change with changes of temperature and pressure. Therefore, **standard temperature and pressure** (abbreviated **STP**) have been chosen of comparing volumes of gases. Standard temperature is *0°C* and standard pressure is *1 atm or 760mm.* Volumes of gases are corrected to STP prior to most stoichiometric calculations by performing a combined Boyle's law-Charles' law calculation.

Example 6.4

A sample of hydrogen gas has a volume of 1.10 litres at –40°C and 0.520 atm. What will be its volume at STP?

Convert the required temperatures to K.

$K = °C + 273° = -40° + 273° = 273K\ (T_1)$

$K = °C + 273° = 0° + 273° = 273K\ (T_2)$

$P_1 = 0.520$ atm $V_1 = 1.10$ litre

$P_2 = 1.00$ atm $V_2 = ?$

Using the combined gas equation

$$\frac{P_1V_1}{T_1} = \frac{P_2V_2}{T_2}$$

$$V_2\text{, Volume of } H_2 \text{ at STP} = (1.10\ \text{litres})\left(\frac{0.520\ \text{atm}}{1.00\ \text{atm}}\right)\left(\frac{273\ \text{K}}{233\ \text{K}}\right) = 0.670\ \text{litre}$$

Molar Volume

The number of moles of gas is required to perform stoichiometric calculation using the mole concept. It has been found experimentally that *one mole of any ideal gas at STP occupies 22.4 liters.* Thus, 1.00 mole H_2 (2.02 grams) at STP occupies 22.4 litres; 0.500 mole H_2 (1.01 gram) at STP occupies 11.2 litres.

Example 6.5

Calculate the number of moles of hydrogen collected in Example 6.4

The volume of H_2 at STP is 0.670 litre. Determine the number of moles of H_2 by dividing the volume by the molar volume of an ideal gas.

$$\text{Number of miles of } H_2 = \frac{0.670 \text{ litre}}{22.4 \text{ litres / mole}} = 0.0299 \text{ mole of } H_2$$

Example 6.6

What volume of ammonia (NH_3) at STP is required to react with excess hydrogen chloride (HCl) to produce 1.78 grams of ammonium chloride (NH_4Cl)?

The balanced chemical equation representing the reaction is

$$NH_{3(g)} + HCl_{(g)} \longrightarrow NH_4Cl_{(g)}$$

Determine the number of moles of NH_4Cl formed by dividing the mass of NH_4Cl formed by the molar mass of NH_4Cl.

$$\text{Number of moles of } NH_4Cl = \frac{1.78 \text{ g}}{53.491 \text{ g / mole}} = 0.0333 \text{ mole of } NH_4Cl$$

From the balanced equation, 1 mole of NH_3 is required for each mole of NH_4Cl formed. Therefore, 0.0333 mole of NH_3 is required to produce 0.0333 mole of NH_4Cl.

Determine the volume of NH_3 at STP required by multiplying the number of moles of NH_3 by the molar volume at STP.

$$\text{Volume of } NH_3 \text{ at STP} = (0.0333 \text{ mole})(22.4 \text{ litres/mole})$$
$$= 0.745 \text{ litre or } 745 \text{ ml}$$

Example 6.7

What mass of zinc must be reacted with excess hydrochloric acid solution to produce 2.00 litres of hydrogen collected over water at 25°C when the atmospheric pressure is 745 mm?

The balanced equation representing the reaction is

$$ZN_{(g)} + 2\,HCl \longrightarrow ZnCl_2 + H_{2(g)}$$

The pressure in the collection vessel is due to both H_2 and water vapor. Water has a vapor pressure of 23.8 mm at 25°C. The partial pressure of H_2 is equal to the atmospheric pressure minus the vapor pressure of water, or 745 mm - 23.8 mm = 721 mm.

Convert the required temperatures to K.

$$K = {}^\circ C + 273^\circ = 25^\circ + 273^\circ = 298 \text{ K } (T_1)$$
$$K = {}^\circ C + 273^\circ = 0^\circ + 273^\circ = 273 \text{ K } (T_2)$$
$$P_1 = 721 \text{ mm}; \quad V_1 = 2.00 \text{ litres}$$
$$P_2 = 760 \text{ mm}; \quad V_2 = ?$$

The volume of dry H_2 (2.00 litres) is then corrected to STP using the expression

$$\frac{P_1V_1}{T} = \frac{P_2V_2}{T_2}$$

gives

$$V_2 = V_1\left(\frac{P_1}{P_2}\right)\left(\frac{T_2}{T_1}\right)$$

Thus :

$$V_2\text{, Volume of dry } H_2 \text{ at STP} = (2.00 \text{ litres})\left(\frac{721 \text{ torr}}{760 \text{ torr}}\right)\left(\frac{273 \text{ K}}{298 \text{ K}}\right)$$

Determine the number of moles of H_2 produced by dividing the volume of dry H_2 at STP by the molar volume at STP.

$$\text{Number of moles of } H_2 = \left(\frac{1.74 \text{ litres}}{22.4 \text{ litres / mole}}\right) = 0.0777 \text{ mole of } H_2$$

From the balanced equation, 1 mole of Zn is required to produce 1 mole of H_2. Therefore, 0.0777 mole of Zn is required to produce 0.0777 mole of H_2.

Determine the mass of Zn needed by multiplying the number of moles of Zn by the molar mass of Zn.

$$\text{Mass of Zn} = (0.0777 \text{ mole})\ (65.38 \text{ g/mole}) = 5.08 \text{ g}$$

LIQUIDS

Like most solids, ice has a highly ordered structure composed of regular pattern of atoms and molecules. Oxygen atoms of neighboring water molecules, attracted by hydrogen bonds to a given central O atom, are located at the corners of a regular tetrahedron around the given central O atom. This basis structure is extended in three dimensions to form a regular crystal lattice with hexagonal channels as in a honeycomb. When ice melts, its highly ordered structure is partially broken down. Liquid water maintains short-range order around any given O atom, but the overall arrangement of tetrahedra is more random and is constantly changing. The structure of the liquid is between the extremes of complete disorder characteristic of gases and precise order characteristic of solids. Structural changes between liquids and solids give rise to differences in properties.

This chapter deals with some of the properties of liquids and solids, how these properties differ from each other and from those of gases, and how these properties can be understood in terms of the kinetic molecular theory.

General Properties of Liquids.

Like gases, *liquids have no characteristic shape.* Liquids assume the shape of whatever volume of a vessel is required to contain the liquid, whereas gases assume the shape of an entire closed container.

Unlike gases, *liquid maintain their volume,* regardless of the shape or size of the container. Whereas a gas occupies the total volume accessible to it, 100 mL of water occupies a 100 mL volume whether it is placed in a mL beaker or 750 mL bottle.

Unlike gases, *liquids are incompressible.* The volume of a gas at constant temperature halves when its pressure is doubled, whereas the volume of a sample of water decreases only $5 \times 10^{-3}\%$ when the pressure on the sample is doubled from 1 atm to 2 atm.

Unlike gases, *liquid show little change in volume with changes in temperature.* The volume of a gas at constant pressure doubles when its absolute temperature is doubled, whereas the volume of a sample of water increases only 4% with an increase in temperature from 273 K to 373 K.

Unlike gases, *liquids held at constant volume show a spectacular change in pressure with changes in temperature.* The pressure of a gas at constant volume doubles when the absolute temperature is doubled, but simply heating mercury contained in a fixed volume by 1° C increases the pressure from 1 atm to 46 atm.

Like gases, *liquids increase in volume as additional moles of the sample are added.* The volume of water at constant temperature and pressure doubles as the number of moles of water is doubled.

Liquids diffuse much more slowly then gases. Unless coffee is stirred when milk is added to it, several minutes are required for the milk to spread spontaneously throughout the cup. No such stirring is required for rapid mixing of two gases.

Liquids exhibit internal resistance to flow called **viscosity**. It is easier to pour water out of a narrow-mouth bottle than to pour glycerine or honey from the same bottle because the later has a higher viscosity.

The surface of a liquid experiences a net inward pull, called **surface tension,** that tends to draw the surface molecules into the body of the liquid. In the absence of gravity and other forces, the stable form of a mass of liquid is a sphere. This is nearly exhibited by a free-falling drop of rain. More commonly visible is the tear-shaped drop acted upon by gravity as it hangs from a medicine dropper.

The magnitude of surface tension is responsible for the phenomenon called **capillary action**, *the rise or fall of liquids in very small diameter tubes.* When a very small diameter tube is placed vertically in a beaker of water, the water rises in the capillary tube well above the level of water in the beaker. This climbing up the walls occurs because the attraction between water and the glass surface is stronger than the surface tension between water molecules. In the same manner, plant roots get water through capillaries in the soil Seven though the water table may be several feet below the roots of the plants. There are extended networks of very fine capillaries in many types of soil. Water molecules have a stronger attraction to the capillary walls of the soil than to other water molecules and thus rise by capillary action well above the water table to plant roots.

Liquids evaporate to form gases. Water placed in an open beaker gradually evaporates so that the beaker will be empty within a few days.

Kinetic Molecular Theory Applied to Liquids

Assumptions 1,4,5 and 6 explaining the behavior of gases, are also valid for liquids. Thus, liquids consist of tiny, discrete atoms or molecules that are in perpetual random motion, that suffer elastic collisions with each other and with container walls, and whose average kinetic energies are directly proportional to absolute temperature. However, explanation of the observed properties of liquids by means of the kinetic molecular theory requires several postulates in direct contrast to assumptions 2 and 3 for gases.

1. *Individual atoms or molecules composing a liquid must be close together, essentially in contact, so that free space between them is very small.* Four general properties can be explained on the basis of this assumption.
   a. Liquids are practically incompressible because, with free space reduced to a minimum, electron clouds of adjacent molecules repel immediately upon attempting to force molecules closer together.
   b. Liquids at constant volume show drastic changes in pressure with changes in temperature because liquid molecules are much closer together than gas molecules and a much greater number of collisions per unit wall area results for liquids than for gases when higher temperature imparts higher velocities and kinetic energies to the molecules.

c. The volume of a liquid increases as additional moles (or molecules) of liquid are added because, with a minimum of free space, more molecules cannot be forced into the same volume.

d. Liquids diffuse more slowly than gases because the average distance between molecules is small, and a molecule suffers great hindrance from billions of collisions with neighboring molecules while migrating from one side of container to another.

2. *Attractive forces between atoms or molecules in a liquid prevent them from separating spontaneously from each other, but they are not strong enough to hold the positions of liquid molecules fixed.* Five general properties can be explained on the basis of this assumption.

a. Liquids have no characteristic shape because the positions of liquid molecules are not fixed, but are free to slip past each other and fill from the bottom of a container upward so as to occupy positions of lowest potential energy.

b. Liquids maintain their volume regardless of the shape or size of the container.

c. Liquids show little change in volume with changes in temperature.

d. Liquids exhibit surface tension and resistance to flow because attraction between molecules are strong enough to keep them clustered together.

e. Liquids evaporate because, in a collection of molecules, a finite number may be near the surface of the liquid and possess kinetic energy great enough to overcome attractive forces and escape into the gas phase.

Evaporation

Evaporation is *the conversion of a liquid to a gas.* Kinetic energy must be applied to break attractive forces between liquid molecules in order that molecules may escape to the gas phase. In the process of boiling kinetic energy is applied in the form of heat; but how do many liquids vaporize at ambient temperatures without the addition of heat?

The average kinetic energy of a collection of liquid molecules is directly proportional to the absolute temperature. Nevertheless, all molecules do not have the same energy; any given molecule may have a high or a low kinetic energy. A collection of molecules has a distribution of kinetic energies at any given temperature as illustrated in Fig 6.5. If the dashed line represents the minimum kinetic energy required to overcome attractive forces of the liquid and escape into the gas phase, it can be seen that even at the lower temperature, a few molecules have the necessary kinetic energy.

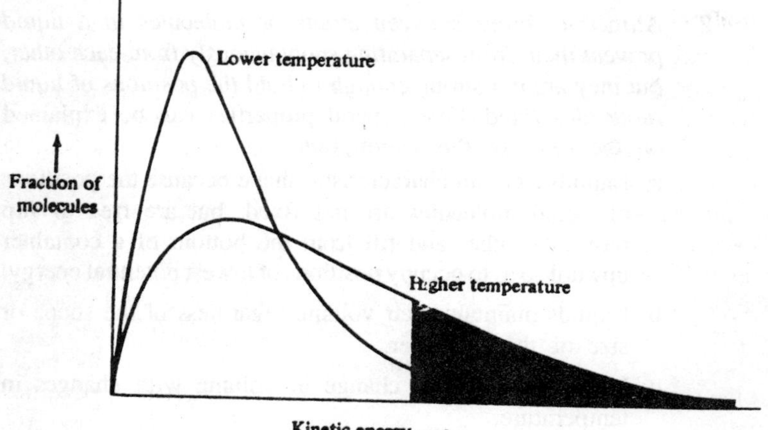

Fig. 6.5 : Kinetic energy distributions of molecules of a liquid at two different temperatures. The dashed line represents the minimum kinetic energy required for molecules of the liquid to overcome attractive forces and escape into the vapor phase

In addition to decreasing the amount of liquid, evaporation causes cooling of the remaining liquid. Cooling occurs because highly energetic molecules overcome attractive forces and pass from the liquid. They leave behind molecules of *lower* average kinetic energy, giving a corresponding decrease in the temperature of the liquid.

The decrease in temperature is actually observed experimentally only when the liquid is held in an *insulated* container. When water evaporates from a *noninsulated* beaker, there is sufficient heat flow from the surroundings to compensate for the loss of energy in the escaping highly energetic molecules. The temperature of the liquid is maintained at or near room temperature even though evaporation continues. However, when water evaporates from an *insulated* open container such as a thermos bottle, the heat flow from the surroundings is not sufficient to compensate for the loss of energy in the highly energetic escaping molecules. Thus, the temperature of the liquid drops, the curve in Fig. 6.5 shifts to the left, the number of highly energetic molecules decreases, and the rate of evaporation is less. The net result is that water evaporates less rapidly from an insulated thermos bottle than from an open beaker. This phenomenon is used to advantage when evaporation of liquid nitrogen (—196°), used as a coolant, is decreased by containing it in a thermos bottle.

Equilibrium Vapor Pressure

Water placed in an open beaker evaporates freely into the air. If a bell jar is sealed over the beaker, all the water does not evaporate. The liquid level drops for a time, but then remains constant. Initially, highly energetic molecules pass into the vapor state, the amount of liquid decreases, heat is absorbed by the water from the surroundings to maintain a nearly constant temperature, and evaporation continues. However, as an increasing number of water molecules convert to water vapor, the probability that gaseous water molecules, diffusing throughout the bell jar in their rapid, random motion, will collide with and condense to the liquid phase, also increases. Eventually condensation and evaporation occur at the same rate; that is, water molecules enter the liquid at the same rate at which water molecules leave the liquid. A condition of *dynamic equilibrium*, in which two changes oppose each other exactly, is attained. Neither process stops; they simply occur at the same rates, with the result that the number of molecules in each phase remains *constant* (but not equal).

The gaseous water molecules, in their rapid, random motion, constantly bombard the walls of the bell jar and thus exert a pressrue. *The pressure exerted by a vapor in equilibrium with its liquid* is called *the* **equilibrium vapor pressure** (frequently termed *vapor pressrue* with equilibrium assumed).

The magnitude of the vapor pressure depends on the nature of the liquid and on its temperature. Liquids with relatively large forces of attraction, such as the dipole-dipole and hydrogen bonding forces in water, have small tendencies to escape into the vapor phase and have low vapor pressures. Liquids with relatively small forces of attraction, such as the weak dispersion forces present in butane, have large tendencies to escape into the vapor phase and have high vapor pressures.

Substances having high vapor pressures at room temperature are said to be **volatile**. As the temperature of any liquid is increased, the average kinetic energy of its molecules and the number of highly energetic molecules increase. Hence, more molecules can escape to the vapor phase, and vapor pressure increases. Conversely, the vapor pressure decreases with decreasing temperature. The vapor pressures of *n*-butane, diethyl ether, ethyl alcohol, and water plotted as a function of temperature in Fig 6.6, illustrate these ideas.

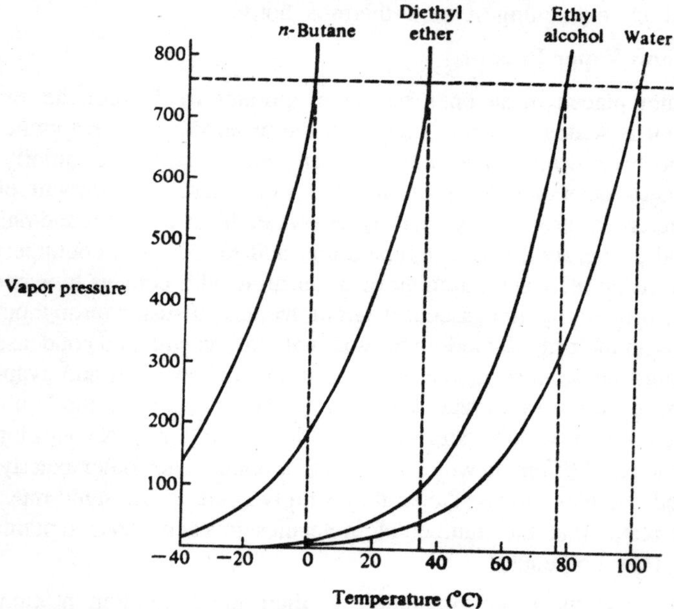

Figure 6.6 : Vapor pressures of n-butane, diethyl ether, ethyl alcohol, and water as a function of temperature. The horizontal dashed line indicates one atmosphere pressure, and the vertical dashed lines indicate normal boiling points of the respective compounds.

Our physical well-being is determined partially by humidity, or the presence of water vapor in the air. Air can hold a great quantity of water at high temperatures, but very little at low temperatures. On cold winter days the vapor pressure of water in the air may be less than 4 torr. Since water vapor from inside condenses on cold windows and since there are air passages to the outside despite our best construction efforts, the vapor pressure inside may also approach this value. This is much too low for comfort and health. Dry air causes rapid evaporation of moisture from mucous membranes in our breathing passages, causing irritation and increasing the change of infection. Thus, humidifiers are used to add water vapor to the air. The reverse problem frequently occurs in summer when very moist air prevents the necessary elimination of excess heat from our bodies by evaporation. Then dehumidifiers are used.

Boiling

Evaporation occurs for highly energetic molecules close to the liquid-vapor interface. **Boiling** is *a special form of evaporation in which conversion from liquid to vapor occurs through formation of bubbles within the liquid.* These bubbles increase markedly the area of the liquid- vapor boundary and allow evaporation even from the interior of the liquid. Boiling occurs when the vapor pressure of a liquid increases with rising temperature until, at some characteristic temperature that is different for different liquids, bubbles from and push the surrounding air aside. The boiling point of a liquid is the temperature at which the vapor pressure of the liquid equals the prevailing atmospheric pressure.

The boiling point of a liquid depends on the nature of the bonding forces in the liquid. Liquids with relatively small forces of attraction between molecules have low boiling points; liquids with relatively large forces of attraction have high boiling points.

The boiling points of a liquid also depends on pressure. The **normal boiling point** is defined as *the temperature at which the vapor pressure of a liquid is equal to 1 atm or 760 mm* (100° for water). The boiling point of water decreases under the lower atmospheric pressures prevalent at higher altitudes, being 97.7° at pressure of 700 mm. On the other hand, water boils at 120.1° in a pressure cooker if the pressrue is 2 atm. The boiling point at any pressure may be determined from vapor pressure curves like those of Fig. 6.2 by finding the temperature that corresponds to the desired pressure. Thus, the boiling point of ethyl alcohol at 400mm is 63°.

The boiling point of a pure liquid is constant at a given pressure. Addition of heat to a liquid at its boiling points only gives more molecules sufficient kinetic energy to escape from the liquid to the gas phase and does not change the temperature of the sample until all of the liquid has vaporized. The boiling point at a given pressure is a commonly used criterion for identifying a liquid and characterizing its purity. A liquid containing impurities boils at a higher temperature than the corresponding pure liquid.

Water

Water is an extraordinary liquid. It covers about three fourths of the earth's surface and is present in the atmosphere and in cloud formations. It is by far the most abundant compound in living organisms, constituting 60% by mass of the human body and about 90% by mass of certain foods such as lettuce, turnips, and apples. Water is of major importance to chemists because of its ability to dissolve a great variety of substances. In addition, it is a reactant or product in many chemical reactions, a coolant in distillations, and a medium for heat transfer or a coolant for maintaining reactions at constant temperature.

The bent molecular structure of water is responsible for its unique properties. Water has polar covalent bonds between a highly electronegative oxygen atom and two hydrogen atoms. It is a polar molecule with a partial negative charge residing on the oxygen and a partial positive charge shared by two hydrogen atoms. The polar nature of the water molecule accounts for its high solvent ability and for a number of properties arising from strong intermolecular hydrogen bonding acting between adjacent H_2O molecules in the liquid.

Water is the most common dissolving medium in nature and in the laboratory, but is not a universal solvent. It dissolves many polar molecules such as hydrogen chloride (HCl) and ionic compounds such as sodium chloride (NaCl). The polar ends of the water molecules interact with dissolved ions or with oppositely charged ends of dissolved molecules. Water does not dissolve nonpolar substances such as oxygen (O_2), sulfur (S_8), bromine (Br_2), carbon tetrachloride (CCl_4), or hydrocarbons. These nonpolar molecules are repelled by water molecules and tend to cluster about each other rather than dispersing in water.

The melting and boiling points of water are unusually high compared to those of similar compounds. Unlike water, the other three are gases at room temperature. This anomalous behaviour of water results from the strong intermolecular hydrogen bonding present in both ice and water. The hydrogen atoms, though covalently bonded to an oxygen in one H_2O molecule, are capable of bridging adjacent water molecules together into three-dimensional aggregates throughout the bulk of the liquid. The energy necessary to break a hydrogen bond (ca 5 kcal/mole) is only about one tenth of the energy required to disrupt a covalent bond. Nevertheless, the cumulative effects of all of the hydrogen bonds within the water structure necessitate the expenditure of a considerable amount of energy and result in the abnormally high melting and boiling points.

Solid water (ice) is less dense than liquid water (0.917 versus 1.00g/ml). Very specific hydrogen bonding in ice gives rise to open hexagonal channels as in a honeycomb, Liquid water has a randomly changing, collapsed structure with less empty space, and thus is more dense.

The density difference between ice and water has important consequences. One good consequence is that ice floats on water and forms on the top surface of a pond in winter. The ice acts as an insulator between cold air and the remaining water below so that the pond does not freeze solid. Furthermore, the insulating ice moderates the temperature of the underlying water so that fish and frogs can survive the rigors of winter. This would not be so if ice were more dense than water and the pond froze from the bottom upward. Bad features that result from ice being less dense than water include the need to drain water pipes and toilet bowls in summer cottages that are closed for the winter and the need to maintain minimal heating in your home when you are away for a winter vacation. Since the same amount of water will occupy a larger volume as a solid than as a liquid, on freezing the increased pressure from ice requiring a larger a volume will burst the pipes and could cause extensive flooding and water damage.

SOLIDS

General Properties of Solids

Unlike gases and liquids, *solids have characteristic shapes, geometric patterns, or crystal structures that are distinctive for particular substances.* Solid sodium chloride forms a lattice of cubes. The plane surfaces or faces within the crystal always intersect at an angle of 90°. When a sodium chloride crystal is broken, or even when ground into a fine powder, it cleaves in preferred directions so that the characteristic faces and angles are retained. Solids also crystallize in the variety of other forms depicted in Fig. 6.7 Furthermore, there are several modifications of some of these forms. Some solids, which form two or more different crystalline modifications, are said to exhibit *polymorphism*. Diamond and graphite are different modifications of solid carbon. Different modifications are favored under different conditions. Thus, graphite can be converted to the more dense diamond under very high pressure.

The actual crystal structure of solid can be determined experimentally by x-ray diffraction. X-rays, like those used for these x-rays, are passed through a solid.

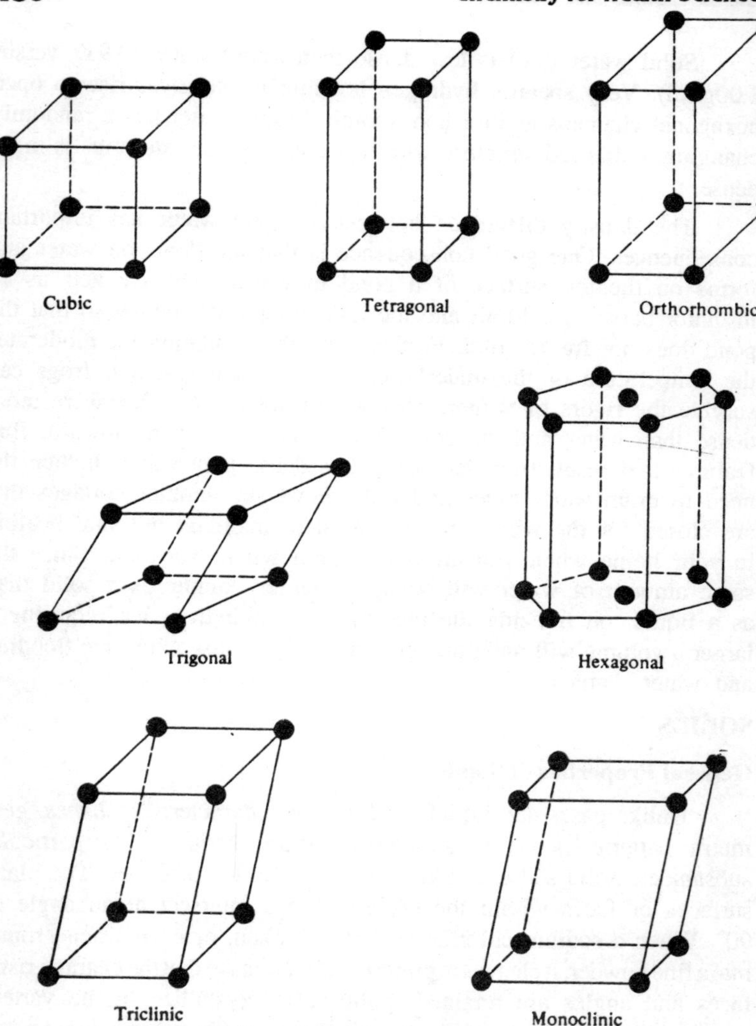

Fig. 6.7 : The simplest units in different crystal systems.

They are deflected by the electron density of atoms or ions in the solid and produce characteristic patterns on a photograph. A detailed analysis of these patterns can yield an exact crystal structure.

Unlike gases, but like liquids, *solids maintain their volume.* A cube of sugar has an identical volume to the same sugar recrystallized in the form of a sphere or some other regular or irregular shape.

Unlike gases, but like liquid, *solids are essentially incompressible.* The volume of an iron sample decreases negligibly when the pressure on the sample is doubled from 1 atm to 2 atm.

Unlike gases, but like liquids, *solids show little change in volume with changes in temperature.* The volume of an iron sample held at constant pressure increases only 0.35% with an increase in temperature from 0° to 100°C.

Unlike gases, but like liquids, *solids held at constant volume show a tremendous change in pressure with changes of temperature.* Expansion joints are used in heavy construction so that solid materials are not forced to remain in a constant volume through temperature fluctuations.

Like gases and liquids, *solids increase in volume as additional moles of a sample are added.* A0.6 carat (a unit of mass equal to 0.2 gram) diamond has twice the number of moles and also twice the volume of a 0.3 carat diamond at the same temperature and pressrue.

Solids diffuse even more slowly than liquids. A gold plate in contact with a lead plate shows only slight evidence for diffusion between the two plates after five years. Rock layers in contact with each other in the earth's crust still retain their boundaries even on a time scale of millions of year.

Under appropriate conditions, *solids melt to form liquids and sublime to form gases.* At room temperature and atmospheric pressure, ice melts to form liquids water whereas dry ice (solid carbon dioxide) sublimes, or converts directly from the solid to CO_2 vapor.

Kinetic Molecular Theory Applied to Solids

Assumptions 1, 4, 5, and 6 explaining the behavior of gas, are valid for solids as well as liquids. Thus, solids consist of tiny discrete atoms, molecules, or ions, that are in perpetual ransom motion, that suffer elastic collisions with each other, and whose average kinetic energies are directly proportional to absolute temperature. However, explanation of the observed properties of solids by means of the kinetic molecular theory requires, as it did for liquids, modification of assumptions 2 and 3 for gases.

1. *Atoms, molecules, or ions in a solid are in contact with each other.* Free space is minimal. Four general properties of solids can be explained on the basis of this assumption.

a. Solids, are essentially incompressible because, with minimal free space, electron clouds of adjacent atoms, molecules, or ions repel immediately on attempting to force them closer together.

b. Solids at constant volume show drastic changes in pressure with changes of temperature because the directly contacting atoms, molecules, or ions cause an even greater number of collisions per unit wall area that those in liquids when higher temperature imparts higher velocities and kinetic energies to the molecules.

c. The volume of a solid increases as additional moles are added because, with minimal free space, more molecules cannot be forced into the same volume.

d. Solids diffuse even more slowly than liquids because the average distance between atoms, molecules, or ions is so small that they suffer great hindrance from billions of collisions with neighboring atoms, molecules, or ions in a crystal.

2. *Attractive forces between atoms, molecules, or ions in a solid are strong enough to hold them in fixed positions.* Three general properties of solids can be explained on the basis of this assumption.

a. Solids have characteristic shapes or geometric patterns because of the strong attractive forces holding atoms, molecules, or ions in well-defined position.

b. Solids maintain their volume and show little change in volume with changes in temperature because the strong attractive forces keep the atoms, molecules, ions clustered together.

c. Solids melt or sublime because a finite number of atoms, molecules, or ions may be near the surface of the solid and possess kinetic energy enough to overcome attractive forces and escape into the liquid or vapor phase.

Melting

Ice melts when the temperature rises above 0°. **Melting** *is the conversion of a solid to a liquid.* Melting occurs when a solid a warmed sufficiently so that atoms, molecules, or ions at the lattice points in a crystal overcome attractive forces and are free to slide past each other and exchange neighbors.

The melting point of solid depends on the nature of the bonding forces in the solid. Solids with relatively small forces of attraction have low melting points; solids with relatively large force of attraction have high melting points.

Melting points also change slightly with pressure. *The* **normal melting points** is defined as *the temperature of the solid-to-liquid transition at a pressure of* 1 *atm*. Melting points increase marginally with increasing pressure for most solids. However, the opposite is true for ice.

The melting point of a pure solid is constant at a given pressure. Addition of heat to a solid at its melting point only gives more molecules sufficient kinetic energy to form the liquid and does not change the temperature of the sample until all of the solid has melted. The melting point is a commonly used criterion for identifying a solid and characterizing its purity. A solid contain in impurities melts at a lower temperature than the corresponding pure solid.

Sublimation

Ice and snow disappear gradually in the winter even when the temperature does not rise above 0°. No melting occurs; instead the ice and snow are converted directly to water vapor. **Sublimation** is *the direct conversion of a solid to a gas* without passing through the liquid state. Just as in a liquid the atoms, molecules, or ions composing a solid have a distribution of kinetic energies at any given temperature. At any given time a few of them have the requisite kinetic energy to escape into the gas phase (to sublime). As with evaporation of liquids, an *equilibrium vapor pressure* arises from sublimation of solids. At higher temperatures more atoms, molecules, or ions have the necessary kinetic energy to sublime. Thus, vapor pressures of solids increase with increasing temperature, having lower magnitudes but the same trends as illustrated for liquids. Solids with high vapor pressures at room temperature are said to be *volatile*.

Vapor pressures of solids depend on the nature of the bonding forces in a solid. Solids with relatively weal forces of attraction have high vapor pressures; solids with relatively strong forces of attraction have low vapor pressures.

Types of Solids

We have indicated how the specific properties of solids are related to the nature of the bonding forces in such solids. Bonding forces in turn depend upon what particles occupy the lattice sites in a crystal. The bonding forces may be very strong covalent forces as between C atoms in the diamond lattice, or may be very weak dispersion forces as between Ar atoms in solid argon.

Six types of solids can be distinguished on the basis of particles that occupy lattice sites. Table 6.1 summarizes these six types in terms of the nature of bonding forces between particles, the general properties which each type exhibits, and specific examples of each type. Many solids can be classified into one of these six general categories.

Polymeric solids have polymers at the lattice points. **Polymers** are *very large molecules*, frequently consisting of long chains of atoms, two-dimensional sheets of atoms, or three-dimensional network of atom. Polymeric solids like Teflon are very important industrially. Others like cellulose and proteins are important biologically.

Amorphous Solids

Many polymeric materials have a particularly strong tendency to cool below their freezing point without crystallizing. Under appropriate conditions, a number of liquids can be made to do this. As the temperature is decreased, such liquids become less mobile (more viscous) and eventually become quite rigid. *Rigid, noncrystalline materials* are called **amorphous solids** *or* **glasses**. Amorphous solids resemble liquids in so far as they have short-rang order rather than the ling-range order characteristic of crystalline solids. However, they differ from liquids and resemble solids in being rigid and immobile.

Ordinary window glass is a good example of an amorphous solid. It does not have the highly ordered structure of quartz (SiO_2), Instead, it has Si and O atoms arranged in *irregular* tetrahedral networks, each Si surrounded by four O atoms and each O atom shared between two Si atoms. Glass has reasonable strength and remains quite rigid except at high temperatures because of strong Si—O bonds. However, if it does break, due to its disorderly structure it will fracture into irregular shell-like chips rather the cleaving with the formation of plane faces.

Various types of glass, rubber, and plastic are popular commercial amorphous solids because they can be fabricated into almost any shape by molding or blowing. Their constituents can be chosen to meet a variety of requirements: reasonable strength per unit mass, good insulation of heat or electricity, good resistance to chemical attack, or good resistance to mechanical abrasion.

Table 6.1 : Types of Solids

Type	*Particles Occupying Lattice Sites*	*Forces Between Particles*	*Properties of Solids*	*Examples*
Atomic solids	Atoms	Covalent bonds between atoms	Very high melting points Very hard Non volatile Insulators	Diamond, $C_{(s)}$ $SiO_{2(s)}$ $Ge_{(s)}$ $B_{(s)}$ $Si_{(s)}$ Boron nitride
Ionic solids	Cations and anions	Electrostatic attraction between oppositely charged ions	High melting points nonvolatile Hard, but shatter Insulators	$NaCl_{(s)}$ $CaF_{2(s)}$ $MgSO_{4(s)}$ $(NH4)_3PO_{4(s)}$ $KBr_{(s)}$ $KNO_{3(s)}$
Polar molecular solids	Polar molecules	Dipole-dipole attraction between molecules	Moderate melting points Moderately volatile and hard Insulators	$HCl_{(s)}$ $H_2O_{(s)}$ $SO_{2(s)}$ $NO_{(s)}$ $NH_{3(s)}$ $HF_{(s)}$ $PCl_{3(s)}$
Nonpolar molecular solids	Nonpolar molecules (or atoms)	Dispersion forces between molecules (or atoms)	Low melting points Volatile Soft Insulators	$I_{2(s)}$ $CO_{2(s)}$ $CH_{4(s)}$ $SiCl_{4(s)}$ $Ar_{(s)}$
Metallic solids	Cations in an electron gas	Attraction between cations and electrons	Variables melting Variable hardness Low volatility Good conductors	$Na_{(s)}$ $Cu_{(s)}$ $Mg_{(s)}$ $Al_{(s)}$ $Fe_{(s)}$ $Cr_{(s)}$ $Ni_{(s)}$ $Zn_{(s)}$
Polymeric solids	Polymers	Dispersion forces and hydrogen bonding	Variable melting Variable hardness Low volatility	Polyethylene Polystyrene Nylon Orlon Starch Protein

SOLUTIONS

Introduction

This chapter is concerned with mixtures of substances. Mixtures can be divided into two general categories: heterogeneous mixtures and homogeneous mixtures. Mixtures in the latter category are called *solutions.* Here, homogeneous means that visually the material appears uniform throughout. Thus, by visual of observation it is not possible to distinguish between solutions and pure substances.

Although we normally think of solutions as liquids, there are gaseous solutions and solid solutions. Air is an example of a gaseous solution consisting primarily of a mixture of elemental nitrogen and oxygen. Certain alloys (such as 14-karat gold, a mixture of gold, silver, and copper) are example of solid solutions.

When a solid is mixed with a liquid to produce a solution, the solid is the *solute* and the liquid, the *solvent.* Similarly, the gas is the *solute* and the liquid, the *solvent* in a solution prepared from mixing a gas with a liquid. When a solution is prepared from two liquids, the distinction between solute and solvent is less clear. Normally, the liquid present in the larger amount is called the solvent and the one in lesser quantity, the solute.

Why do Substances Dissolve

Some materials dissolve in certain solvents and others do not dissolve. For example, when a spoonful of sugar is put into coffee or tea, the sugar dissolves but the spoon does not ! Common experience also shows us that grease does not dissolve in water; instead, it floats on the surface of the water. Grease dissolves in cleaning fluid, and it appears to dissolve in water when soap or detergent is present. Why do these substances behave in this fashion? What is different about cleaning fluids that enables them to dissolve grease while water will not? In order to answer these questions we need to consider the molecular structure, bonding characteristics, and forces of attraction between the molecules of both the potential solutes and solvents.

We know that the strong electrical attractions between oppositely charged ions caused them to gather in an orderly arrangement. These strong forces allow little movement of the ions, with the result that ionic compounds are high-melting solids.

Some Covalent compounds are composed of polar molecules and other of nonpolar molecules. Polar molecules are attracted to one another. The attraction between polar molecules can be accounted for by attraction between dipoles or by hydrogen bonding. These forces are weaker than the forces between ions and the molecules may move past one another more easily than ions may move. Consequently, polar molecules are generally liquids or low-melting solids.

Inter molecular attractive forces between nonpolar molecules must exist between nonpolar molecules. Otherwise, why would nonpolar molecules come together to form liquids and solids? In the absence of any intermolecular forces, the molecules would not be expected to come together to form a liquid or a solid. The attractive forces between nonpolar molecules are called *London forces* or *dispersion forces.* In nonpolar compounds, the London forces reveal that they gradually increase as the size and molecular weight of the molecules increase.

Generally speaking, except when hydrogen bonding is involved, the London forces also account for the bulk of the attractive forces between polar molecules. Thus, in the absence of hydrogen bonding, the London forces are mainly responsible for inter molecular attraction in both nonpolar and polar compounds. When hydrogen bonding is involved, it is primarily responsible for intermolecular attraction. Now let's see how this discussion helps answer questions about solution formation.

Experimentally it is found that carbon tetrachloride dissolves in hexane but is insoluble in water. Both carbon tetrachloride and hexane are nonpolar substances. Thus, London forces account for the for attraction between CCl_4 molecules. Similarly, London forces hold the hexane molecules together in the liquid state. Since London forces exist between all molecules and only slowly increase with molecular size, we might expect that the attraction between a CCl_4 molecular and a hexane molecule would be about the same as the attraction between two CCl_4 molecules or between two hexane molecules.

Following this reasoning, the sum of the attractive forces between the molecules in a mixture carbon tetrachloride and hexane would be approximately the same as the sum of the attractive forces in pure carbon tetrachloride and those in pure hexane. In agreement with the experimental results we predict that carbon tetrachloride will dissolve in hexane since there is little or no change in the attractive forces upon mixing. These arguments hold true for mixtures of most nonpolar compounds and we may formulate a general rule: *Nonpolar compounds dissolve in nonpolar solvents.*

Since carbon tetrachloride is insoluble in water, it must be concluded by the arguments in the paragraphs above that the sum of the attractive forces in a mixture of carbon tetrachloride and water would be less than the sum of those in pure carbon tetrachloride plus those in pure water. Water possesses a high molecular polarity and that hydrogen bonding occurs between water molecules. Both of these factors lead to large attractive forces between water molecules. Carbon tetrachloride is nonpolar; consequently, there should be little attraction between CCl_4 and H_2O molecules. Under these circumstances, water molecules will associate with each other in preference to associating with CCl_4 molecules. Thus, water and carbon tetrachloride are mutually insoluble because of the lack of association between the molecules of these two compounds. Thus we can generalise that *nonpolar compounds are insoluble in water because the attractive forces between nonpolar molecules and water molecules are less than the attractive forces between water molecules.*

The Solubility of Ionic Compounds

Crystalline ionic compounds are collections of ions held together by the attractive forces between the oppositely charged cations and anions. When an ionic compounds is dissolved, a breakdown of his orderly arrangement of ions occurs. Interionic forces must be overcome during the dissolving process. Thus, if an ionic compound is dissolved, it seems reasonable to assume that the attractive forces between the ions and the solvent are greater than the interionic forces in the crystal. Chemists have found that, in general, ionic compounds are insoluble in most liquids except water. This observation suggests that an electrical attraction between the charged ions and partially charged atoms on polar molecules is responsible for the dissolving of ionic compounds.

Consider a crystal of sodium chloride made up of alternation sodium and chloride ions. When the salt crystal is placed into water, the water molecules will turn so as to place the partially positively charged hydrogen atoms of nearby water molecules near negatively charged chloride ions on the surface of the salt crystal. Other water molecules will be oriented so as to place the partially negatively charged oxygen atom near the positive sodium ions on the crystal surface. This process is called *solvation*; when water is the solvent, it may be called *hydration*. The hydration process reduces the attraction between the oppositely charged ions in the crystal. The result of this reduced attraction between ions is a breaking away of ions from the crystal surface. As each ion is hydrated and breaks away, another ion in the crystal is exposed to the water and the hydration-breaking away process may be repeated. Thus, the crystal is literally picked apart and the ions dispersed among the water molecules as the compound dissolves Fig 6.8.

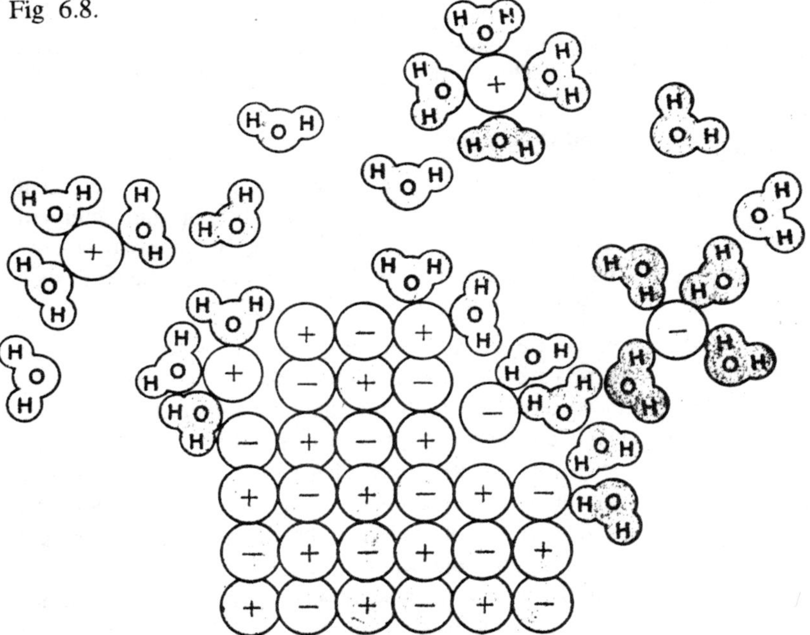

Fig. 6.8 : The dissolving of an ionic compound in water.

However all ionic compounds are not water soluble. Thus should not be surprising if it is remembered that there are two opposing forces that influence the solubility of ionic compounds: (1) the attraction between oppositely charged ions, tending to keep the ions in the crystal (2) the attraction of the polar solvent molecules for the ions, tending to pull the ions away from the crystal. Since both forces originate from the electrical attraction between the ions in the crystal (such as a larger ionic charge) will also increase the attraction between the ions and the water molecules. Consequently, there are no simple rules enabling us to predict the solubility of ionic compounds. It has been possible, however, to compile some useful guidelines to the solubility of ionic compounds. These guidelines are :

1. Ionic compounds of the alkali metals and ammonium ion are soluble.
2. All ionic compounds containing acetate and nitrate ions are soluble.
3. All metal chlorides are soluble except those of Ag^+, Hg_2^{2+}, and Pb^{2+}
4. All carbonates and phosphates, except these of the alkali metals and ammonium ion, are insoluble.
5. Most metal hydroxides are insoluble except those of the alkali metals and alkaline earths. Hydroxides of the latter are not slightly soluble.

Action of Soaps and Detergents

From everyday experience we know that a soap or detergent will apparently cause grease to dissolve in water. Soap and detergent molecules consist of a relatively large portion that is nonpolar or only slightly polar and an ionic group as the other portion. Several such molecules are shown in Fig 6.9.

The reason grease or fats and oils don't dissolve in water is because they are relatively nonpolar molecules. The very polar water molecules prefer to associate with one another rather than with the nonpolar molecules. Consider the structure of soap and detergent molecules; we find that one portion of the molecules would prefer to associate with the nonpolar fat or oil molecules and the other portion with the water molecules. The result of this is to break up grease globs and to disperse them throughout the water as shown in Fig 6.10. Notice in this figure that the grease is not dispersed throughout the water as individual molecules; microscopic droplets of a mixture of grease and soap are dispersed in the water. These droplets are called *micelles* (pronounced "my-cells"). The micelles disperse in the water because of the attraction between the ionic portion of the soap molecules and the polar water molecules. Also, notice that the micelles are negatively charged, causing them to repel one another, which prevents them from clumping together again. Because of the presence of these microscopic micelles, a true solution has not been formed; instead an *emulsion* or *suspension* has been produced. For this reason, soaps and detergents are classified, along with other molecules possessing similar types of structural features, as *emulsifying agents* or *emulsifiers.*

The difference between a soap and a detergent.

In a soap the ionic portion of the molecule is always either of

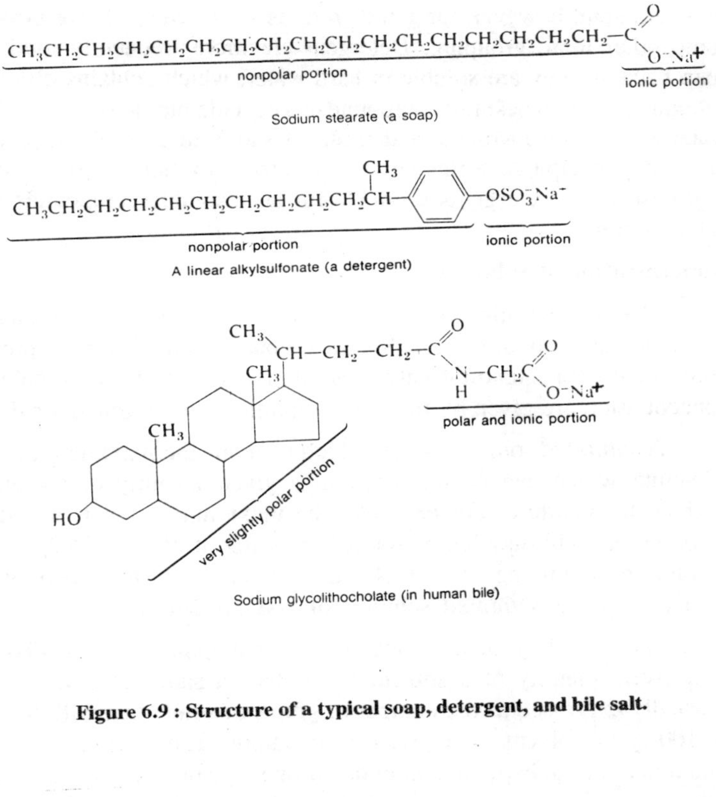

$CH_3CH_2CH_2CH_2CH_2CH_2CH_2CH_2CH_2CH_2CH_2CH_2CH_2CH_2CH_2CH_2CH_2{-}C({=}O)O^-Na^+$

nonpolar portion — ionic portion

Sodium stearate (a soap)

$CH_3CH_2CH_2CH_2CH_2CH_2CH_2CH_2CH_2CH_2CH(CH_3){-}C_6H_4{-}OSO_3^-Na^+$

nonpolar portion — ionic portion

A linear alkylsulfonate (a detergent)

CH_3, $CH{-}CH_2{-}CH_2{-}C({=}O){-}N(H){-}CH_2C({=}O)O^-Na^+$, CH_3, CH_3, HO

very slightly polar portion — polar and ionic portion

Sodium glycolithocholate (in human bile)

Figure 6.9 : Structure of a typical soap, detergent, and bile salt.

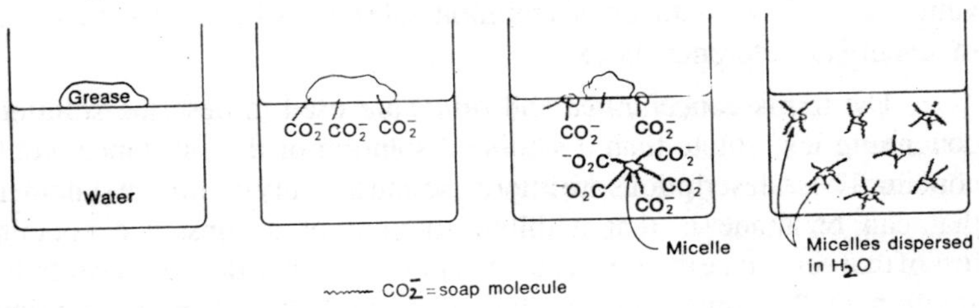

Fig. 6.10 : The process by which a soap disperses grease in water.

these groups:

$$-C\overset{\displaystyle O}{\underset{\displaystyle O^{-}Na^{+}}{\Large\lessgtr}} \quad \text{or} \quad -C\overset{\displaystyle O}{\underset{\displaystyle O^{-}K^{+}}{\Large\lessgtr}}$$

In a detergent, however, the ionic group is —O—SO_3^- Na^+ or possibly some other ionic grouping. Detergent are more frequently used that soap because they are soluble in hard water, which contains dissolved calcium and magnesium compounds. The calcium ions in the hard water will combine with the soap molecules to form an insoluble precipitate. This precipitate forms curds that adhere to clothes, giving them a grayish cast. Detergents do not form an insoluble compounds with calcium ions.

Concentration of Solutions

The concentration of a solution is a measure of the amount of solute in a quantity of solvent. Solution concentration can be expressed qualitatively or quantitatively. *Qualitative* expressions of solution concentration are given by the terms *saturated, concentrated,* or *dilute.*

A *saturated* solution is one which contains the maximum amount of solute which can be dissolved in a given quantity of solvent at a given temperature. For example, the maximum amount of sodium chloride that will dissolve in 100 mL of water at 25°C is 36.2g. Thus, a solution containing 36.2 g of NaCl dissolved in 100 mℓ of water is said of be a *saturated* solution of sodium chloride.

The *solubility* of a compound is the amount that will dissolve in a given quantity of a solvent to produce a saturated solution and is usually expressed as the number of grams of solute that will dissolve in 100 g of solvent at a given temperature. Thus the solubility of a compound is an experimentally determined quantity and the solubility values for a large number of common substances have been tabulated in chemical reference books.

The terms *concentrated* and *dilute* are used to describe solutions containing less solute than a saturated solution of the substance would contain. These descriptions are imprecise and the only definitive statement that can be made is that a dilute solution of a substance contains less of that substance in a given quantity of solvent than does a concentrated solution in the same solvent. Both concentrated and dilute solutions of a substance are said to be *unsaturated* solutions since they contain less solute that is required to produce a saturated solution.

Quantitative Aspects of Solutions

Qualitative descriptions of solutions may at times be useful but in general are unsatisfactory. In chemistry, medicine, and other fields it is often necessary to know exactly how much of a substance is contained in a given amount of a solution. Medicinal dosages generally require the administration of a prescribed amount of a therapeutic agent of known concentration. A large number of ways for quantitatively expressing concentrations of solutions have been devised. We shall limit out study to several of the most common methods of expressing concentration.

Percent concentration

The definition of percent concentration we shall use is the so-called weight-volume percent, defined as the *grams of solute per 100 mL of solution.*

$$\text{Percent concentration} = \frac{\text{grams of solute}}{\text{100 mL of solution}}$$

A 5 percent glucose solution is one containing 5 g of glucose per 100 ml of solution. To prepare 100 mℓ of 5 percent glucose solution. To prepare 100 mL of 5 percent glucose solution, weigh 5 g of glucose and add to it enough water to give a *total volume of solution* of 100 mL. Notice that is the volume of the solution and *not* the volume of water added that must be 100 mL.

***Example* 6.8**

How would one prepare 6 L of 5 percent glucose solution?

amount of glucose = (6L)(1000 mL/1 L)(5g glucose/100 mL solution)
= 300 g of glucose

Weigh out 300 g of glucose and add enough water to make a total volume of 6 L.

***Example* 6.9**

How many milliliters of 5 percent glucose solution should be used to have 10 g of glucose?

(10 g glucose) (100 mL solution/ 5 g glucose) = 200 mL

Molarity

Chemicals react on mole-to-mole ratios. Since many reactants are in solution, the number of moles of solute in a given quantity of solution should be a useful way expressing concentration.

The molarity of a solution is defined as the number of moles of solute per liter of solution. A solution containing 1 mole of solute per litre of solution is said to be *1 molar,* abbreviated *1 M.*

$$\text{Molarity} = \frac{\text{No. of moles of solute}}{\text{Volume of solution in litres}}$$

A *1 M* solution of glucose contains 1 mole of glucose per litre. Since the gram formula weight of glucose of 180g, 1 L of 1 *M* glucose solution could be prepared by weighing 180 g of glucose and adding enough water to yield 1 L of solution. Greater or lesser amounts of solution can be prepared by using different quantities of solute and solvent.

***Example* 6.10**

100 mL of 1 m glucose can be prepared by using 18 g of glucose and enough water to give a solution of 100 mL. The amount of glucose required is calculated in the following manner.

$$\text{Grams glucose required} = (0.1\text{L})(1\ \cancel{\text{mole}}/\text{L})(180\text{ g glucose}/1\ \cancel{\text{mole}}\text{ glucose})$$
$$= 18\text{ g glucose}$$

***Example* 6.11**

Solution concentrations may be greater or less than 1 *M*. 500 mL of 0.3 *M* glucose can be prepared by dissolving 27 g of glucose in enough water to yield 500 mL of solution. The grams of glucose needed are calculated:

$$\text{grams glucose required} = (0.5\text{ L})(0.3\ \cancel{\text{mole}}/\text{L})(180\text{ g glucose}/\cancel{\text{mole}}\text{ glucose})$$
$$= 27\text{ g glucose required}$$

A general equation to calculate the number of grams of solute required is

grams solute required = (amount, in litres, of solution desired)(molarity of desired solution)(gram formula weight of solute)

Frequently we will find that a solution has already been prepared and we need to calculate *the volume of solution to use in order to obtain a specified number of moles of solute.* This calculation is easily performed via the equation

$$\text{volume of solution needed} = \frac{\text{moles solute desired}}{\text{solution molarity}}$$

Another useful calculation involves determining *the number of moles of solute contained in a given quantity of a solution of know concentration.*

The equation

moles of solute = (solution molarity) (volume, in litres)

allows this calculation to be made.

Equivalents Per Litre

Just as the concentration of solutions can be expressed in moles per liter, the concentration of an ion in a solution can be expressed in equivalents per liter. To determine the number of equivalents of an ion in a given quantity of a solution of known concentration, we multiply the concentration expressed in equivalents per liter by the volume in liters.

***Example* 6.12**

Calculate the number of equivalents of chloride ion in 500 mL of solution containing 0.3 eq/L of chloride ion.

$$\text{eq of } Cl^- = (0.3\ \text{eq}/\cancel{L})\ (0.5\ \cancel{L}) = 0.15 \text{ eq of } Cl^-$$

***Example* 6.13**

Calculate the number of equivalents of aluminium in 500 mL of 0.05 *M* $Al_2(SO_4)_3$.

$$500\ \cancel{mL}\ (1\ L/1000\ \cancel{mL})[0.05\ \cancel{\text{moles } Al_2(SO_4)_3}/L]$$

$$[2\ \cancel{\text{moles } Al^{3+}}/\cancel{\text{mole } Al_2(SO_4)_3}](3 \text{ eq } Al^{3+}/\cancel{\text{mole } Al^{3+}}) = 0.15 \text{ eq } Al^{3+}$$

Since the concentration of ions in body fluids is relatively low, it is often given in milliequivalents per litre. There are 1000 milliequivalents (meq) in 1 eq. A solution containing 0.01 eq/L of Cl^- contains

$$(0.01\ \cancel{eq}/L)\ (1000\ meq/\cancel{eq}) = 10\ meq/L$$

The number of milliequivalents of an ion contained in a given quantity of a solution is calculated by multiplying the *concentration of the ion expressed in equivalents per litre by the volume in millilitres.*

Dilutions

Laboratory supply rooms commonly stock concentrated solutions that can be diluted to yield solutions of lesser concentration. When a solution is diluted, the solute/solution ratio is decreased. To accomplish this, all we have to do is mix more solvent with the concentrated solution. After addition of more solvent we end up with a larger volume of solution of lesser concentration. For example, if we have 1 L of 1 *M* glucose and we add enough water to give a volume 2 L, the resulting solution must be 0.5 *M*. Remember that 1 L of 1 *M* glucose contains 1 mole of glucose. After we add more water, the same mole of glucose is dissolved in 2 L of solution. The molarity must now be

$$1 \text{ mole glucose}/2 \cancel{L} = 0.5 \text{ mole glucose}/1 \cancel{L} = 0.5\ M \text{ glucose}$$

The total number of moles of glucose in the concentrated solution equals the number of moles in the diluted solution since no glucose was added or removed. *In the general case, the moles of solute in the concentrated solution equals the moles of solute in the diluted solution,* expressed mathematically by Eq. (7.1)

$$\text{moles of solute}_{(\text{solution 1})} = \text{moles of solute}_{(\text{solution 2})} \qquad (7.1)$$

Remember for any solution concentration expressed in molarity that Eq(7.2) must be true.

$$\text{moles of solute} = (\text{molarity})(\text{volume in litres}) \qquad (7.2)$$

Combination of Eq. (7.1) and Eq (7.2) yields Eq (7.3)

$$[\text{molarity}_{(\text{solution 1})}][\text{volume in litres}_{(\text{solution 1})}] = [\text{molarity}_{(\text{solution 2})}][\text{volume in litres}_{(\text{solution 2})}]$$

or $$M_1V_1 = M_2V_2 \qquad (7.3)$$

***Example* 6.14**

Suppose we need 500 mL of 0.5 *M* glucose and all that is available is a 2.0 *M* glucose solution. Applying Eq. 7.3,

$$M_1 = 2.0\ \text{M}$$
$$V_1 = ?$$
$$M_2 = 0.5\ \text{M}$$
$$V_2 = 0.500\ \text{L}$$
$$(2.0\ M)\ (V_1) = (0.5000\ \text{L})\ (0.5M)$$

$$V_1 = \frac{(0.500\ L)(0.5\ \cancel{M})}{(2.0\ \cancel{M})} = 0.125\ \text{L} = 125\ \text{mL}$$

Thus, 125 mL of the 2.0 *M* glucose solution should be diluted to 500 mL to have a 0.5 *M* glucose solution. (*Do not add 500 mL of water; dilute the given solution of 125 mL to 500 mL).*

In medical laboratories and hospital, percent concentration is used much more frequently than molarity is; yet there remains the need to carry out dilutions of solutions. Since percent concentration is the grams of solute per 100 mL of solution, an equation similar to Eq.(7.3) that can be used in dilution problems involving percent concentration is Eq. (7.4).

$$[\%\ \text{concentration}_{(\text{solution 1})}][\text{volume in millilitres}_{(\text{solution 1})}] = [\%\ \text{concentration}_{(\text{solution 2})}][\text{volume in millilitres}_{(\text{solution 2})}]$$

***Example* 6.15**

Suppose the doctor prescribes injection of 1 cc of 1.25 percent chloropromazine and the only solution available is 2.5 percent chlorpromazine. How should one proceed to administer the drug?

$$(2.5\%)(\text{volume in milliliters}) = (1.25\%)(1\ \text{mL})$$

$$\text{volume} = \frac{(1.25\%)(1\ \text{mL})}{(2.5\%)} = 0.5\ \text{mL}$$

Thus, the correct procedure would be to use 0.5 mL of 2.5 percent chlorpormazine solution that has been diluted to 1.0 mL for the injection.

Titrations

Titrations are used to analyze a solution to determine the concentration of solute. When a titration is used to analyze a solution, it is necessary to know the balanced equation for the reaction that occurs.

Suppose that we have a solution of sodium chloride of unknown concentration and it is necessary to determine its concentration. It is known that a solution of sodium chloride will react with a solution of silver nitrate to produce insoluble silver chloride according to the equation

$$NaCl + AgNO_3 \longrightarrow AgCI + NaNO_3$$

In order to find the concentration of the sodium chloride solution we would measure out a known volume of the NaCl solution and titrate it with (add to it) a solution of known concentration of $AgNO_3$ until the formation of the insoluble AgCl ceases. The apparatus shown in Fig. 6.11 is used for this procedure. When the formation of AgCl ceases, all the NaCl must have reacted and the end point of the reaction has been reached. From the data collected the molarity of NaCl can be determined.

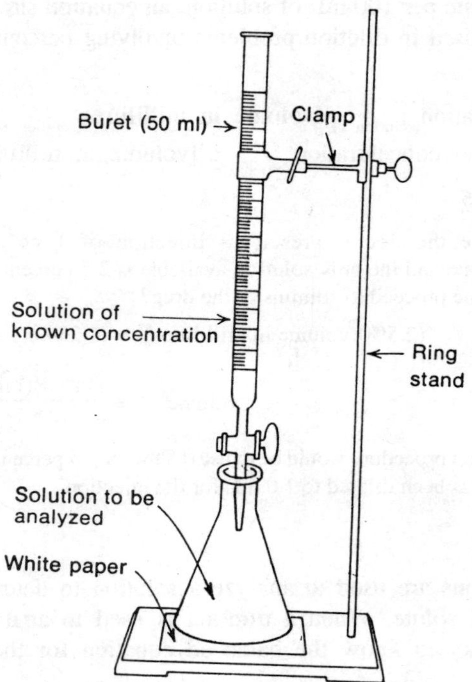

Fig. 6.11 : Apparatus for titrations.

***Example* 6.16**

In an experiment, 37.2 mL of 0.115 *M* $AgNo_3$ was required to titrate 25.0 mL of a NaCl solution to the end point. What is the molarity of the NaCl solution?

The balanced equation tells us that one mole of $AgNO_3$ reacts with one mole of NaCl. Using the equation $M_1V_1 = M_2V_2$

we have

$$\text{Molarity of NaCl } (M_1) = \frac{(0.115\ \text{M})(0.0372\ \text{L})}{(0.025\text{L})} = 0.171\ M$$

Many tests performed in hospital laboratories employ titrations. One example is the determination of the concentration of oxalate ions in urine. In one procedure, calcium chloride is added to a urine sample to precipitate the insoluble calcium oxalate. The calcium oxalate is redissolved in sulfuric acid and titrated with a potassium permanganate solution. The balanced equation for the reaction between calcium oxalate and potassium permanganates is

$$8H_2SO_4 + 5CaC_2O_4 + 2KMnO_4 \longrightarrow 10CO_2 + 8H_2O + 5CaSO_4 + 2MnSO_4 + K_2SO_4$$

***Example* 6.17**

In a typical oxalate determination, 10 mL of urine was treated with $CaCl_2$ and the calcium oxalate formed redissolved in 1 *M* sulfuric acid solution. To reach the titration end point, 8.7 mL of 0.00102 *M* $KMnO_4$ solution was required. What was the concentration of oxalate ions in the urine sample?

The chemical equation tells us that 5 moles of CaC_2O_4 reacts with 2 moles of $KMnO_4$. Therefore at the end point the moles of CaC_2O_4 in the sample must equal $\frac{5}{2}$ the moles of $KMnO_4$ added. Thus,

$$\text{moles } CaC_2O_4 = \frac{5}{2}(0.0087\,\cancel{L}\ \text{KMnO4 added})\left(0.00102\frac{\text{mole } KMnO_4}{\cancel{L}}\right)$$

$$\text{moles } CaC_2O_4 = 0.000008974 = 8.974 \times 10^{-6} \text{ mole } CaC_2O_4$$

Since 1 mole of oxalate ions ($C_2O_4^{2-}$) is contained in 1 mole of CaC_2O_4, the urine sample contained 8.974×10^{-6} mole of $C_2O_4^{2-}$

Therefore the concentration of $C_2O_4^{2-}$ in the urine sample is

$$\frac{8.974 \times 10^{-6} \text{ mole}}{0.010\,\cancel{L}} = 8.974 \times 10^{-4} \text{ mole}/\cancel{L}$$

To calculate the concentration in moles per liter, we must divide the moles of soluble in the sample by the volume (in liters) of the sample.

7

Chemical Equilibrium

REACTION REVERSIBILITY

The solution-making process, osmosis, and dialysis are all examples of a dynamic situation. Each involves a reaction in one direction in competition with a reverse reaction. In such cases, the forward *and* reverse reactions occur *simultaneously*. There are some chemical reactions, however, that proceed only in the forward direction with reactants being converted to products. Such reactions are called *irreversible reactions*. In effect, the products are not able to chemically *regroup* and to be reconverted to reactants by the reverse reaction. Because of this irreversibility, it is not possible for an irreversible reaction to establish an equilibrium between reactants and products. The thermal decomposition of sugar (sucrose) exemplifies an irreversible reaction:

$$C_{12}H_{22}O_{11}(s) \xrightarrow{\Delta} 12C(s) + 11H_2O(g) \qquad \Delta \text{ signifies heat}$$

No conditions are known under which elemental carbon can directly react with water to form sucrose. The reaction above cannot be an equilibrium reaction. (If it could equilibrate, rain could convert piles of coal to sugar). Baking a cake, frying meat and boiling an egg are examples of irreversible reactions.

Reversible reactions are those in which the reverse reaction can compete significantly with the forward reaction. Equilibrium is reached when the *rates* of forward and reverse reactions become equal. At equilibrium there is no *apparent* change in the reaction mixture as no **net** loss or gain of material occurs. At equilibrium the concentrations of reactants and products do not have to be equal; they usually are not.

CHEMICAL EQUILIBRIUM

Let's first look at the establishment of an equilibrium by using a generalized reversible reaction in which substances A and B are converted chemically into substances C and D according to the equation

$$A + B \rightleftharpoons C + D$$

A and B represent reactants
C and D represent products

Before reaction, no product C and D has formed and their concentration is zero. As the reaction begins, the rate of the forward reaction is greater than the reverse reaction rate since the concentration of reactants is large compared to the concentration of products. As the reactions continues, reactants are converted to products, reactant's concentration decreases, and the rate of the *forward* reaction slow (recall that a change in concentration affects reaction rates). As products is produced, its concentration increases and the reverse reaction's rate increases. When the forward and reverse *rates* become equal, equilibrium is established and no net change in the concentrations of reactants and products occurs as long as neither reactants nor products is lost from the reaction mixture. This is shown in Fig 7.1.

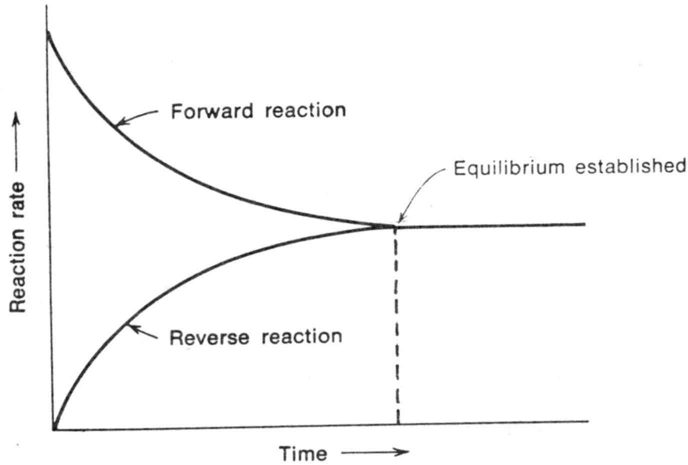

Figure 7.1 : Relationship of reaction rates for a reversible reaction and establishment of equilibrium.

At equilibrium, how do the concentrations of reactants and products compare? We find there are only three possible situations for all equilibria that can be represented by reactants $\rightleftharpoons$ products.

Situation 1. The concentration of products is approximately equal to the concentration of reactants at equilibrium: [products] = [reactants].

Situation 2. At equilibrium, the concentration of products concentration of reactants : [Products] > [Reactants]. This situation is described by saying the equilibrium favors products formation, *i.e.*, the forward reaction.

Situation 3. The concentration of products *less than* the concentration of reactants: [products] < [reactants] at equilibrium. In such a condition, the equilibrium favors the reactants, *i.e.*, the reverse reaction.

In most chemical reactions that reach equilibrium, situation 2 or situation 3 apply: [products] ≠ [reactants] at equilibrium. Either products formation (forward reaction) or reactants are favored depending on reaction conditions.

Many physiologically and biochemically significant processes involve reversible reactions that may establish equilibrium. In the blood, the normal range of glucose concentration is 65 to 100 mL of blood. This level is adjusted in several ways, three of which are by (1) dietary intake (eating), (2) exercise, and (3) the use of stored glucose called glycogen. Under normal conditions, dietary intake provides sufficient glucose to maintain this level and any excess glucose ingested can be chemically converted to glycogen and stored in the muscles and liver. This is to say that an equilibrium exists between glucose formation and glycogen formation:

$$x \text{ glucose} \underset{\text{Reactants}}{\overset{\text{Product}}{\rightleftarrows}} xH_2O + \text{glycogen}$$

At equilibrium,

rate of forward reaction (glycogen formation) =
rate of reverse reaction (glucose formation)

If the rates of these two opposing reactions are not equal, the system is not at equilibrium. The direction (forward reaction or reverse reaction) *from which equilibrium is approached does not alter (determine) the concentrations of products and reactants at equilibrium.*

Le Chatelier's Principle

Let's now think about three situations in which the position of the glycogen/glucose equilibrium is altered by placing a chemical demand (stress) on it.

Situation 1. *Cause*: Not enough glucose is taken in by eating. *Result:* Insufficient glucose in present for the reverse rate to equal the forward rate. The forward rate is less than the reverse rate. This imbalance causes a net conversion of glycogen to glucose that continues until equilibrium is reestablished.

Situation 2. *Cause*: An excess of glucose is ingested by a person. *Result*: The glucose concentration is increased by the addition of glucose. The forward reaction rate increases converting the excess glucose to glycogen.

Situation 3. *Cause*: A person participates in strenuous exercise. *Result:* Glucose is metabolized to furnish energy required during exercise. This lowers the concentration of glucose. The forward reaction rate on longer equals the reverse reaction rate and glycogen is converted into glucose until equilibrium is re-established.

Each situation represents a condition in which the status quo of the glycogen/glucose system is changed by the applied stress. The stress causes the system to shift temporarily from an equilibrium to a nonequilibrium condition and the system responds until equilibrium is reestablished. After studying a broad range of chemical equilibria, the French chemist Henri Le Chatelier (1888) pointed out the effect a stress would have on a chemical system at equilibrium: *When a change (stress) is placed on a chemical system that is at equilibrium, the system responds so as to counteract the imposed change and to reestablish equilibrium.*

In the lungs, oxygen combines with hemoglobin (HHb) to form oxyhemoglobin:

$$HHb + O_2 \rightleftharpoons HHbO_2$$

In the lungs, the relatively high concentration of oxygen causes the forward reaction to be favored. The oxyhemoglobin is transported by the blood to the cells. Within cells that require oxygen, the oxygen concentration is low. As oxyhemoglobin is carried in a capillary adjacent to the cell, the low oxygen concentration creates a stress on the equilibrium above and oxygen is liberated. It diffuses across the cell membrane into cell.

For a reversible reaction to achieve equilibrium, both reactants and products must be present. Several situations exist that reversibility of reaction:

The applied stress comes from outside the chemical system.

1. Reactions involving gaseous reactants and/or products can achieve equilibrium only in a closed container that prevents loss of the gaseous species.
2. Reaction involving aqueous *ions* are driven to completion, i.e., no reverse reaction occurs when the formation of an insoluble product (precipitate) occurs.

A catalyst speeds up a reaction by lowering the activation energy. Since the activation energy is lowered for both the forward and reverse reactions, *the presence of the catalyst does **not** affect the equilibrium position.*

Homeostasis

In living cells, very few individual reactions ever reach equilibrium because these reaction are often coupled with each other. This sequencing of reactions places a continuous stress preventing equilibrium from being reached in any previous reactions. We would be dead if all cells in our body achieved equilibrium.

Consider that our body temperature severs as a useful guide to our physiological condition. The normal 37.0 °C (98.6 °F) value and its relative constancy results from the *net* metabolic changes in our body even though our physical activities are not constant. We eat, sleep work, walk, sit, sing—all different activities—and somehow our body accommodates these diverse functions without a drastic change in body temperature even in summer's heat and winter's cold.

Thus, homeostasis involves a total system whose internal environment is constantly compromising by shifting materials from site to site in response to stresses placed upon it so that no **net** change in the environment occurs. Water enters or leaves cells depending on osmotic pressure changes; chloride ions enter cells and bicarbonate ions leave; oxygen and carbon dioxide mutually exchange cellular occupancy: All these processes occur because a stimulus causes an organism to initiate a sequence of metabolic actions attempting to restore normalcy to the system.

Types of Solute Solvent Mixtures

We have described solutions as uniform solute/solvent mixtures and have assumed that the solute would remain uniformly dispersed

---

Reactions are driven to completion because a continuously applied stress prevents equilibrium from being restored.

throughout the solvent. A *true solution* may be described as one in which particles (ions and/ or molecules) with diameters of 0.5 to 2.5 Å are dispersed in a solvent. These solutions are homogeneous and transparent to light. The component particles are too small to be removed by filtration or to be viewed with electron microscopes. To appreciate the size of these particles, consider that approximately 100 million particles, each 2.5 Å in diameter, would form a now only 2.5 cm long when placed side by side.

Opaque or translucent dispersions of solute particles in a solvent are called *suspensions*. Particles in a suspension have diameters greater than 1000 Å, are filterable, and have masses great enough to settle under the force of gravity.

Muddy water is typical example of a suspension.

Is it possible to have a dispersion with particles of average diameters less than 1000 Å but larger than those in a true solution? Yes , it's possible and particles in this size range are called *colloids*. Their dispersions in a solvent are called *colloidal dispersions*. Some common materials present in colloidal form are milk, fruit jellies, gelatin desserts, whipped cream, clouds, and many components of body fluids, including protoplasm. Despite containing larger-sized particles than those in true solutions, colloidal dispersions do **NOT** settle out upon standing.

Solubility and Equilibrium

Experimental data show that a solution is the net result of two processes: (1) the rate of solute dissolving in a solvent and (2) the rate of solute coming out of solution (crystallization).

$$\text{solute} + \text{solvent} \rightleftharpoons \text{solution}$$

Before any solute is added, only pure solvent is present. When a small amount of solute is added, the rate of dissolving is far greater than rate of crystallization. As more solute dissolves in the fixed amount of solvent, the *difference* between the rates of dissolving and crystallization continuously decreases. Eventually, the amount of dissolved solute is such that the rate of solute going into solution (rate 1) equals the rate of solute coming out of solution (rate 2). When this occurs, a saturated solution exists and under these conditions the solute is said to be at equilibrium with the solution.

---

1 Å = 10^{-8} cm = 10^{-10} m

A system at equilibrium is one in which the rate of a process in one direction is equal to the rate of a process in the reverse direction; i.e.,

rate *of forward reaction =* ***rate*** *of reverse reaction*

The symbolism $\rightleftharpoons$ is often used to denote equilibrium. For example, the equation representing the hypothetical reaction A + B $\rightleftharpoons$ C + D is interpreted to mean that **at equilibrium** (*when equilibrium is reached*) the rate of conversion of A and B to C and D (forward reaction) is the same as the rate of conversion of C and D to A and B (reverse reaction). Both reactions are occurring but there is no net change in the system since the rates of these opposing reactions are equal.

Osmosis

The phenomenon of osmosis, a familiar term from a biology course. Osmosis is a mechanism by which water flows in an attempt to equalize solution concentrations between a cell and its environment. Osmosis is the flow of water through an osmotic membrane from a region of pure water or a region of lower solute concentration (greater water concentration) to a region of higher solute concentration (lesser water concentration). An osmotic membrane is one that allows the passage of water molecules but not solute particles.

Consider a U-tube separated into two compartment C and S by an osmotic membrane, o.m. (see Fig 7.2). Initially, compartment C has a higher solute concentration than compartment S. Osmosis will

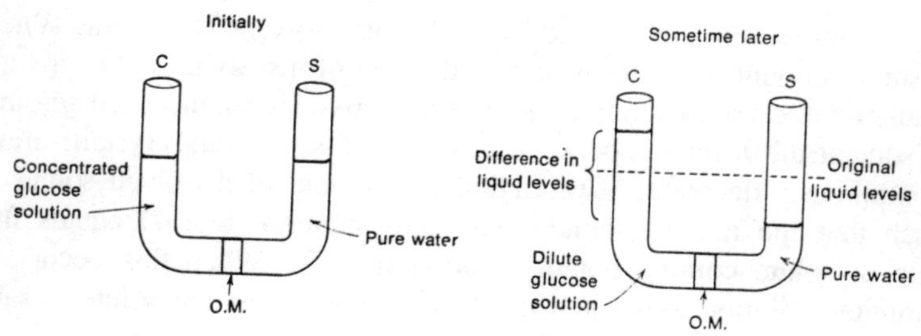

Fig. 7.2 : Solution concentrations and osmosis.

occur and *water* will flow through the osmotic membrane from the region of lower solute concentration (compartment S) to the region of higher solute concentration (compartment C). The flow continues until the water level has risen to a height in tube C great enough so that its weight exerts a sufficient downward pressure on the membrane so that no net transfer of water occurs across the membrane. This downward pressure is called the *osmotic pressure* of the solution. Theoretically the osmotic pressure is controlled solely by the concentration of solute particles present. As the solute concentration changes by water transfer, a point is reached where the rate of water flow in one direction equals the rate of flow in the reverse direction and equilibrium is established. At the point, because the rates of opposing water passage are equal, there is no **net** transfer of water across the membrane.

The osmotic pressure of two solutions, A and B, with respect to each other may be described by one of the following conditions:

Condition 1:

$$\text{osmotic pressure}_A = \text{osmotic pressure}_B$$

Solution are *isotonic.*

Condition 2:

$$\text{osmotic pressure}_A > \text{osmotic pressure}_B$$

Solution A is *hypertonic* to solution B.

Condition 3:

$$\text{osmotic pressure}_A < \text{osmotic pressure}_B$$

Solution A is *hypotonic* to solution B.

Normal saline solution is a 0.9 percent sodium chloride solution, which is isotonic with red blood cells. Water is hypotonic to normal saline and red blood cells placed into pure water swell and rupture, an event called *hemolysis.* Osmosis will occur with a flow of water from the region of lower solute concentration (pure water) across the red blood cell membrane into the region of higher solute concentration (red blood cell). This massive influx of water distends and then ruptures the cell membrane.

Dialysis

An osmotic membrane is permeable to water molecules but not solute particles. A cell restricted to operating under these conditions could neither take in nutrients nor liberate waste materials. Living systems must obviously be capable of more than just the transfer of water molecules. Membranes that are permeable to flow in either direction of true solution-sized particles, including solvent molecules, are called *dialyzing membranes.* When functioning properly, dialyzing membranes are not permeable to particles of colloid or suspension size such as starch or high molecular weight proteins.

If an aqueous mixture of colloidal, suspension, and true solution-sized particles is placed on one side of a dialyzing membrane and pure water on the other side, the true solution-sized particles will pass through the dialyzing membrane into the pure water. This process is called *dialysis.* Schematically, this is shown in Fig. 7.3.

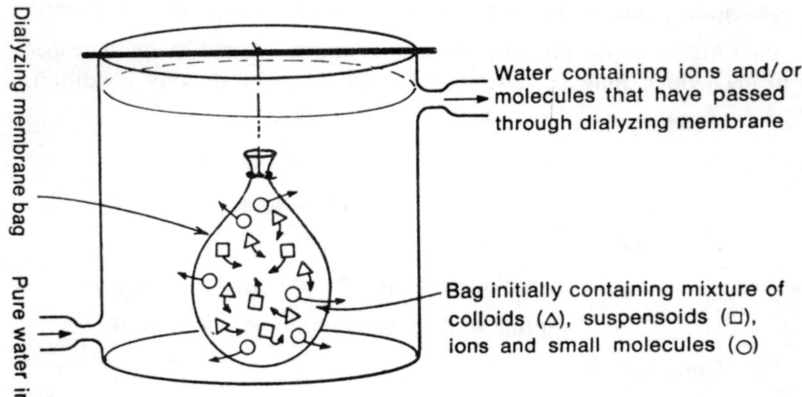

Fig. 7.3 : Dialysis Apparatus

A practical use of this phenomenon is the purification of proteins and vaccines. The impure material is placed into a dialysis bag in setup similar to that in Fig. 7.3. The dialyzable impurities pass through the bag leaving the purified protein or vaccine inside the bag. Milk used in low salt (sodium) diets can also be obtained by this method.

Kidney Function and Dialysis

Physiologically, nearly all body membranes function as dialyzing membranes. The most dramatic physiological example of dialysis is the maintenance of the blood's solute and electrolyte balance through kidney functions. The following data may give you an appreciation of the magnitude of the task handled routinely by the kidneys during normal *daily* operation. The total body water of a 70-kg adult (about 150 lb) is approximately 40 liters (57 percent of the total body mass). Normal fluid intake including fluid synthesized by the body is about

2400 *mL/day*. Of this amount, an average 1500 *mL* (63 percent) is lost in the urine; perspiration, feces, diffusion through the skin and evaporation through the lungs account for the remainder.

In generating the 1.5-L average daily volume of urine, the kidneys filter blood at a rate averaging 125 mL/min, equivalent to about 180 liters daily. The blood is brought to the kidneys by the large renal arteries and after filtration it leaves via the large renal veins. Over 99 percent of the fluid is reabsorbed by kidney action. Less than 1 percent of the fluid is not reabsorbed and is eliminated as urine — an aqueous solution of dissolved electrolytes and organic materials. Thus the kidneys serve *both* to control the concentration of body fluid constituents and to provide a means for excretion of metabolic waste products.

8

Reaction Kinetics

INTRODUCTION

Information about reaction rates and how reactions take place is sought though a subspeciality of chemistry called *reaction kinetics.* A reaction rate tells how much material reacts in a given amount of time.

Exothermic reactions are energy yielding reactions and are termed exothermic. *Endothermic reactions absorb energy and since the energy source is heat, they are called* endothermic *reactions.* Since nature tends to move toward lower energy, we would expect the energetics of a reaction to determine the tendency of a reaction to occur. Chemists have found that most exothermic reactions have a tendency to occur, while endothermic reactions have little tendency to occur. These tendencies are not necessarily related to how rapidly a reaction takes place, however. For example, wood, paper, and gasoline, among other things, will all react with oxygen to evolve heat in a process we call *burning*. The combustion of these materials is an exothermic reaction and has a great tendency to occur. On the other hand, the reaction of these materials with oxygen at normal temperatures does not take place very rapidly. This is very fortunate since most of us would find it inconvenient to sit in a burning chair reading a book that is going up in flames! Thus, although there may be a great tendency for a reaction to occur, this does not necessarily lead to a rapid reaction. There seems to be a barrier preventing these reactions form occurring at a rapid rate at ordinary temperatures.

To ignite combustible materials we heat them in the presence of oxygen. Heat is a form of energy. The barrier to exergonic reactions can be overcome by adding energy to the reactants. In recognition of this fact, the barrier is called an *energy barrier* and the energy required to overcome the barrier is known as the *activation energy,* abbreviated E_a.

During a chemical change, reactant molecules are transformed into product molecules. Bonds in the reactant molecules are broken and new bonds make in forming product molecules during a reaction. Wood and paper are primarily cellulose, which can be represented by the formula $(C_6H_{10}O_5)_n$, where *n* is a very large number. The reaction for the combustion of cellulose can be written

$$(C_6H_{10}O_5)_n + 6nO_2 \longrightarrow 6nCO_2 + 5nH_2O$$

From examination of this equation it is obvious that an oxygen–oxygen bond in the oxygen molecule must be broken so that the oxygen atoms can form new bonds with either carbon or hydrogen to produce the products. Additionally, carbon-hydrogen bonds, carbon–carbon bonds, and perhaps carbon–oxygen or oxygen–hydrogen bonds in cellulose must be broken in order for new bonds to be formed to produce CO_2 and H_2O.

Chemical bonds are formed because bond formation is accompanied by a decrease in potential energy. This energy may be released as heat and light. The lowering of potential energy results in a more stable system. This means that to break a bond, energy must be put into the system in order to increase the potential energy (Fig 8.1).

The necessity to break bonds during the course of a reaction means that energy is required to start a reaction even if that reaction is exothermic. This energy requirement constitutes a barrier to the reaction taking place. It is not possible for chemists to calculate the activation energy for a reaction from knowledge of the energy required to break certain bonds. Activation energies* for reaction must be determined experimentally for each reaction.

The following points are significants :

1. The activation energy for some reaction is relatively small, while for others it is quite large.

---

* The activation energy is the difference in energy between the top of the energy barrier and the reactants.

2. *For exerthermic reactions the activation energy is unrelated to the overall energy change for the reaction.*
3. *The activation energy and not the overall energy change determines the rate of reaction.*

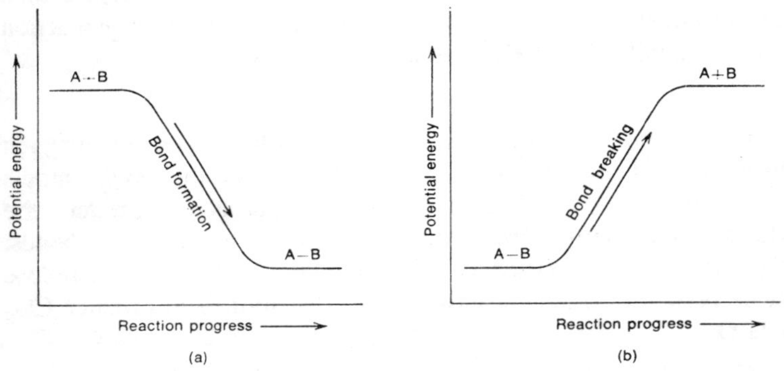

Fig. 8.1 : Change in potential energy during bond-making and bond-breaking. (a) Potential energy decreases during bond formation between atoms A and B. (b) Potential energy increases during breaking of the A—B bond.

The relationships among the activation energy, the overall energy change, and the energy of the reactants and products can be represented diagrammatically as shown in Fig. 8.2. For an exergonic reaction that the energy of the products is less than that of the reactants (Fig 8.2(a). Consequently, the products of the reaction are more stable than the reactants since there is an overall decrease in energy during reaction. Conversely, the products of the endergonic reaction shown in Fig 8.2(b) are less stable than reactants as a result of the increase in energy during reaction. Also notice *the activation energy of the endothermic reaction is larger than the overall energy change for the reaction.* This relationship between the activation energy and the total energy change is true for all endergonic reactions.

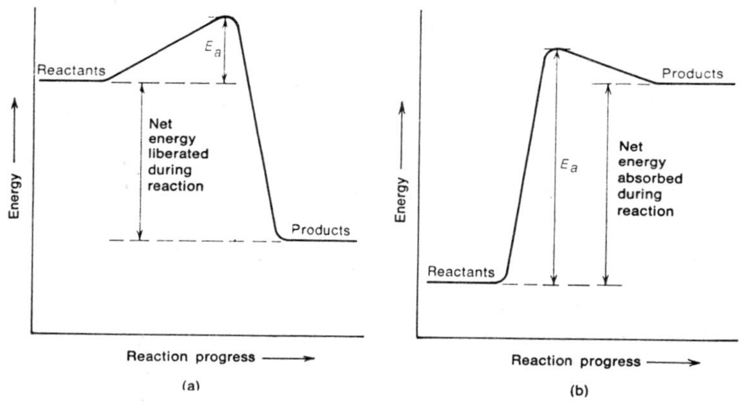

Fig. 8.2 : (a) Reaction energy diagram for an exothermic reaction. (b) Reaction energy diagram for an endothermic reaction.

How Does the Activation Energy Determine Reaction Rates?

The key to understanding how the activation energy determines the rate of reaction lies in remembering that while we tend to think of large amounts of substances reacting in a given time span, the reaction occurs on a molecular basis. For this reason, it is not necessary that all the reactant molecules acquire sufficient energy to overcome the energy barrier at the same time. Instead, the reaction rate depends on the fraction of the total number of molecules that at any given instant possesses enough energy to react.

Consider burning a piece of paper. It isn't necessary to heat the entire piece of paper to give all the molecules enough energy to react with oxygen. We need only to heat a corner of the paper to give just a small fraction of the total number of molecules in the paper enough energy to react. As the paper begins to react, the rate of burning is relatively slow. Since this is an exothermic reaction, more energy is given off than is absorbed. As this energy is given off, it can be used to supply the activation energy to other reactant molecules, causing the reaction to become self-sustaining. Not only is the reaction self-sustaining, but the reaction rate will continually increase (as long as there is a sufficient supply of unreacted molecules) since with the passage of time more and more energy is available to give an increasingly larger fraction of the molecules the required activation energy.

The self-sustaining nature of combustion reactions is a property of all exothermic reactions. As a result, it is necessary to control endothermic reactions either by limiting the amount of reactant present or by providing a means to dissipate energy as it released during reaction. Endothermic reactions on the other hand are more easily controlled since there is a net absorption of energy.

Kinetic Molecular Theory

From everyday experience we know that as heat is applied to a substance, it becomes hotter and the temperature of the substance rises. Therefore, *increasing the temperature at which a reaction is occurring must increase the rate of reaction.* Conversely, a decrease in temperature results in a decreased reaction rate.

The kinetic molecular theory relates temperature to the kinetic energy of molecules and aids in developing a fuller understanding of how heat affects reaction rates. Kinetic molecular theory states that *the average kinetic energy of a collection of molecules is directly proportional to temperature.*

Molecular are in constant motion and the higher the temperature, the faster the molecules are moving. Also, according to the theory, the motion of the molecules is completely random.

While kinetic molecular theory tells us something about the average kinetic energy of a collection of molecules, it says nothing about the kinetic energy of any particular molecule. This means that some molecules may have large kinetic energies, while others posses little kinetic energy. The distribution of kinetic energies among a large number of molecules can be described mathematically. The easiest way of understanding this description is by a graph obtained from plotting the number of molecules versus their kinetic energy. The kinetic energy distribution for a large collection of molecules at two different temperatures is shown in Fig. 8.3. Notice the considerable range of kinetic energies of the molecules at each temperature and that the average kinetic energy is higher at 225°C than at 25°C.

Suppose we place a line on the energy distribution graph at the energy equal to the activation energy of a reaction. The result is shown in Fig. 8.4.

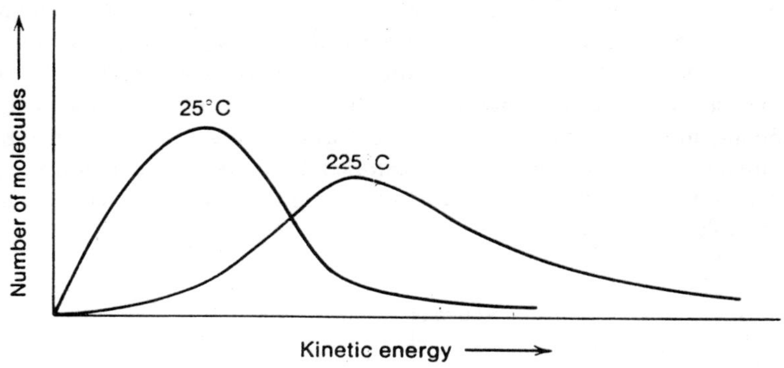

Fig. 8.3 : The distribution of molecular kinetic energies at room temperature (25°C) and 225°C.

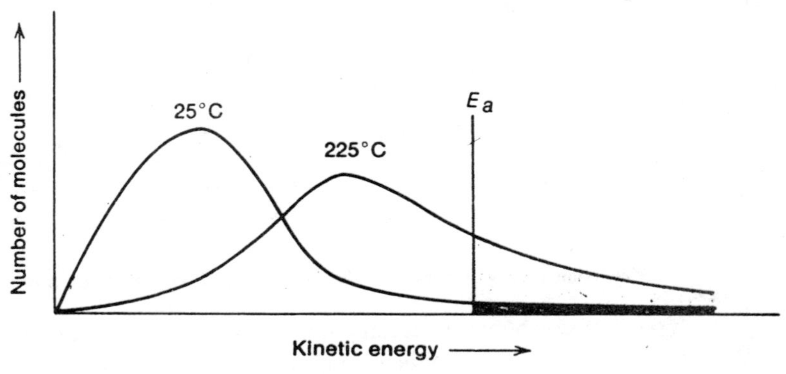

Fig. 8.4 : The number of molecules possessing kinetic energy equal to or greater than the activation energy of reaction at each temperature is proportional to the shaded areas.

Obviously, the fraction of the total number of molecules possessing the activation energy for reaction is smaller at 25°C than at 225°C Hence the rate of reaction should be less at 25°C than at 225°C.

If two molecules are to react, they must come in contact with one another. Kinetic molecular theory tells us that molecules are in constant random motion. The random motion of the molecules means that they are constantly bumping into one another.

During the collision of two objects, their kinetic energy is momentarily converted to potential energy; this potential energy is reconverted to kinetic energy as the objects bounce away from each other.

When molecules collide with one another, their kinetic energy is momentarily converted to potential energy. If the collision is between molecules with high kinetic energies, the potential energy at impact will equal or exceed the activation energy for reaction. In this case the reactant molecules may stick together and react instead of immediately bouncing apart. The species formed during collision of these energetic molecules is called an *activated complex*. Figure 8.5 is a reaction energy diagram for an exergonic reaction, showing the position of the activated complex.

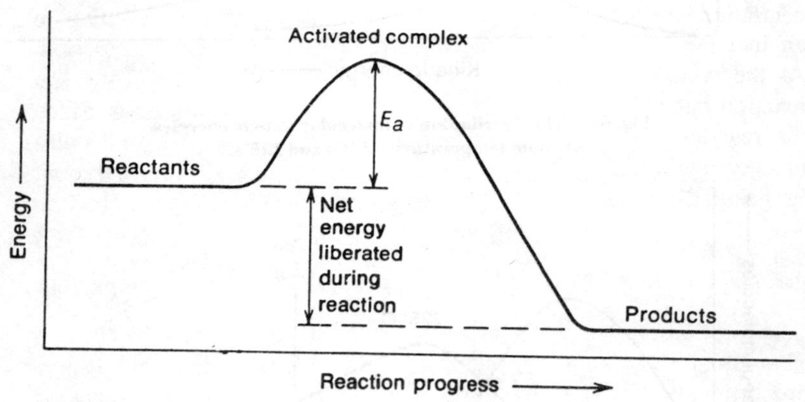

Fig. 8.5 : A reaction energy diagram showing the energy of the activated complex.

If the collision between reactant molecules is not energetic enough, the activated complex will not be formed and the molecules will simply bounce away from each other without reaction.

Concentration and the Rate Law

We have just seen that the reaction rate is dependent on the collision energy of the reactant molecules. The energy of the collision depends solely on the temperature. Consequently, one way of controlling reaction rates is by controlling the temperature at which the reaction occurs. There are many instances, however, in which it is not possible to control the temperature at which reaction occurs or else it is undesirable to change the reaction temperature. Such is the case with the reactions occurring in our bodies. Body temperature is normally 37.0°C If control is to be maintained over body reactions, there must exist other means for controlling reaction rates.

In a situation in which it is not possible to control the energy of collisions, we might try to control the number of collisions per time interval. If we can increase the number of collision occurring between energetic molecules per item interval, the reaction rate should increase. Correspondingly, a decrease in the number of collisions should decrease the rate.

The likelihood of collisions between molecules can be controlled by changing the concentrations of the reactants. An increase in concentration crowds more molecules into a given volume, resulting in an increase in the number of collisions during a given time period. As the collision rate increases, the reaction rate increases. Thus, the reaction rate is proportional to the concentration of the reactants. Since the reaction rate is dependent on the number of reactant molecules in a given volume, molarity is the best unit for expressing concentration. In general, for any single-step reaction of the type

$$aA + bB + ... \longrightarrow zZ + yY + ...$$

the reaction rate at any given temperature will obey the equation

$$\text{reaction rate} = k[A]^a\ [B]^b$$

This equation is known as the rate law for the reaction and k is a proportionality constant called the *specific rate constant.** The brackets [ ] in the rate equation indicate that the concentration of the reactants is expressed as molarity.

Sucrose (cane sugar) will react with water to produce glucose plus fructose according to the equation

$$\underset{\text{sucrose}}{C_{12}H_{22}O_{11}} + H_2O \longrightarrow \underset{\text{glucose}}{C_6H_{12}O_6} + \underset{\text{fructose}}{C_6H_{12}O_6}$$

Accordingly, the rate equation for this reaction is written

$$\text{rate} = k[\text{sucrose}][H_2O]$$

* The specific rate constant varies with temperature changes since the reaction rate changes with temperature and the rate equation as written does not include a temperature term. Consequently. It is necessary to measure reaction rates at a constant temperature.

Suppose we measure the rate of this reaction when the sucrose concentration is 0.1 *M* and the water concentration* is 0.1 *M*. If in another experiment the concentration of sucrose is doubled (0.2 *M*) while the water concentration remains 0.1 *M*, the reaction rate must double.

During reaction the reactants are used up. As a result, their concentration is decreasing with the passage of time, causing the rate of reaction to decrease with time. This feature of reaction rates has significant biological implications. The body must be able to conduct a wide variety of reactions and control the rate at which these reactions take place. The control mechanisms are very complex; yet they must be able to respond very rapidly to changes in a person's activities; regardless of activities, the body must conduct these reactions at some minimal constant rate necessary to sustain life. One means by which the reactions can be controlled is by controlling the concentrations of reactants.

Because we do not ingest food continuously, the body must have the means to store nutrients when they are in abundant supply and release them from storage when required. If the body did not have this capability, the reactions might occur too rapidly shortly after a meal. Several hours later, with depletion of the nutrient concentration, the reaction rates would become too low for good health. Two examples of substances that are stored and released as needed are glucose, which can be stored as glycogen, and fat, which can be stored in fat depots.

The normal range of values for glucose concentration in the blood of healthy individuals in 0.065 to 0.095 percent. If the glucose concentration exceeds this value, the person is said to be *hyperglycemic*; if below this level, *hypoglycemic*. Glucose concentration within normal range provides a smooth flow of glucose into body tissue for use in cells for energy production. The movement of glucose from the bloodstream into the cells maintains cells concentration of glucose at a level the will ensure a glucose reaction rate necessary to meat the energy demands for the cells. **If after ingestion of a meal the blood glucose levels tend to exceed the normal concentration range, glucose is removed from the blood stream and stored as glycogen in the liver. Later as the blood glucose level is diminished by cell utilization, the liver glycogen is broken down to glucose and released into the bloodstream to maintain normal concentration.

---

* This reaction is being conducted in a solvent other than water.

** Cell concentrations of glucose are controlled by both the blood glucose concentration and by insulin concentration.

Catalysis

The addition of small amounts of certain substances a reaction mixture increase the rate of reaction. These substances are called *catalysts.* A catalyst is defined as *a substance that changes the rate of a reaction and can be recovered essentially unaltered.* Substances decreasing reaction rats are called *inhibitors.*

The catalyst (or inhibitor) is not shown in a balanced chemical equation as a reactant or products since it is recovered unchanged after reaction. Nevertheless, it is necessary to mention that a catalyst or inhibitor has been used in the reaction. This information is usually given by writing the catalyst (or inhibitor) above or below the arrow in the balanced equation.

Catalysts and inhibitors are found in, or used in, the manufacture of many materials. Biologically, catalysts and inhibitors are of great importance. Nature employs catalysts in almost all biological reactions. These biological catalysts are given the special name ***enzymes.*** Many drugs used in medical treatment of disease are inhibitors.

Catalysts take part in the reaction in such a way they are regenerated during reaction. Their role in a reaction is to lower the activation energy for a reaction by providng an alternate way for reaction to occur. In many cases the lowering of the activaton energy is accomplished; by combination of the catalyst with one of the reactants to form an intermediate or activated complex.

Consider the general reaction represented by the equation

$$A + B \xrightarrow{\text{catalyst}} Z + Y$$

The catalyst combines with A to give an activated complex :

$$A + \text{catalyst} \longrightarrow [\text{catalyst} — A]$$

This activated complex then interacts with the other reactant to give the products and regenerated the catalyst:

$$[\text{catalyst} — A] + B \longrightarrow Z + Y + \text{catalyst}$$

The activation energy for the formation of the activated complex between catalyst and reactant is lower than that for the activated complex formed between the reactants. A reaction energy diagram for an uncatalyzed and catalyzed reaction is given in Fig. 8.6.

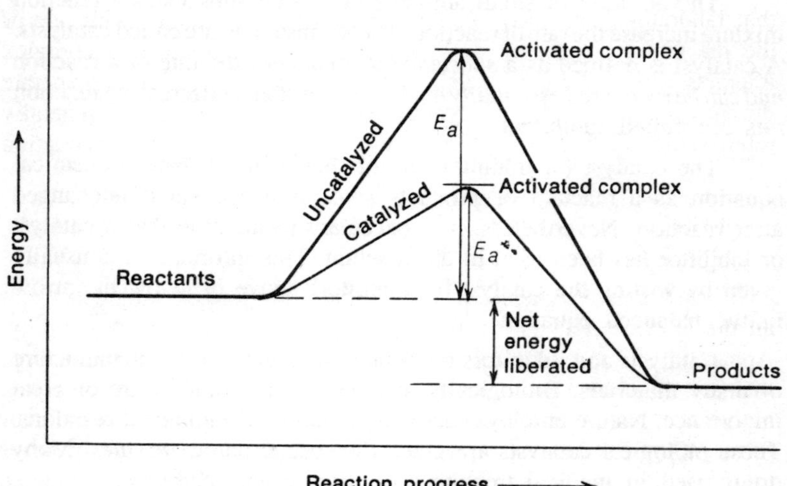

Fig. 8.6 : Comparison of reaction energies for an uncatalyzed and catalyzed reaction.

Enzymes as Catalysts

Enzymes are the catalysts for the chemical reactions of life. In many respects these are the most remarkable compounds known. An enzyme is a very complex molecule consisting of a high molecular weight protein molecule. In addition to the protein part, the enzyme may also consist of another smaller portion. This smaller portion may be either a covalent molecule or a metal ion. The protein portion, called the ***apoenzyme***, must combine with the smaller portion known as a ***cofactor*** to produce the active enzyme. To distinguish between the two types of cofactors, the cofactor is called a ***coenzyme*** if it is an organic molecules and an ***activator*** if it is a metal ion. Minerals are the body's source of metal ions to act as enzyme activators .Among the required metal ions are K^+, Mg^{2+}, Ca^{2+}, Zn^{2+}, Mn^{2+}, Co^{2+}, Fe^{2+}, and Cu^{2+}. Several of the vitamins serve as coenzymes in the chemical machinery of the cell. Thus, the necessity of including minerals and vitamins in the daily diet can be understood. Without these substances, the body's enzymes would be inactive and reaction rates would be too slow to maintain proper growth and maintenance.

Basically, an enzyme functions as a catalyst in the same way that laboratory catalysts function. The enzyme combines with one of the reactants to form an activated complex. This activated complex interacts with the other reactant to produce the products and regenerate the enzyme. In the case of enzymatic catalysis, the reactants are generally referred to as ***substrates***. The sequence can be represented by the equation

$$\underset{\text{enzyme}}{E} + \underset{\text{substrate}}{S} \longrightarrow \underset{\substack{\text{enzyme—substrate}\\ \text{activated complex}}}{E\text{—}S} \longrightarrow \underset{\text{enzyme}}{E} + \underset{\text{products}}{P}$$

Although enzymes are very large and complex molecules, not all the enzyme molecule is directly involved in the catalytic activity. Apparently there exists several relatively small areas of the enzyme that are important to its catalytic activity. One of these areas is the *active site* that actually catalyzes the reaction. Generally, the cofactor forms a portion of this active site. The relationship of the cofactor to the active site is shown in Fig 8.7. Other areas of the enzyme molecule serve as binding sites. These binding sites serve to hold the substrate on the enzyme in the most favorable position for reaction to occur on the *active site*. The binding of the substrate to the enzyme occurs primarily via dipolar forces and hydrogen bonding. These features of enzymes are represented schematically in Fig. 8.8 for a reaction involving breakage of a bond.

The presence of binding sites on the enzymes conveys the property of specificity to enzymes. *Specificity* means the enzyme will catalyze some reactions but will not catalyze other very closely related reactions. For example, the enzyme urease will catalyze the hydrolysis of urea to carbon dioxide and ammonia.

$$\underset{\text{urea}}{H_2N\text{—}\overset{\overset{\displaystyle O}{\|}}{C}\text{—}NH_2} + H_2O \xrightarrow{\text{urease}} 2NH_3 + CO_2$$

The compound biuret can also be chemically hydrolyzed to ammonia and carbon dioxide.

$$\underset{\text{biuret}}{H_2N\text{—}\overset{\overset{\displaystyle O}{\|}}{C}\text{—}\overset{\overset{\displaystyle H}{|}}{N}\text{—}\overset{\overset{\displaystyle O}{\|}}{C}\text{—}NH_2} + 2H_2O \longrightarrow 2CO_2 + 3NH_3$$

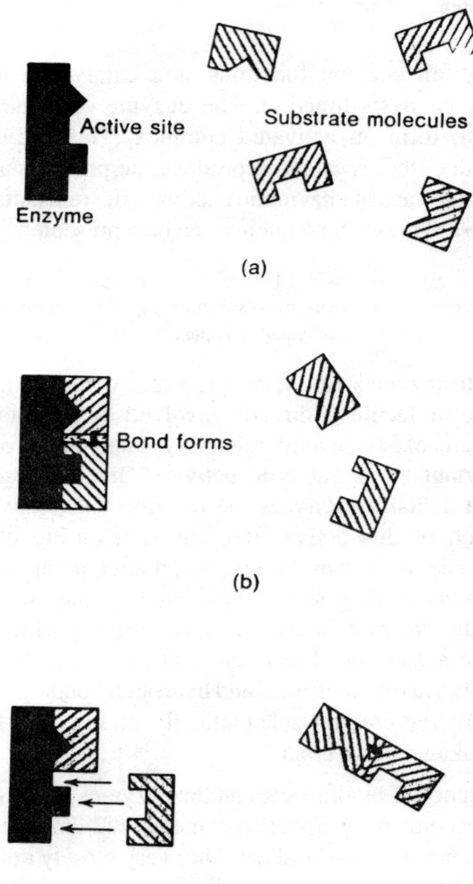

Fig. 8.7 : (a) The active site of the enzyme molecule catalyzes the reaction. (b) Molecules react while attached to the surface of the enzyme. (c) The same enzyme may be used again.

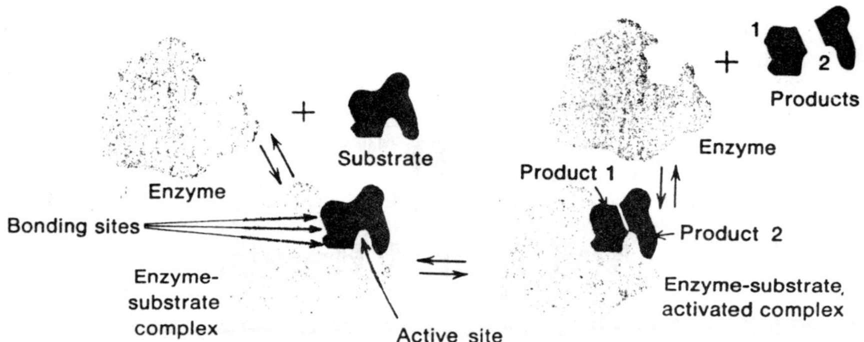

Fig. 8.8 Enzyme Catalysis.

Both urea and biuret belong to the general class of organic compounds known as *amides*. Recall the general formula of amides is

$$\mathrm{R{-}\overset{\overset{\displaystyle O}{\|}}{C}{-}NH_2}.$$

Although urease will catalyze the hydrolysis of urea, it will not catalyze the hydrolysis of biuret or any other amid. For this reason, urease is said to possess *absolute specificity.*

Other enzymes are less specific. They may possess *reaction specificity* or *group specificity*. Dehydrogenases and esterases are examples of reaction specific enzymes. Dehydrogenases are enzymes catalyzing the removal of hydrogen atoms from the substrate and are important in the metabolism of organic acids resulting from the breakdown of fats. Esterases catalyze the hydrolysis of esters. The enzymatic hydrolysis of an ester is shown in Fig. 8.9.

$$\underset{\text{an ester}}{\mathrm{R{-}C}\!\begin{array}{l}{/\!\!/}\mathrm{O}\\ \backslash\,\mathrm{O{-}R}\end{array}} + \mathrm{H_2O} \xrightarrow{\text{esterase}} \mathrm{R{-}C}\!\begin{array}{l}{/\!\!/}\mathrm{O}\\ \backslash\,\mathrm{OH}\end{array} + \mathrm{ROH}$$

Fig. 8.9 : Enzymatic hydrolysis of an ester.

9

Acids and Bases

THE ARRHENIUS THEORY

In 1680 Robert Boyle noted that acids (1) dissolve many substances. (2) change the colors of some natural dyes (indicators), and (3) lose their characteristic properties when mixed with alkalis (bases). By 1814 Gay-Lussac concluded that acids *neutralize* bases and that the two classes of substances can be defined only in terms of their reaction with each other.

In 1884 Svante Arrhenius presented his theory of electrolytic dissociation, which resulted in the Arrhenius theory of acid-base reactions. In his view, an **acid** is substance that contains hydrogen and produces H^+ in aqueous solution. A **base** is substance that contains the OH group and produces hydroxide ions. OH^-, in aqueous solution. **Neutralization** is defined as the combination of H^+ ions with OH^- ions to form H_2O molecules:

$$H^+ \text{ (aq)} + OH^- \text{ (aq)} \longrightarrow H_2O(L) \quad \text{(neutralization)}$$

The Arrhenius theory of acid-base behavior satisfactorily explained reactions of **protonic acids** (those containing acidic hydrogen atoms) with metal hydroxides (hydroxy bases). It represented a significant contribution to chemical thought and theory in the latter part of the nineteenth century. We used this theory in introducing acids and bases and discussing some of their reactions. The Arrhenius model of acids and bases, although limited in scope, led to the development of other general theories of acid-base behavior. These are considered in later sections.

THE HYDRATED HYDROGEN ION

Although Arrhenius described H^+ ions in water as bare ions. We now know that they are hydrated in aqueous solution and exist as $H^+(H_2O)_n$ in which n is some small integer. This is due to the attraction of the H^+ ions, or protons, for the oxygen end (δ—) of water molecules. While we do not know the extent of hydration of H^+ in most solution. We often represent the hydrated hydrogen ion as the hydronium ion. H_3O^+, $H^+(H_2O)_n$ in which $n = 1$. The hydrated hydrogen ion is the species that gives aqueous solutions of acids their characteristic acidic properties. Whether we use the designation H^+ (aq) or H_3O^+, we are always referring to the hydrated hydrogen ion.

$$H^+ + :\ddot{\underset{\cdot\cdot}{O}}:H \longrightarrow H:\ddot{\underset{\cdot\cdot}{O}}:H^+$$
$$\qquad\quad H \qquad\qquad\qquad\quad H$$

THE BRØNSTED-LOWRY THEORY

In 1923 Brønsted and Lowry independently presented logical extensions of the Arrhenius theory. Brønsted's contribution was more thorough than Lowry's, and the result is known as the **Brønsted theory** or the **Brønsted-Lowry theory.**

An **acid** is defined in this theory as a *proton donor*, H^+, and a **base** is defined as a *proton acceptor*. The definitions are sufficiently broad that any hydrogen-containing molecule or ion capable of releasing a proton, H^+, is an acid, while any molecule or ion that can accept a proton is base. According to the Bronsted-Lowry theory.

an acid-base reaction is the transfer of a proton from an acid to a base.

Thus the complete ionization of hydrogen chloride, HCl, a *strong* acid, in water is an acid-base reaction in which water acts as a base, a proton acceptor.

Step 1:	HCl(aq)	$\longrightarrow$	H^+ (aq) + Cl (aq)	(Arrhenius description)
Step 2:	H^+ (aq) + H_2O (L)	$\longrightarrow$	H_3O^+	
Overall :	HCl (aq) + H_2O (L)	$\longrightarrow$	$H_3O^+ + Cl^-$ (aq)	(Brønsted-Lowry description)

$$\text{(H)}:\ddot{\underset{\cdot\cdot}{Cl}}: + H:\ddot{\underset{\cdot\cdot}{O}}: \longrightarrow H:\ddot{\underset{\cdot\cdot}{O}}: + :\ddot{\underset{\cdot\cdot}{Cl}}:$$

(with H below the O of water; H above and below the O of the product)

acid$_1$ base$_2$ acid$_2$ base$_1$

The ionization of hydrogen fluoride, a weak acid, is similar, but it occurs to only a slight extend. So we use double arrows to indicate that it is reversible.

$$HF(g) + H_2O(\ell) \rightleftharpoons H_3O\ (aq) + F^-\ (aq)$$

acid$_1$	base$_2$	acid$_2$	base$_1$

$$\text{(H)}:\ddot{\underset{..}{F}}: + H:\ddot{\underset{\underset{H}{..}}{O}}: \rightleftharpoons H:\overset{\overset{H}{..}}{\underset{\underset{H}{..}}{O}}:^+ + :\ddot{\underset{..}{F}}:$$

We can describe Bronsted-Lowry acid-base reactions in terms of *conjugate acid-base pairs.* **Conjugate acid-base pairs** are species that differ by a proton. In the preceding equation. HF (acid$_1$) and F (base$_1$) are one conjugate acid-base pair and H_2O (base$_2$) and H_3O^+ (acid$_2$) are the other pair. The members of each conjugate pair are designated by the same numerical subscript. In the forward reaction. HF and H_2O act as acid and base, respectively. In the reverse reaction. H_3O^+ acts as the acid, or proton donor, and F^- acts as the base, or proton acceptor.

When HF is dissolved in water, the HF molecules give up some H^+ ions that can be accepted by either of tow bases. F^- and H_2O. The fact that HF is only slightly ionized tells us that F is a stronger base than H_2O. When HCl (a strong acid) is dissolve din water, the HCl molecules give up H^+ ions that can be accepted by either of two bases. Cl^- and H_2O. The fact that HCl is completely ionized in dilute aqueous solution tells us that Cl^- is a weaker base than H_2O. Thus the stronger acid. HCl, has the weaker conjugate base, Cl^-. The weaker acid. HF, has the stronger conjugate base. F^-. We can generalize:

The weaker an acid is, the greater is the base strength of its conjugate base. Likewise, the weaker a base is, the stronger is its conjugate acid.

Strong and weak, like many other adjectives, are used in a relative sense. We do not mean to imply that the fluoride ion, F^-, is a strong base compared with species such as the hydroxide ion. OH. We mean that *relative to the anions of strong acids*, which are very weak bases. F^- is a much stronger base.

Ammonia acts as a weak Bronsted-Lowry base, and water acts as an acid in the ionization of aqueous ammonia:

$$\underset{\text{base}_1}{NH_3(aq)} + \underset{\text{acid}_2}{H_2O\ (\ell)} \rightleftharpoons \underset{\text{acid}_1}{NH_4^+(aq)} + \underset{\text{base}_2}{OH\ (aq)}$$

$$\begin{array}{c} \text{H} \\ \text{H}:\ddot{\text{N}}: \\ \text{H} \end{array} + \begin{array}{c} \\ \text{(H)}:\ddot{\underset{..}{\text{O}}}: \\ \text{H} \end{array} \rightleftharpoons \begin{array}{c} \text{H} \\ \text{H}:\ddot{\underset{..}{\text{N}}}:\text{H} \\ \text{H} \end{array} + \begin{array}{c} \\ :\ddot{\underset{..}{\text{O}}}: \\ \text{H} \end{array}$$

As we see in the reverse reaction, ammonium ion, NH_4^+, is the conjugate acid of NH_3. The hydroxide ion, OH^+, is the conjugate base of water.

Water acts as an acid (H^+ donor) in its reaction with NH_3, whereas it acts as a base (H^+ acceptor) in its reaction with HCl and HF. Whether water acts as an acid or a base depends on the other species present.

Careful measurements show that pure water ionizes ever so slightly to produce equal numbers of hydrated hydrogen ions and hydroxide ions:

$$H_2O + H_2O \rightleftharpoons H_3O^+ + OH^-$$

$$\begin{array}{cccc} \text{Base}_1 & \text{Acid}_2 & \text{Acid}_1 & \text{Base}_2 \end{array}$$

$$\begin{array}{c} \text{H}:\ddot{\underset{..}{\text{O}}}: \\ \text{H} \end{array} + \begin{array}{c} \text{H}:\ddot{\underset{..}{\text{O}}}: \\ \text{H} \end{array} \rightleftharpoons \left[\begin{array}{c} \text{H} \\ \text{H}:\ddot{\underset{..}{\text{O}}}: \\ \text{H} \end{array}\right]^+ + \left[\begin{array}{c} :\ddot{\underset{..}{\text{O}}}: \\ \text{H} \end{array}\right]^-$$

The **autoinization** (self-ionization) of water is an acid-base reaction according to the Bronsted-Lowry theory. One H_2O molecule (the acid) donates a proton to another H_2O molecule (the base). The H_2O molecule that donates a proton becomes an OH^- ion, the conjugate base of water. The H_2O molecule that accepts a proton becomes an H_3O^+ ion. Examination of the reverse reaction (right-to-left) shows that H_3O^+ (an acid) donates a proton to OH^- (a base) to form two H_2O molecules. One H_2O molecule behaves as an acid while the other acts as a base in the autoinization of water. Water is said to be **amphiprotic**, that is, H_2O molecules can accept and donate protons.

The combination of H_3O^+ and OH^- ions to form nonionized water molecules is the reaction that occurs when strong acids and strong soluble bases react to form soluble salts and water. The reverse reaction, the autoinoization of water, occurs only slightly, as expected.

PROPERTIES OF AQUEOUS SOLUTIONS OF ACIDS AND BASES

Aqueous solutions of most protonic acids exhibit certain properties, which are properties of hydrated hydrogen ions.

1. They have a sour taste. Pickles are usually preserved in vinegar, a 5% solution of acetic acid. Many pickled condiments contains large amounts of sugar so that the taste of acetic acid isn't so pronounced. Lemons contain citric acid, which is responsible for their characteristic sour taste.
2. They change the colors of many indicators (highly colored dyes). Acids turn blue litmus red, and bromothymol blue changes from blue to yellow in acids.
3. Nonoxidizing acids react with metals above hydrogen in the activity series to liberate hydrogen gas, H_2.
4. They react with (neutralize) metal oxides and metal hydroxides to form salts and water.

WATER

The water making up about 57 percent of an adult's body weight performs a variety of functions, some of which are a solvent to dissolve and transport substances necessary for metabolism so that new cells may be formed, a solvent to dissolve and transport metabolic waste products from the body, a heat-exchange liquid that can help to maintain a stable body temperature, an osmotic fluid whose release and intake maintains the concentration of solute materials within normal limits, and a reactant in many of the body's chemical processes.

Autonization of water

Water undergoes slight ionization into hydrogen ions H_3O^+ and hydroxide ions (OH^-) via the reversible reaction

$$H_2O + H_2O \rightleftharpoons H_3O^+ + OH^-$$

In pure water, the concentration of hydrogen ions, hydroxide ions, and *undissociated* water are experimentally found to be 1×10^{-7} *M*, 1×10^{-7} *M*, and 55.5 *M*, respectively. Note from the equation above that upon dissociation (unization) each water molecules forms one hydrogen ions and one hydroxide ion. Thus, *in pure water the concentrations of hydrogen ions and hydroxide ions are equal.*

$$[H_3O^+] = [OH^-] = 1 \times 10^{-7}\ M$$

Because the $[H_3O^+]$ equals the $[OH^-]$ in *pure* water, we say water is *neutral. Any aqueous solution* in which

$$[H_3O^+] = [OH^-] = 1 \times 10^{-7}\ M$$

is said to be a *neutral* solution.

In any aqueous solution (not pure water), the mathematical product of the hydrogen and hydroxide concentrations equals a constant, K_w, whose value is 1×10^{-14}.

$$K_w = [H_3O^+] \times [OH^-] = 1 \times 10^{-14}$$

The value of K_w is a constant for a *given* temperature. The value of 1×10^{-14} is for 25°C. K_w is called the *dissociation* (ionization) *constant of water.*

Table 9.1 K_w at Some Different Temperatures

TEMP (°C)	K_w
0	1.13×10^{-15}
10	2.92×10^{-15}
25	1.00×10^{-14}
37*	2.38×10^{-14}
45	4.02×10^{-14}
60	9.61×10^{-14}

* Normal human body temperature

Acid Solutions

It is possible to have an aqueous solution where hydrogen ion concentration does *not* equal hydroxide ion concentration. Consider what happens when some HCl is added to a liter of water. HCl dissociates to furnish H_3O^+ and Cl^- ions:

$$H_2O(\ell) + HCl \longrightarrow H_3O^+ (aq) + Cl^-(aq)$$

The added substance (HCl) release H_3O^+ ions into the water. Thus, the dissociation of water is not the sole source of hydrogen ions. Table 9.2 indicates experimental data determined for various amounts of HCl.

Table 9.2 (H_3O^+) and (OH^-) In aqueous HCl Solution

Total Amount HCl Added	(H_3O^+)	$(OH^-)=\frac{K_w}{H_3O+}=\frac{1\times10^{-14}}{H_3O+}$
0.000 mol/L	$1\times10^{-7}\,M$	$\frac{1\times10^{-14}}{1\times10^{-7}}=1\times10^{-7}\,M$
0.001 mole/L	$1\times10^{-3}\,M$	$\frac{1\times10^{-14}}{1\times10^{-3}}=1\times10^{-11}\,M$
0.010 mol/L	$1\times10^{-2}\,M$	$\frac{1\times10^{-14}}{1\times10^{-2}}=1\times10^{-12}\,M$
0.100 mol/L	$1\times10^{-1}\,M$	$\frac{1\times10^{-14}}{1\times10^{-1}}=1\times10^{-13}\,M$

Two conclusions may be drawn from these data:

1. Because the [H^+] increased upon addition of HCl, the HCl must have donated hydrogen ions to the water. *Substances that donate hydrogen ions in aqueous solution are defined as acids.* Table 9.3 is a list of several common laboratory acids.

Table 9.3 : Common Laboratory Acids

Name	*Formula*	*Dissociation*
Hydrochloric	HCl	$HCl \xrightarrow{H_2O} H_3O^+(aq)+Cl^-(aq)$
Hydrobromic	HBr	$HBr \xrightarrow{H_2O} H_3O^+(aq)+Br^-(aq)$
Hydroiodic	HI	$HI \xrightarrow{H_2O} H_3O^+(aq)+I^-(aq)$
Perchloric	$HClO_4$	$HClO_4 \xrightarrow{H_2O} H_3O^+(aq)+ClO_4^-(aq)$
Nitric	HNO_3	$HNO_3 \xrightarrow{H_2O} H_3O^+(aq)+NO_3^-(aq)$
Sulphuric	H_2SO_4	$H_2SO_4 \xrightarrow{H_2O} H_3O^+(aq)+SO_4^{2-}(aq)$

Acids that liberate 1 mole of hydrogen ion per mole of acid are called monoprotic; those that libterate 2 moles of hydrogen ions per mole acid per diprotic.

2. The hydrogen and hydroxide ion concentrations are no longer equal. The [H_3O^+] is greater than the [OH^-]. *Any aqueous solution in which the hydrogen ion concentration is greater than the hydroxide ion concentration is called an acidic solution.*

Recall that the equilibrium constant K_w is the product of the hydrogen and hydroxide ion concentrations. Although the product of these two terms must equal 1×10^{-14}, *the H_3O^+ and OH^- concentrations are not necessarily equal.* Notice that the addition of HCl results in an increase in $[H_3O^+]$, causing a stress on the water dissociation equilibrium. In response to this stress, a decrease in $[OH^-]$ occurs. Le Chatelier's principle allow us to predict this result.

So far then if

$[H_3O^+] = [OH^-]$ solution is *neutral*

if

$[H_3O^+] > [OH^-]$ solution is *acidic*

Basic Solutions

Sodium hydroxide is soluble in water. It will dissolve, dissociating into Na^+ (aq) and OH^- (aq). Let's add solid NaOH to a liter of water and measure the $[H_3O^+]$ that results as NaOH is added. Table 9.4 contains the experimental data.

Table 9.4 (H^+) and (OH^-) in aqueous NaOH Solution

Total Amount NAOH(s) Added	(H_3O^+)	$(OH^-) = \frac{K_w}{H_3O^+} = \frac{1\times10^{-14}}{H_3O^+}$
0.000 mol/L	1×10^{-7}	$\frac{1 \times 10^{-14}}{1 \times 10^{-7}} = 1 \times 10^{-7}$
0.001 mol/L	1×10^{-11}	$\frac{1 \times 10^{-14}}{1 \times 10^{-11}} = 1 \times 10^{-3}$
0.010 mol/L	1×10^{-12}	$\frac{1 \times 10^{-14}}{1 \times 10^{-12}} = 1 \times 10^{-2}$
0.100 mol/L	1×10^{-13}	$\frac{1 \times 10^{-14}}{1 \times 10^{-13}} = 1 \times 10^{-1}$

In each case, upon addition of NaOH, the $[H_3O^+]$ is less than the $[OH^-]$. *Aqueous solutions in which the hydrogen ion concentration is less than the hydroxide concentration are called basic (alkaline) solutions.* Substances that *liberate hydroxide ions* in aqueous solution are called bases. Table 9.5 lists several common laboratory bases.

Table 9.5 : Common Laboratory Bases

Name	*Formula*	*Aqueous Dissociation*
Sodium hydroxide	NaOH	$NaOH \xrightarrow{H_2O} Na^+(aq) + OH^-(aq)$
Potassium hydroxide	KOH	$KOH \xrightarrow{H_2O} K^+(aq) + OH^-(aq)$
Barium hydroxide	$Ba(OH)_2$	$Ba(OH)_2 \xrightarrow{H_2O} Ba^{2+}(aq) + 2OH^-(aq)$

In acidic solutions the H_3O^+ concentration is always greater than the OH^- concentration. We should not conclude that acidic solutions contain no OH^- ions. Rather, the $[OH^-]$ is always less than 1.0×10^{-7} *M* in such solutions. The reverse is true for basic solutions, in which the $[OH^-]$ is always greater than 1.0×10^{-7} *M*. By definition. "neutral" aqueous solutions at 25°C are solutions in which $[H_3O^+] = [OH^-] = 1.0 \times 10^{-7}$ *M*.

Solution	*General Condition*	*at 25°C*
Acidic	$[H_3O^+] > [OH^-]$	$[H_3O^+] > 1.0 \times 10^{-7}$; $[OH^-] < 1.0 \times 10^{-7}$
Neutral	$[H_3O^+] = [OH^-]$	$[H_3O^+] = 1.0 \times 10^{-7}$; $[OH^-] = 1.0 \times 10^{-7}$
Basic	$[H_3O^+] < [OH^-]$	$[H_3O^+] < 1.0 \times 10^{-7}$; $[OH^-] > 1.0 \times 10^{-7}$

THE pH SCALE

The pH scale provides a convenient way to express the acidity and basicity of dilute aqueous solutions. The **pH** of a solution is defined as

$$pH = \log \frac{1}{[H_3O^+]} \quad \text{or} \quad pH = -\log[H_3O^+]$$

Note that we use pH rather than pH_3O. At the time pH concept was developed. H_3O^+ was represented as H^+ . Various "p" terms are used. In general, a lower-case "**p**" before a symbol is read "negative" logarithm of the symbol." Thus. pH is the negative logarithm of H_3O^+ concentration. **pOH** is the negative logarithm of OH^- concentration, and **p*K*** refers to the negative logarithm of an equilibrium constant.

$$pH = -\log [H_3O^+] \text{ or } [H_3O^+] = 10^{-pH}$$
$$pOH = -\log [OH^-] \text{ or } [OH^-] = 10^{-pOH}$$

Example 9.1

Calculate the pH of a solution in which the H_3O^+ concentration is 0.050 mol/L.

Solution

We are given $[H_3O^+] = 0.050\ M$ or $[H_3O^+] = 5.0 \times 10^{-2}\ M$.

$$pH = — \log [H_3O^+] = — \log(5.0 \times 10^{-2}) = — (—1.30) = 1.30$$

The "1" in 1.30 is *not* a significant figure it comes from the exponential term.

Example 9.2

The pH of a solution is 3.301. What is the concentration of H_3O^+ in this solution?

Solution

Recall that pH = — log $[H_3O^+]$. Therefore we can write —log$[H_3O^+]$ = 3.301. Multiplying through by —1 given log $[H_3O^+]$ = —3.301. Taking the inverse logarithm of both sides of the equation gives

$$[H_3O^+] = 10^{-3.30}\ [H_3O^+] = 5.00 \times 10^{-4}\ M$$

A convenient relationship between pH and pOH in *all dilute solutions at 25°C* can be derived easily:

$$[H_3O^+][OH^-] = 1.0 \times 10^{-14}$$

Taking the logarithm of both sides of this equation gives

$$\log [H_3O^+] + \log [OH^-] = \log (1.0 \times 10^{-14})$$

Multiplying both sides of this equation by —1 gives

$$(—\log [H_3O^+]) + (—\log [OH^-]) = —\log (1.0 \times 10^{-14})$$

or

$$pH + pOH = 14$$

We can now relate $[H_3O^+]$ and $[OH^-]$ as well as pH and pOH:

$$[H_3O^+][OH^-] = 1.0 \times 10^{-14} \text{ and } pH + pOH = 14 \quad (\text{at } 25°C)$$

From this relationship, we see that pH and pOH can *both* be positive only if *both* are less than 14. If either pH or pOH is greater than 14, the other is obviously negative.

Example 9.3

Calculate $[H_3O^+]$. pH. $[OH^-]$, and pOH for 0.015 M HNO_3 solution.

Solution

$$HNO_3 + H_2O \longrightarrow H_3O^+ + NO_3^-$$

Because nitric acid is a strong acid (it ionizes completely), we know that

$$[H_3O^+] = 0.015\ M$$

$$pH = —log[H_3O^+] = —log(0.015) = —(—1.82) = 1.82$$

We also know that pH + pOH = 14. Therefore.

$$pOH = 14 — pH = 14 — 1.82 = 12.18$$

Because $[H_3O^+][OH^-] = 1.0 \times 10^{-14}$.$[OH^-]$ is easily calculated.

$$[OH^-] = \frac{1 \times 10^{-14}}{[H_3O^+]} = \frac{1.0 \times 10^{-14}}{0.015} = 6.7 \times 10^{-13}\ M$$

To develop familiarity with the pH and pOH scales, consider a series of solutions in which $[H_3O^+]$ varies from 10 *M* to 1.0×10^{-15} *M*. Obviously, $[OH^-]$ will vary from 1.0×10^{-15} *M* to 10 *M* in these solutions. Table 9.6 summarizes these scales.

Table 9.6 : Relationships Among (H_3O+), pH, pOH, and (OH^-)

(H3O+)	pH	pOH	(OH^-)	
10^{-15}	15	–1	10^{1}	Increasing Basicity
10^{-14}	14	0	1	
10^{-13}	13	1	10^{-1}	
10^{-12}	12	2	10^{-2}	
10^{-11}	11	3	10^{-3}	
10^{-10}	10	4	10^{-4}	
10^{-9}	9	5	10^{-5}	
10^{-8}	8	6	10^{-6}	
10^{-7}	7	7	10^{-7}	Neutral
10^{-6}	6	8	10^{-8}	Increasing Activity
10^{-5}	5	9	10^{-9}	
10^{-4}	4	10	10^{-10}	
10^{-3}	3	11	10^{-11}	
10^{-2}	2	12	10^{-12}	
10^{-1}	1	13	10^{-13}	
1	0	14	10^{-14}	
10^{1}	–1	15	10^{-15}	

(H_3O^+ Concentration; OH^- Concentration)

The pH of an aqueous solution may be determined by using indicators or a pH meter. In the indicator method, we prepare a series of solutions of known pH standard solution. We add a mixture of indicators, known as a universal indicator, to each so that solutions with different pH have different colors. We then add the same universal indicator to the unknown solution and compare its color to those of the standard solutions. Solutions with the same pH have the same color.

Universal indicator papers may also be used to determine pH. A piece of the paper is dipped into a solution. The color of the paper is then compared with a color chart on the container to establish the pH of the solution.

pH Range for a Few Common Substances

Substances	*pH Range*	
Gastric contents (human)	1.6–3.0	↑ more acidic
Soft drinks	2.0–4.0	
Lemons	2.2–2.4	
Vinegar	2.4–3.4	
Tomatoes	4.0–5.0	
Beer	4.0–5.0	
Urine (human)	4.8–8.4	
Milk (cow's)	6.3–6.6	
Saliva (human)	6.5–7.5	
Blood plasma (human)	7.3–7.5	
Egg white	7.6–8.0	
Milk of magnesia	10.5	
Household ammonia	11–12	more basic ↓

IONIZATION CONSTANTS FOR WEAK MONOPROTIC ACIDS AND BASES

We have discussed solutions of strong acids and strong soluble bases. There are relatively few of these. Weak acids are much more numerous than strong acids.

Weak acids ionize only slightly in dilute aqueous solution. Our classification of acids as strong or weak is based only on the *extent to which they ionize in dilute aqueous solution.* You may know that hydrofluoric acid dissolves glass.

Table 9.7 : Ionization Constants for Some Weak Monoprotic Acids

Acid	*Ionization Reaction*	K_w *At 25°C*
Hydrofluoric acid	$HF + H_2O \rightleftharpoons H_3O^+ + F^-$	7.2×10^{-4}
Nitrous acid	$HNO_2 + H_2O \rightleftharpoons H_3O^+ + NO_2^-$	4.5×10^{-4}
Acetic acid	$CH_3COOH + H_2O \rightleftharpoons H_3O^+ + CH_3COO^-$	1.8×10^{-5}
Hypochlorous acid	$HClO + H_2O \rightleftharpoons H_3O^+ + ClO^-$	3.5×10^{-8}
Hydrocyanic acid	$HCN + H_2O \rightleftharpoons H_3O^+ + CN^-$	4.0×10^{-10}

But HF is *not* a strong acid. The reaction of glass with hydrofluoric acid occurs because silicates react with HF to produce silicon tetrafluoride,

SiF_4, a very volatile compound. This reaction tells us nothing about the acid strength of hydrofluoric acid.

Several weak acids are familiar to us. Vinegar is a 5% solution of acetic acid. CH_3COOH. Carbonated beverages are saturated solutions of carbon dioxide in water. This produces carbonic acid ($CO_2 + H_2O \rightleftharpoons H_2CO_3$). Citrus fruits contain citric acid. $C_3H_5O(COOH)_3$. Some ointments and powders used for medicinal purposes contain boric acid. H_3BO_3 or $B(OH)_3$. These everyday uses of weak acids indicate that there is a significant difference between strong and weak acids. The difference is that *strong acids ionize completely in dilute aqueous solution. Whereas weak acids ionize only slightly.*

Let us consider the reaction that occurs when a weak acid, such as acetic acid, is dissolved in water. The equation for the ionisation of acetic acid is

$$CH_3COOH + H_2O \rightleftharpoons H_3O^+ + CH_3COO^-$$

The equilibrium constant for this reaction could be represented as

$$K_c = \frac{[H_3O^+][CH_3COO^-]}{[CH_3COOH][H_2O]}$$

This expression contains the concentration of water. If we restrict our discussion to *dilute* aqueous solutions, the concentration of water is very high. There are 55.5 moles of water in one liter of pure water. In dilute aqueous solutions, the concentration of water is *essentially* constant: if we assume that it is constant we can rearrange the preceding expression to give

$$K_c[H_2O] = \frac{[H_3O^+][CH_3COO^-]}{[CH_3COOH]}$$

Because both K_c and $[H_2O]$ are constant (at a specified temperature) the ionization constant of a weak acid is the product $K_c[H_2O]$; i.e., $K_a = K_c[H_2O]$.

For acetic acid. $$K_a = \frac{[H_3O^+][CH_3COO^-]}{[CH_3COOH]} = 1.8 \times 10^{-5}$$

This expression tells us that in dilute aqueous solutions of acetic acid, the concentration of H_3O^+ multiplied by the concentration of CH_3COO^- and then divided by the concentration of *nonionized* acetic acid is equal to 1.8×10^{-5}.

Ionization constants for weak acids (and bases) must be calculated from experimentally determined data. Conductivity, depression of freezing point, or measurements of pH provide data from which these constants can be calculated.

Neutralization

Acids (hydrogen ion donors) and bases (hydroxide ion donors) react to form water and a class of compounds called *salts*. This reaction is called a *neutralization* reaction. Let's see what happens when equal volumes of 0.1 *M* hydrochloric acid and sodium hydroxide solutions react:

In the HCl solution,

$$HCl \longrightarrow H^+(aq) + Cl^-(aq)$$

In the NaOH solution,

$$NaOH(s) \xrightarrow{H_2O} Na^+(aq) + OH^-(aq)$$

Upon mixing the two solutions, we don't anticipate reaction between the species of like charge, i.e. Na^+ with H^+. We have already described the reaction occurring between hydrogen ions and hydroxide ions, however;

$$H^+(aq) + OH^-(aq) \longrightarrow H_2O$$

Therefore,

$$H^+(aq) + Cl^-(aq) + Na^+(aq) + OH^-(aq) \longrightarrow H_2O + Na^+(aq) + Cl^-(aq)$$

Thus, the acidic component of the solution (H^+) has been neutralized by the presence of the basic component (OH^-) to produce the neutral species H_2O. The salt sodium chloride, being soluble, remains dissolved.

The net reaction for neutralization may be generalized as follows:

$$\underbrace{H^+(aq) + X^-(aq)}_{\text{an acid}} + \underbrace{M^+(aq) + OH^-(aq)}_{\substack{\text{a base} \\ \text{(metal hydroxide)}}} \rightarrow \underbrace{HOH}_{\text{water}} + \underbrace{M^+(aq) + X^-(aq)}_{\text{salt}}$$

Note that the particular salt formed depends on the cation of the base and the anion of the acid used. The salt's solubility depends on it component ions. The solid crystalline salt is obtained upon evaporation of the water.

The salt sodium chloride is made by the neutralization reaction between sodium hydroxide and hydrochloric acid. Let's set up an experiment that allows us to determine when all the acid has been neutralized by the controlled addition of base. The most common way to carry out this experiment is to perform a titration.* We know that 1 mole of hydroxide ions neutralizes 1 mole of hydrogen ions. By using a sodium hydroxide solution whose concentration is accurately known (a standardized solution), the experiment will also enable us to calculate the concentration of acid in the sample.

A carefully measured volume of hydrochloric acid of unknown concentration is placed into a flask. A few drops of an indicator, a substance that changes color when all the acid is neutralized, is also added. A standardized sodium hydroxide solution is placed in a buret. The base is slowly added to the flask. Upon initial addition of base, hydroxide ions react with H^+ ions in the acid solution to produce water. Sodium and chloride ions remain dissolved. As additional base is added, more hydrogen ions are neutralized. Thus, as the titration proceeds, the concentration of H^+ is continuously decreasing from its original value. Eventually, when sufficient base is added to neutralize all the hydrogen ions present from the dissociation of HCl in the acid sample, the indicator changes color. The point at which the number of moles of hydroxide ion added equals the number of moles of H^+ liberated by HCl in the original acid sample is called the *end point* of the titration.

***Example* 9.4**

If 25 mL of 0.100 *M* NaOH was required to titrate 50.0 mL of an HCl solution to the end point, calculate the concentration of the HCl solution. At the end point we know that

No. of moles OH^- added = No. of moles H^+ originally present in HCl

No. of moles OH- = volume of base × molarity of base =

$$0.025\ \ell \times 0.100\ \text{mole}/\ell = 0.0025\ \text{mole OH}^-$$

No. of moles H^+ = No. of moles OH^- = 0.0025 mole H^+ in original sample

Therefore, 0.0025 mole of H^+ were present in 0.050 ℓ of HCl and the original HCl concentration was 0.0025 mole/0.050 ℓ or 0.050 mole/ℓ (0.050 *M*).

A generalized equation can be written for the neutralization of **monoprotic** acids:

$$(\text{volume of acid, liters}) \times (\text{molarity of acid}) = (\text{volume of base, liters}) \times (\text{molarity of base})$$

By using an experimentally determined volume of standardize base to neutralize a known volume of acid, the concentration of the acid can be calculated.

---

* Acid-base titrations are also performed by adding standardized acid to a solution of base whose concentration is to be determined.

Weak Acids

Acetylsalicylic acid, commonly called aspirin, contains a carboxyl group

$$-\overset{\overset{\displaystyle O}{\|}}{C}-OH$$

Acids containing this group are called *carboxylic* acids. In aqueous solution, a carboxyl acid can act as a hydrogen ion donor (see Table 9.8):

$$R-C\begin{matrix}\nearrow O \\ \searrow OH\end{matrix} + H_2O \longrightarrow H_3O^+ + R-C\begin{matrix}\nearrow O \\ \searrow O^{(-)}\end{matrix}$$

When a 0.10 *M* solution of hydrochloric acid and a 0.10 *M* solution of acetylsalicylic acid are prepared, their acidities differ. Experimentally, the *hydrogen ion* concentration of the hydrochloric acid is measured as 1.0×10^{-1} M; that of the acetylsalicylic acid solution is 5.6×10^{-3} *M*. Both solutions are definitely acidic, indicating that both acids release hydrogen ions into solution.

$$\underset{0.10\,M}{HCl} \longrightarrow \underset{0.10\,M}{H^+(aq)} + \underset{0.10\,M}{Cl^-(aq)}$$

$$C_6H_4(COOH)-O-\overset{\overset{\displaystyle O}{\|}}{C}-CH_3 \longrightarrow H^+(aq) + C_6H_4(COO^{(-)})-O-\overset{\overset{\displaystyle O}{\|}}{C}-CH_3(aq)$$

acetylsalicylic acid 0.10 *M* | 0.0056 *M* | acetylsalicylate ion 0.0056 *M*

The fact that the hydrogen ion concentration in the acetylsalicylic acid is considerably less than that in the hydrochloric acid solution indicates acetylsalicylic acid is a much poorer hydrogen ion donor than hydrochloric acid.

Table 9.8 : Some Carboxylic Acids

Name	*Comments*	*Structural Formula*
Lactic	Waste product of carbohydrate metabolism	$CH_3-\underset{H}{\overset{OH}{C}}-C(=O)-OH$
Pyruvic	Compound vital in metabolism	$CH_3-\overset{O}{\overset{\Vert}{C}}-C(=O)-OH$
Carbonic	Found in carbonated beverages	$HO-\overset{O}{\overset{\Vert}{C}}-OH$
Acetysalicylic	Active ingredient in aspirin	$C_6H_4(COOH)-O-\overset{O}{\overset{\Vert}{C}}-CH_3$
Salicylic	Formerly used as an analgesic	$C_6H_4(COOH)-OH$
Acetic	Found in vinegar and poor wines	$CH_3-C(=O)-OH$
Citric	Found in citrus fruits; important intermediate in carbohydrate metabolism	$HO-C(=O)-\underset{H}{\overset{H}{C}}-\underset{C(=O)OH}{\overset{OH}{C}}-\underset{H}{\overset{H}{C}}-C(=O)-OH$
Ascorbic	Vitamin C	$C_6H_8O_6$
Cholic	Most abundant bile acid	$C_{22}H_{40}O_5$
Acetoacetic	Metabolic breakdown products of fats	$CH_3-\overset{O}{\overset{\Vert}{C}}-CH_2-C(=O)-OH$
β-Hydroxy-butyric	Metabolic breakdown product of fats	$CH_3-\underset{H}{\overset{OH}{C}}-CH_2-C(=O)-OH$

Contd..

Table 9.8 : (Contd...)

Name	*Comments*	*Structural Formula*
Hippuric	Formed in the liver	$C_9H_9O_3N$
Formic	Found in bees and ants	H—C(=O)—OH

The hydrogen ion concentration in the hydrochloric acid solution results from the essentially *complete* dissociation of 0.1 *M* HCl into 0.1 mole of hydrogen ions and 0.1 mole of chloride ions per liter. The much lower concentration of hydrogen ions in the acetylsalicylic acid solution is due to the *incomplete* dissociation of acetylsalicylic acid into ions. This means that a significant amount of *undissociated* acetylsalicylic acid molecules are present in solution. The term *weak acid* is used to describe *an acid that incompletely dissociates in aqueous solution.* Most acids are of this type including those of physiological significance. *An acid that is essentially completely dissociated is termed a strong acid.* A strong acid has a *strong tendency* to liberate hydrogen ions. Although both solutions in the example above are 0.10 M, they contain differing hydrogen ion concentrations because HCl is a strong acid, whereas acetylsalicylic acid is a weak acid. Be careful not to confuse the terms *strong* and *concentrated*; they are **not** synonymous.

Because undissociated acetylsalicylic acid molecules along with hydrogen and acetylsalicylate ions are present in solution, an equilibrium is established.

$$H_2O + C_6H_4(COOH)\text{—O—}\overset{O}{\overset{\|}{C}}\text{—}CH_3 \rightleftharpoons H_3O^+(aq) + C_6H_4(COO^{(-)})\text{—O—}\overset{O}{\overset{\|}{C}}\text{—}CH_3(aq)$$

acetylsalicylic acid   acetylsalicylate ion

In the forward reaction, *acetylsalicylic* acid functions as an acid by donating hydrogen ions. In the reverse reaction, *acetylsalicylate* ions function as a base by combining with hydrogen ions. An acid- base pair, such as acetylsalicylic acid- acetylsalicylate ion, is called a *conjugate acid- base pair*. In this example, acetylsalicylate ion is the conjugate base of the acetylsalicylic acid. This can be generalized symbolically:

$$\underset{}{H_2O} + \underset{\text{acid}}{HB} \rightleftharpoons \underset{\substack{\text{conjugate}\\\text{base}}}{B^-} + H_3O^+$$

In the dissociation of a *weak* acid (**HB**) the conjugate base (**B⁻**) is such a powerful hydrogen ion *acceptor* that the *reverse* reaction is highly favored. Thus the acid remains largely undissociated.

Let's look at another example of a conjugate acid-base pair relationship. Carbonic acid

$$\text{H—O—C(=O)—O—H}$$

dissociates into hydrogen and bicarbonate ions:

$$H_2O + \underset{\substack{\text{carbonic acid}\\(H_2CO_3)}}{\text{H—O—C(=O)—O—H}} \rightleftharpoons H_3O^+(aq) + \underset{\substack{\text{bicarbonate}\\\text{ion }(HCO_3^-)}}{\text{H—O—C(=O)—O}^{(-)}(aq)}$$

Thus, bicarbonate ion is the conjugate base of carbonic acid. This relationship will be taken up again as it applies to the control of acidity and alkalinity.

Notice in the examples above that water behaves as a base and hydronium ion is its conjugate acid.

Buffers—pH Control Agents

Aqueous solutions in the body generally have a very narrow range of hydrogen ion concentration within which the body function best and a limited hydrogen ion concentration range in which the organism can function at all. One reason for this is that enzymes have an optimum pH for maximum efficiency.

The extracellular fluid, including blood plasma, is normally regulated at a hydrogen concentration of approximately 4×10^{-8} *M*, corresponding to a pH of 7.4. Cellular metabolic waste products that function as acids (e.g., lactic acid) or bases (e.g., bicarbonate ion), upon entering the extracellular fluid, can alter its pH. The [H^+] can vary from a lower limit of 1.6×10^{-8} *M* (pH = 7.8) to an upper limit of 1.2×10^{-7} (pH = 6.9) but only for brief periods or else death occurs. The normal pH of arterial blood is 7.4. *Whenever the pH falls within the range of 7.4-7.0,* a person is said to have *acidosis* caused by an *increase* in hydrogen ion concentration. *Alkalosis* occurs *within the range 7.4—7.8* due to a *decrease* in hydrogen ion concentration. How does the body manage to prevent acidosis or alkalosis under normal conditions? What we need is something to neutralize the acidic and/or alkaline waste products of metabolism.

The body three primary methods to accomplish this: (1) If the blood hydrogen ion concentration changes from its normal range, the kidneys either *reabsorb fewer* hydrogen ions causing an *acidic urine* or *reabsorb additional* hydrogen ions resulting in an alkaline urine. Thus, the fluctuation in urine pH reflects its role in maintaining blood pH within the very narrow normal range. (2) Pulmonary ventilation is altered, changing the rate at which carbon dioxide is removed form body fluids. The change in CO_2 concentration helps return hydrogen ion concentration to normal limits. (3) Buffers, a special system of hydrogen ion donors and hydrogen ion acceptors, are present in body fluids.

The regulation of hydrogen ion concentration in body fluids is frequently called *acid-base balance.* The hydrogen ion concentration in a solution of an acid depends on the concentrations of undissociated acid and conjugate base present. Let's examine this relationship using a solution containing citric acid (HCit) and sodium citrate. Citric acid is a weak acid and sodium citrate (NaCit) is a *soluble* salt containing citrate ion, the conjugate base of citric acid.

$$\text{HCit} \rightleftharpoons \text{H}^+(aq) + \text{Cit}^-(aq)$$

$$\text{NaCit}(s) \xrightarrow{H_2O} \text{Na}^+(aq) + \text{Cit}^-(aq)$$

Let's begin with a 1.00 *M* citric acid solution with a hydrogen ion concentration of 3×10^{-2} M due to partial dissociation of citric acid. To this solution we add a small amount of sodium citrate. Sodium citrate is an ionic compound that dissociates *completely,* releasing sodium and citrate ions.

What effect does addition of sodium citrate have on the citric acid dissociation equilibrium? Adding citrate ions to the solution has placed a stress on this equilibrium. Le Chatelier's principle tells us the equilibrium will shift to offset this stress. In attempting to offset the addition of citrate ions, hydrogen ions combine with citrate ions to produce undissociated citric acid. In doing so they cause the concentration of hydrogen ions to decrease; hence, the solution should be less acidic and have a hydrogen ion concentration less than 3×10^{-2} M. Experimentally the $[H^+]$ is determined to be 1×10^{-5} M.

Adding more sodium citrate further reduces the hydrogen ion concentration. After addition, the $[H^+]$ is 1×10^{-6} *M*. This means that citrate ions, regardless of their source, can function as a base by combining with hydrogen ions. A solution of citric acid and sodium citrate contains a hydrogen ion donor, HCit, and a hydrogen ion acceptor, Cit^-. *By adjusting how much of each is present, i.e., the HCit/Cit⁻ ratio, the solution's hydrogen ion concentration can be regulated.* This is just the kind of system needed for acid-base balance in body fluids.

Such a *weak acid/conjugate base pair* is called a *buffer pair. A buffer solution contains a weak acid/ conjugate base pair.* Presence of the weak acid and its conjugate base allows the pH of a solution to remain relatively constant despite addition of limited quantities of acids or bases from other sources.

Generalizing, we could say

$$B^- + H^+ \text{ added to solution} \longrightarrow HB$$

and $$HB \rightleftharpoons H^+ + B^-$$

$$H^+ + \text{Base added to solution} \longrightarrow \text{HBase}$$

A note of caution: A buffer system's ability to offset addition of acid or base without a significant change in pH is limited. There is a point at which addition of a huge excess of either acid or base overwhelms the buffer's capacity for offsetting this addition. At that point the pH changes sharply.

Body Fluid Buffers

Body fluids contain three major buffer systems: the carbonic acid/ bicarbonate buffer; the $H_2PO_4^-/HPO_4^{2-}$ buffer, and the protein buffer. The body uses each of these to carry out necessary buffering action.

Dissolved carbon dioxide gas from metabolism combines with water in blood plasma to form unstable carbonic acid (H_2CO_3).

$$CO_2(g) + H_2O(l) \rightleftharpoons H_2CO_3(aq)$$

This equation indicates the amount of H_2CO_3 present is related to the amount of CO_2 present. Recall that adjustment of pulmonary ventilation adjusts CO_2 Concentration.

Carbonic acid is a weak acid that dissociates:

$$\underset{\text{(acid)}}{H_2CO_3} \rightleftharpoons H+(aq) + \underset{\substack{\text{bicarbonate ion} \\ \text{(conjugate base)}}}{HCO_3^-(aq)}$$

Since the amount of CO_2 present governs the amount of H_2CO_3, it also affects the concentration of hydrogen and bicarbonate ions.

Recall the [citric acid[/[citrate] ratio's effect on [H^+] described previously. Likewise, the undissociated acid-to-conjugate base ratio for carbonic acid and bicarbonate, i.e., [H_2CO_3]/[HCO_3^-], governs the [H^+] of a solution in which they are present. Normally, 20 times as much bicarbonte ion is present as carbonic acid. Carbonic acid is capable of neutralizing base added to the solution (reaction1): bicarbonate ion can neutralize hydrogen ions from an acid added to the solution (reaction 2), e.g. lactic acid.

$$H_2CO_3 + \text{Base} \longrightarrow \text{HBase} + HCO_3^- \quad (1)$$

$$HCO_3^- + H^+ \longrightarrow H_2CO_3 \xrightarrow{\text{carbonic anhydrase}} CO_2 + H_2O \quad (2)$$

The enzyme carbonic anhydrase catalyzes the breakdown of the carbonic acid formed into water and carbon dioxide. The latter is expelled by the lungs. Because of their ability of neutralize hydrogen ions, bicarbonate ions are occasionally called the *alkaline reserve* of the body fluids. Bicarbonate is present in higher concentrations than other buffer conjugate bases.

The dihydrogen phosphate ($H_2PO_4^-$)/monohydrogen phosphate (HPO_4^{2-}) buffer pair is unusual in that an anion functions as an acid:

$$\underset{\substack{\text{dihydrogen} \\ \text{phosphate ion} \\ \text{(acid)}}}{HO-\overset{\overset{O}{\|}}{\underset{\underset{O^{(-)}}{|}}{P}}-O-H} \rightleftharpoons H^+ + \underset{\substack{\text{monohydrogen} \\ \text{phosphate ion} \\ \text{(conjugate base)}}}{H-O-\overset{\overset{O}{\|}}{\underset{\underset{O^{(-)}}{|}}{P}}-O^{(-)}} \qquad K_a = 6.2 \times 10^{-8}$$

Functioning as a weak acid, $H_2PO_4^-$ releases hydrogen ion to neutralize base added from an external source. The conjugate base HPO_4^{2-} accepts H^+ added from some external source. The plasma concentration of monohydrogen phosphate is normally four times greater than that of dihydrogen phosphate.

pH and Oxygen Transport

To met normal metabolic demands, the human body requires an enormous amount of oxygen on a continuous basis. The solubility of oxygen in blood would be very slight if not for the presence of the oxygen carrier, hemoglobin. Hemoglobin functions as a reversible oxygen carrier, picking up oxygen in the lungs and releasing it in tissues for use in cellular metabolism. The maintenance of a relative constancy of 7.4 of blood pH is vital for oxygen transport.

Inhaled oxygen form the lungs diffuses into the capillaries due to a difference in oxygen concentration between the lungs and capillaries. Upon entering a capillary, oxygen diffuses into the red blood cells where it combines with hemoglobin (HHb) to form oxyhemoglobin ($HHbO_2$), Combination with oxygen promotes the loss of a hydrogen ion form histidine in hemoglobin. The oxygenation reactions may be summarized;

$$O_2 + HHb \rightleftharpoons HHbO_2 \rightleftharpoons H+ + HbO_2^- \qquad (3)$$

From our very experience of being alive we know that oxygen uptake and release occurs; but what causes the equilibria to be shifted in our favor as needed? Recalling Le chatelier's principle, high oxygen concentration in the lungs favors the forward reaction. Additionally, bicarbonate ions in the red blood cells combine with hydrogen ions forming carbonic acid (H_2CO_3):

$$HCO_3^- + H^+ \rightleftharpoons H_2CO_3$$

Both of these factors favor oxyhemoglobin formation.

Oxyhemoglobin is then carried to the tissues where carbon dioxide is being produced by cellular metabolism. The CO_2 diffuses into the red blood cells where it forms carbonic acid by combining with water. Carbonic acid dissociation provides bicarbonate and hydrogen ions:

$$H_2O + CO_2 \rightleftharpoons H_2CO_3 \rightleftharpoons H+ + HCO_3^- \qquad (4)$$

What does release of hydrogen ions do to the equilibria in Eq. (3)? It places a stress on the equilibria; oxyhemoglobin combines with hydrogen ion forming $HHbO_2$, which splits into oxygen and hemoglobin. The oxygen diffuses into the cells for use in cellular metabolism. Thus oxygen transport is a pH controlled process.

According to Eq. (4) as hydrogen ions combine with oxyhemoglobin, bicarbonte ion concentration in the red blood cells increases. When this occurs. bicarbonate ions diffuse out of the red blood cell into the plasma. To balance the loss of negative ions by the red blood cell, chloride ions migrate from the plasma into the cell, a process called the *chloride* shift.

Respiratory Acidosis and Alkalosis

From the preceding discussion, carbonic acid and bicarbonate levels in the blood, if uncompensated, can change the blood's pH. By decreasing the amount of breathing (hypoventilation), insufficient carbon dioxide is released. Holding your breath will do this. *The increased level of carbon dioxide in the blood leads to increased carbonic acid and hydrogen ion levels and respiratory acidosis results.* Fainting can occur from mild acidosis: coma can be the result of severe acidosis. Alternatively, by hyperventilating (overbreathing), an excessive amount of carbon dioxide is expelled. Equation 4 indicates that excessive loss of carbon dioxide shift the equilibrium to compensate for CO_2 loss. Hydrogen ions and bicarbonate ions combine to form carbonic acid, which decomposes to water and carbon dioxide. *The reduction in hydrogen ion concentration leads to respiratory alkalosis.*

Metabolic Acidosis*—Diabetes Mellitus

Normal carbohydrate metabolism results in the formation of acidic by-products, which can be termed *metabolic acids.* Lactic, citric, and pyruvic acids are examples of metabolic acids. Their concentration is ordinarily such that normal body buffer action prevents continuing acidosis. In diabetes mellitus, *abnormal* carbohydrate metabolism leads to acidosis. Lack of insulin secretion by the pancreas prevents glucose from being metabolized. In its continuing need for energy, the body instead metabolizes stored fats. Two of the products of fat metabolism are

$$CH_3{-}\overset{\overset{O}{\|}}{C}{-}CH_2{-}C\overset{\nearrow O}{\searrow OH} \qquad \text{and} \qquad CH_3{-}\underset{\underset{H}{|}}{\overset{\overset{OH}{|}}{C}}{-}CH_2{-}C\overset{\nearrow O}{\searrow OH}$$

acetoacetic acid — β-hydroxybutyric acid

* Metabolic alkalosis is not nearly so common as metabolic acidosis. Ingestion of excessive alkaline drugs such as sodium bicarbonate or excessive loss of stomach hydrochloric acid through excessive vomiting of gastric contents are two causes of occasional metabolic alkalosis.

Because these acids are formed continuously by a person suffering from untreated diabetes mellitus, their dissociation increases the hydrogen ion concentration in extracellular fluids metabolic acidosis. If untreated, serve acidosis leads to coma and even death.

10

Energy and Biochemical Reactions

INTRODUCTION

In general, the synthesis of cellular components is endergonic (energy consuming). This chapter describes how catabolism makes energy available to cells. The roles of oxidation/reduction reactions, ATP, several coenzymes, and the respiratory chain are featured and their interdependence discussed. Catabolism is the biochemical breakdown of materials; anabolism is the synthesis of new materials.

ENERGY TRANSFER

Because many cellular activities are endergonic, they must be *coupled* with exergonic (energy producing) reactions that provide the energy necessary for the endergonic processes.

Exothermic

$$\underset{\text{(at higher energy)}}{\text{reactants}} \longrightarrow \underset{\text{(at lower energy)}}{\text{products}} + \text{energy}$$

Endothrmic

$$\text{Energy} + \underset{\text{(at lower energy)}}{\text{reactants}} \longrightarrow \underset{\text{(at higher energy)}}{\text{products}}$$

To be coupled successfully, the net energy evolved must be greater than the energy required for endothermic the reaction. Photosynthesis, an endergonic process, consumes 686 kcal of radiant energy for each mole of glucose produced.

$$\underset{(686\ \text{kcal})}{\text{radiant energy}} + 6CO_2 + 6H_2O \xrightarrow{\text{chlorophyll}} \underset{\text{glucose}}{C_6H_{12}O_6} + 6O_2$$

Therefore, energy absorbed during photosynthesis is stored in the bonded atoms in glucose molecules.

When animals eat green plants, the glucose stored in the plants is broken down to carbon dioxide and water with the liberation of 686 kcal/mol^{-1} of glucose (Fig. 10.1).

$$C_6H_{12}O_6 + 6O_2 \longrightarrow 6CO_2 + 6H_2O + 686\ \text{kcal/mol}^{-1}\ \text{glucose}$$

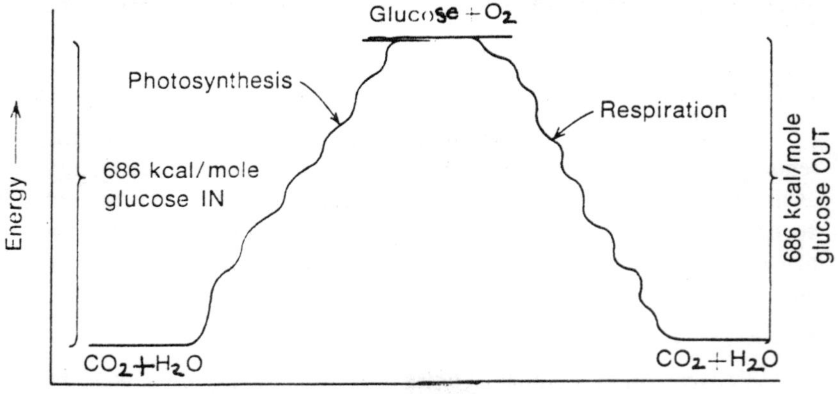

Fig. 10.1 : Relationship of Photosynthesis to Respiration.

This exothermic conversion supplied the body with energy to be coupled with many endothermic processes such as muscle and nerve actions, cellular growth and repair, and maintenance of a relatively constant body temperature.

ATP, Hydrolysis, And Energy

In all living systems, the principal substance used for the exchange of intracellular energy is adenosine triphosphate (ATP), present in cellular fluid at a concentration of 0.5—2.5 mg/mL. Adenosine triphosphate is a substance composed of three main components: (1) adenine; (2) the monosaccharide, ribose; and; (3) phosphate groups. The structural formula for ATP is shown in Fig. 10.2. The combination of adenine and ribose is called *adenosine*.

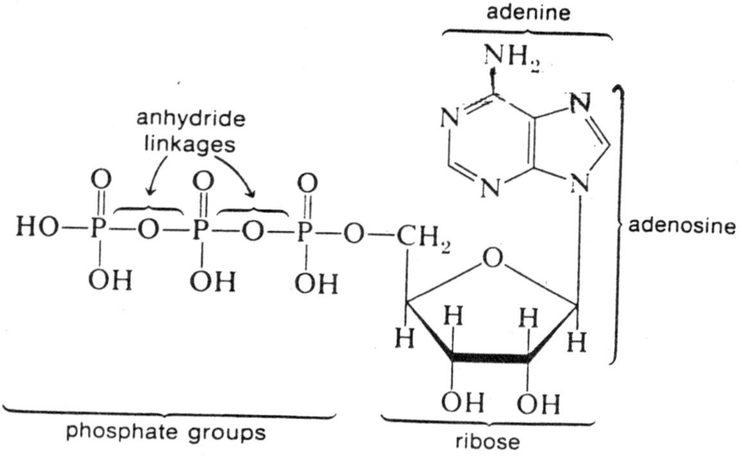

Fig. 10.2 : Adenosine triphosphate, ATP

The phosphate groups are joined by the anhydride linkage.

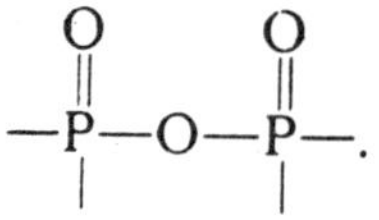

$$-\overset{\overset{\displaystyle O}{\|}}{\underset{|}{P}}-O-\overset{\overset{\displaystyle O}{\|}}{\underset{|}{P}}-.$$

energy + adenosine + phosphates ⟶ ATP + water

ATP is frequently described as an *energy-rich* compound with bonds called *high-energy bonds. A high-energy bond is one that liberates a large amount of energy when* ***hydrolyzed*** *(>5 kcal mole)*. Two such high-energy groups exist in the ATP anhydride linkage.

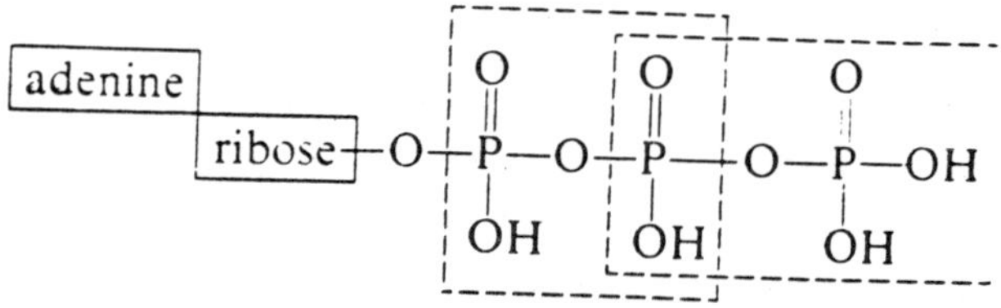

Hydrolysis of phosphate anhydride bonds liberates energy. Anhydrides can be hydrolyzed. Upon hydrolysis of 1 mole of ATP to adenosine diphosphate (ADP), about 8000 cal are liberated under the operating conditions of the cell.

adenine ribose—O—P(=O)(OH)—O—P(=O)(OH)—O—P(=O)(OH)—OH + HOH ⇌

adenosine triphosphate (ATP) + water ⇌

adenine ribose—O—P(=O)(OH)—O—P(=O)(OH)—OH + P(=O)(HO)(OH)(OH) + energy(8000 cal/mol^{-1})

adenosine diphosphate (ADP) + P_i + energy

Pi indicates inorganic phosphate, *i.e.*, all forms of phosphate not combined with a carbon compound.

ATP can also hydrolyze exergonically to produce adenosine monophosphate (AMP) and pyrophosphate (PPi).

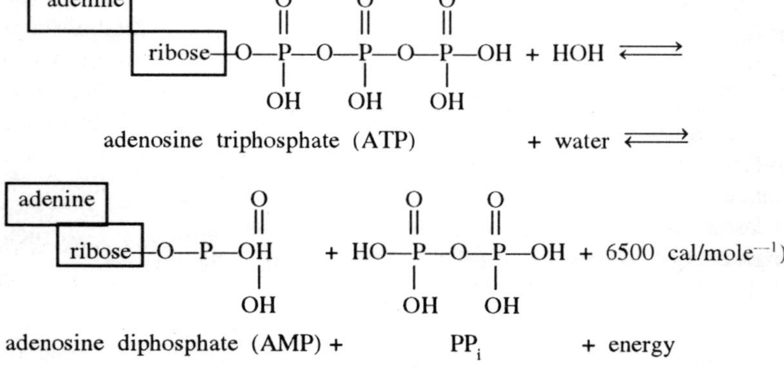

Thus, by exergonically hydrolyzing ATP to ADP or AMP, a cell liberates energy that can be used to power endergonic processes.

The hydrolysis of ADP to AMP is exergonic but does not frequently occur in cellular metabolism.

ATP Synthesis

Energy is released upon ATP hydrolysis. This means that the reverse process is endergonic and energy is stored in ATP during its formation (Fig. 10.3).

$$\text{energy} + \text{ADP} + \text{P}_i \rightleftharpoons \text{ATP} + \text{HOH}$$

$$\text{energy} + \text{AMP} + \text{PP}_i \rightleftharpoons \text{ATP} + \text{HOH}$$

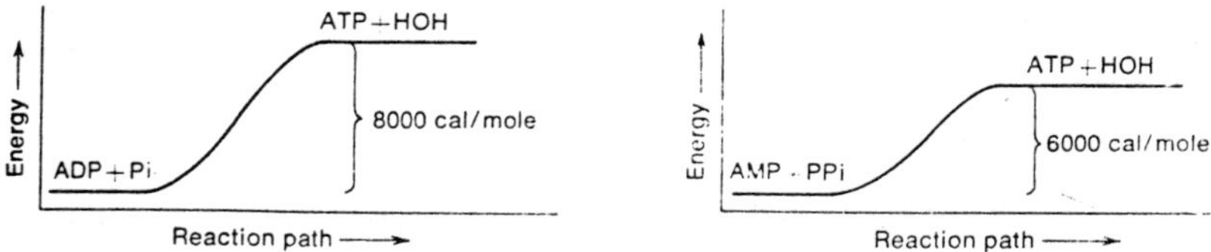

Fig. 10.3 : ATP formation–an endergonic process.

A logical question is: Where does the energy come from that's used in ATP formation? It arises during catabolism in which cells chemically break down large metabolite molecules into simpler substances of *lower* energy. A net release of energy occurs and some of the energy is stored by conversion of AMP and ADP to ATP. The formation of ATP by capturing energy liberated during catabolism is called *oxidative phosphorylation.* The majority of ATP is produced by this process. Oxidative phosphorylation occurs through the action of the *respiratory chain,* also known as the cytochrome chain or the electron transport system (ETS). The components of the *cytochrome chain* or the *electron transport system* (ETS). The components of the ETS are found on the inner membrane surface of mitochondria. Mitochondria are rod-shaped structures occurring in most living cells, with their number varying from about 50 to 5000 per cell depending on cell type and function.

It has been established that the enzymes of the ETS function as a sequence in which hydrogens are passed along carries. Ultimately, the hydrogens combine with oxygen to produce water.

Hydrogen atoms can be imagined to consist of component hydrogen ions and electrons.

$$H\cdot + H\cdot = 2e^- + 2H^+$$

The formation of water involves the reaction

$$2H^+ + 2e^- + \frac{1}{2}O_2 \longrightarrow H_2O$$

Therefore, a biological requirement exists for hydrogen ion and electron carries.

Two kinds of carriers are present in the electron transport systems: (1) coenzymes, which are electron and hydrogen ion carriers, and (2) cytochromes, which are exclusively electron carriers. The coenzymes presently known to participate in the ETS are (1) nicotinamide adenine dinucleotides (NAD and NADP), (2) flavin nucleotides (FAD and FMN), and (3) coenzyme Q (CoQ).

Nicotinamide Nucleotide Coenzymes

A nucleotide contains a nitrogen base bonded to a sugar that is bonded to a phosphate group. A dinucleotide such as NAD or NADP contains two nucleotide units joined through their phosphate groups (Fig. 10-4). Substitution of a phosphate group for a hydrogen of an OH group on the starred ribose in Fig. 10.4. forms NADP from NAD.

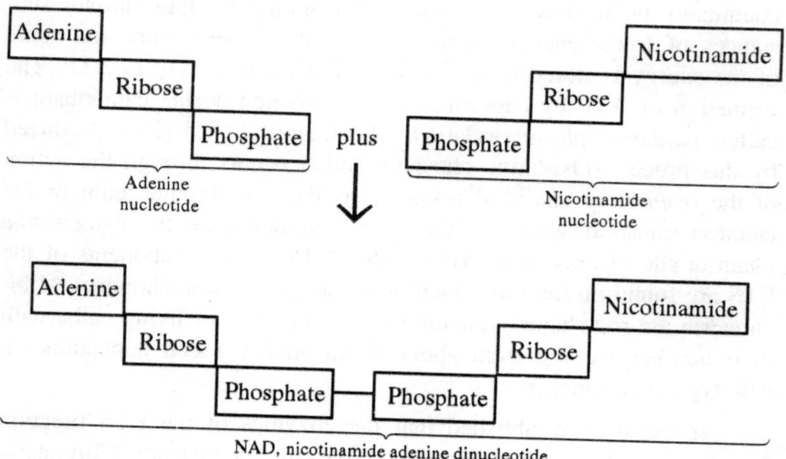

Fig. 10.4 : Combination of nucleotides to form NAD.

The coenzyme (NAD or NADP) is bonded to an apoenzyme to form a complete enzyme. The *reversible* acceptance and release of hydrogens (hydrogen ions and electrons) occur in the nicotinamide portion of the dinucleotide coenzyme. Thus, the coenzyme undergoes two types of reactions: (1) reduction (gain of hydrogen) and (2) oxidation (loss of hydrogen). The reactions are the reverse of each other.

Oxidation of $NADH_2$: $NADH_2 \longrightarrow \underset{\text{oxidized form}}{NAD} + 2H\cdot$

Reduction of NAD : $NAD + \underset{(2H^+ + 2e^-)}{2H\cdot} \longrightarrow \underset{\text{reduced form}}{NADH_2}$

NAD and NADP act as coenzymes for dehydrogenases enzymes that catalyze transfer of hydrogens during oxidation-reduction reactions. The hydrogens are acquired from a metabolite MH_2 undergoing oxidation. Since oxidation and reduction occur simultaneously, two half reactions may be written:

Oxidation : $MH_2 \longrightarrow 2H\cdot + M$

Reduction : $NAD + 2H\cdot \longrightarrow NADH_2$

Net reaction : $MH_2 + NAD \longrightarrow M + NADH_2$

A curved double arrow representation of these coupled half reactions is commonly used. Note that both arrows go in the *same* direction. The fate of a particular reactant is indicated by following the path of the arrow for that half reaction. For the reactions above, the diagram

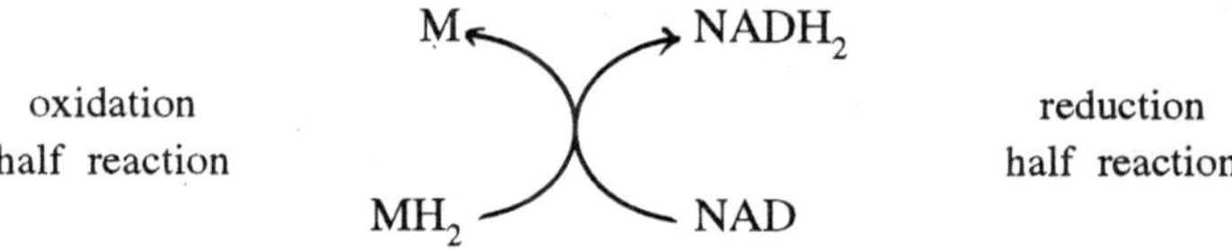

is interpreted to indicate loss of hydrogen from the metabolite (oxidation) and the simultaneous gain of hydrogen by NAD (reduction).

NAD and NADP function primarily during reactions involving carbon-to-oxygen double bonds. For example, lactic acid dehydrogenase catalyzes the conversion of lactic acid to pyruvic acid. NAD is required as the coenzyme.

$$CH_3-\underset{H}{\overset{OH}{C}}-C(=O)OH \xrightarrow[\text{lactic acid dehydrogenase}]{NAD \;\rightarrow\; NADH_2} CH_3-\overset{O}{\overset{\|}{C}}-C(=O)OH$$

lactic acid → pyruvic acid

Flavin Nucleotides

There are two types of flavin nucleotides that take part in the respiratory chain: flavin mononucleotide (FMN) and flavin adenine dinucleotide (FAD). FMN consists of phosphate, ribitol, and flavin bonded together See Fig. 10.5). The combination of ribitol and flavin forms riboflavin (vitamin B_2). FAD is the combination of FMN and AMP joined by an anhydride linkage.

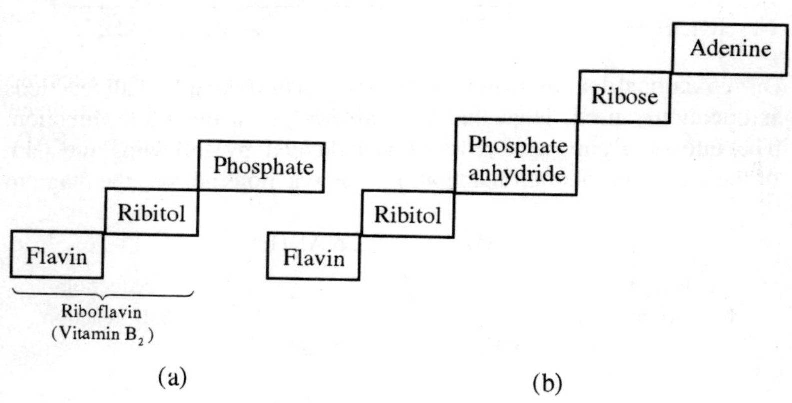

Fig. 10.5 : Flavin nucleotides. (a) Flavin mononucleotide, FMN. (b) Flavine adenine dinucleotide, FAD.

Hydrogens are picked up and released by the flavin portion of the molecule to form oxidized and reduced forms of the coenzymes:

Oxidation of $FMNH_2$: $FMNH_2 \longrightarrow FMN + 2H\cdot$

Reduction of FMN : $FMN + 2H\cdot \longrightarrow FMNH_2$

Oxidation of FADH : $FADH_2 \longrightarrow FAD + 2H\cdot$

Reduction of FAD : $FAD + 2H\cdot \longrightarrow FADH_2$

In the electron transport system. FMN removes hydrogens from $NADH_2$. This makes NAD available to oxidize additional metabolites.

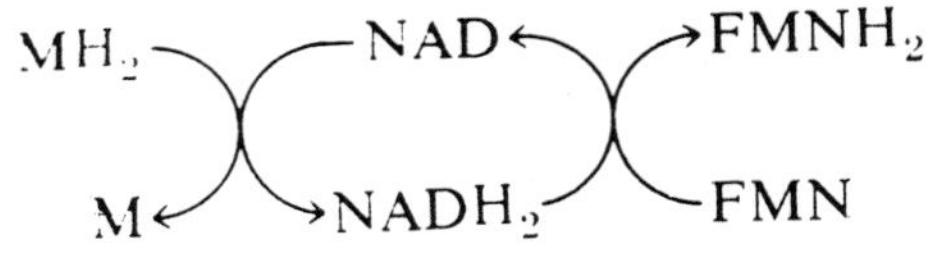

MH_2 → M ; NAD → $NADH_2$; $NADH_2$ → NAD ; FMN → $FMNH_2$

A metabolite may react directly with FAD. The coenzyme is used primarily during reactions in which hydrogens are removed from adjacent carbon atoms during conversion of carbon-carbon single bonds to double bonds.

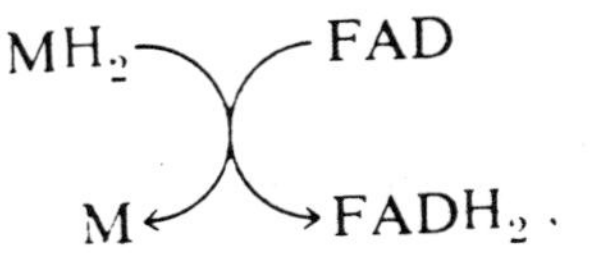

MH_2 → M ; FAD → $FADH_2$.

During carbohydrate and lipid metabolism, the conversion of succinic acid to fumaric acid involves such a change. The reaction is catalyzed by succinic dehydrogenase through the action of its coenzyme, FAD

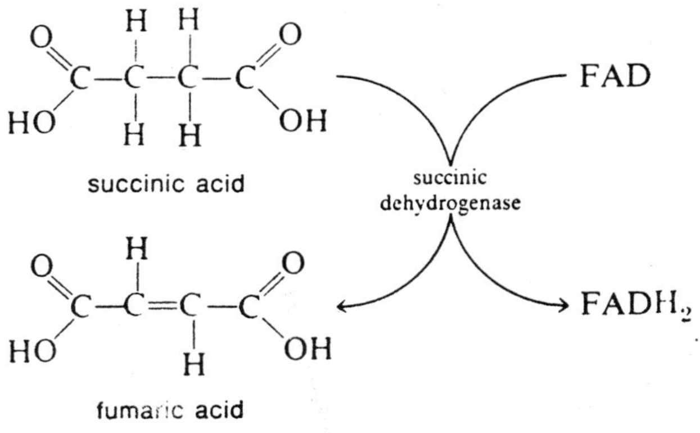

HOOC–CH₂–CH₂–COOH (succinic acid) + FAD —succinic dehydrogenase→ HOOC–CH=CH–COOH (fumaric acid) + $FADH_2$

succinic acid

fumaric acid

Both FMN and FAD are coenzymes for dehydrogenases. If the metabolite is FAD dependent, the metabolite loses hydrogens to FAD directly. A metabolite which is NAD dependent initially loses hydrogens to NAD which passes them on to FMN.

NAD-dependent dehydrogenase sequence : $MH_2 \rightarrow M$; $NAD \rightarrow NADH_2$; $NADH_2 \rightarrow NAD$; $FMN \rightarrow FMNH_2$; $FMNH_2 \rightarrow FMN$ → to coenzyme Q step

FAD-dependent dehydrogenase sequence : $MH_2 \rightarrow M$; $FAD \rightarrow FADH_2$; $FADH_2 \rightarrow FAD$ → to coenzyme Q step

In either case, the reduced flavin nucleotide interacts with coenzyme Q, the next carrier in the respiratory chain.

Coenzyme Q

Coenzyme Q is a lipid material whose role has only recently been completely established. The coenzyme can exist in oxidized and reduced forms, a role consistent with its activity in an oxidation-reduction sequence.

Reduction: $CoQ + H \longrightarrow CoQH_2$ (reduced form)

Oxidation: $CoQH_2 \longrightarrow 2H\cdot + CoQ$ (oxidized form)

CoQ accepts hydrogens from $FADH_2$ or $FMNH_2$ causing oxidation of the $FADH_2$ and reduction of CoQ.

$FADH_2 \rightarrow FAD$; $CoQ \rightarrow CoQH_2$

$FMNH_2 \rightarrow FMN$; $CoQ \rightarrow CoQH_2$

The Cytochromes

These enzymes were probably the first substances for which experimental evidence indicated involvement in biochemical electron transfer reactions. Sequentially in the ETS these are the cytochromes designated as b, c_1, c, a, and a_3. All cytochromes contain a protein bonded to a coenzyme similar to heme. Although subtle differences exist among the cytochromes due to (1) their protein composition, (2) the bonding between the coenzyme and protein, and (3) variations in side chains attached to the coenzyme, each cytochrome possesses a central iron ion in the coenzyme. Unlike the central iron ion in hemoglobin, *the iron ion in a cytochrome is able to readily undergo reversible oxidation and reduction.*

Oxidation: $Fe^{2+} \longrightarrow e^- + Fe^{3+}$

Reduction: $Fe^{3+} + e^- \longrightarrow Fe^{2+}$

An electron is accepted from one cytochrome molecule and passed on to the next one in the sequence. This is only a one-electron change. Because two electrons are removed from the metabolite MH_2 and passed on from NAD to FMN to CoQ, two cytochrome molecules are required to accept the two electrons from $CoQH_2$.

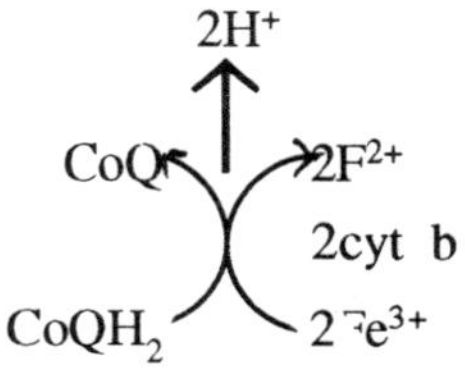

$$\begin{array}{lrl} \text{Oxidation :} & CoQH_2 & \longrightarrow CoQ + 2H+ + 2e^- \\ \text{Reduction :} & 2 \text{ cyt } b_{Fe^{3+}} + 2e^- & \longrightarrow 2 \text{ cyt } b_{Fe^{2+}} \\ \hline \text{Net Reaction :} & 2 \text{ cyt } b_{Fe^{3+}} + CoQH_2 & \longrightarrow 2 \text{cyt } b_{Fe^{2+}} + CoQ + 2H^- \end{array}$$

Cytochromes are not hydrogen ion carriers. The hydrogen ions liberated during oxidation of $CoQH_2$ are released into the intracellular environment. They are used later in the final step of the respiratory chain where they combine with electrons and oxygen to form water.

The cytochrome sequence of the electron transport system can be represented as

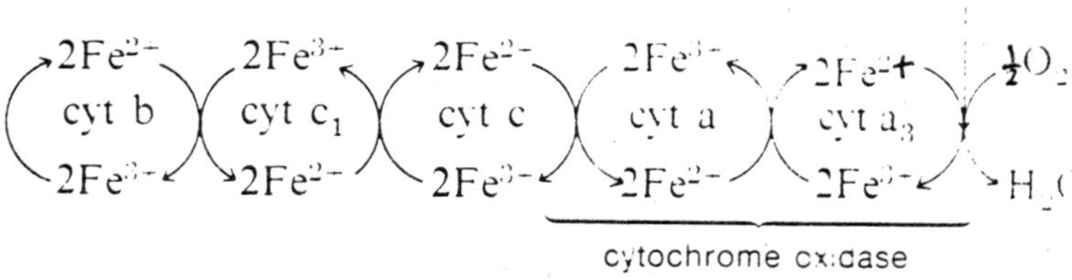

Cytochromes a and a_3 operate as a unit. This unit is the only cytochrome in the ETS that transfers electrons directly to oxygen and is called *cytochrome oxidase.* The vast majority of the oxygen used by cells is involved in this final step of the cytochrome chain. The overall sequence for the electron transport system is shown in Fig 10.6.

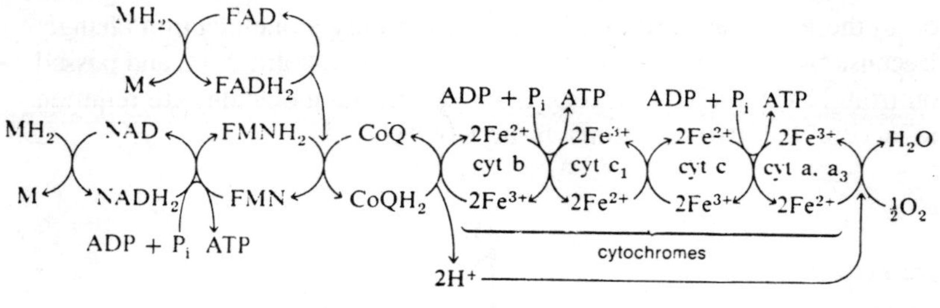

Fig. 10.6 : The electron transport system (respiratory chain). Note the sites of ATP production.

Oxidative Phosphorylation

During the operation of the respiratory chain, metabolites are oxidized, energy is liberated, and water is produced. The driving force for the respiratory chain is the *net* effect of hydrogen reacting with oxygen to produce water. Released energy cab be stored in chemical bonds.

The operation of the respiratory chain efficiently provides the energy for the endergonic conversion of ADP to ATP. At three points in the respiratory chain processes, sufficient energy (at least 8000 cal. mole) is liberated for the endergonic formation of ATP (see Fig 10.6). The coupling of the exergonic respiratory chain processes with the endergonic formation of ATP is termed *oxidative phosphorylation.* Thus, the energy liberated by the oxidation of food molecules during catabolism is stored in ATP. The energy cab subsequently be released by hydrolysis of ATP.

Note from Fig 10.6 that in the respiratory chain an NAD-dependent metabolite liberates 3 moles of ATP per mole of $NADH_2$; an FAD-dependent metabolite liberates 2 moles of ATP per mole of $FADH_2$.

11

Introduction to Organic Compounds

HYDROCARBONS

The simplest carbon-containing molecules exhibiting isomerism are those of carbon and hydrogen collectively called *hydrocarbons*. The hydrocarbons are generally divided into four subclasses depending on particular structural features. These subclasses, alkanes, alkenes, alkynes, and aromatics, are shown in Fig 11.1.

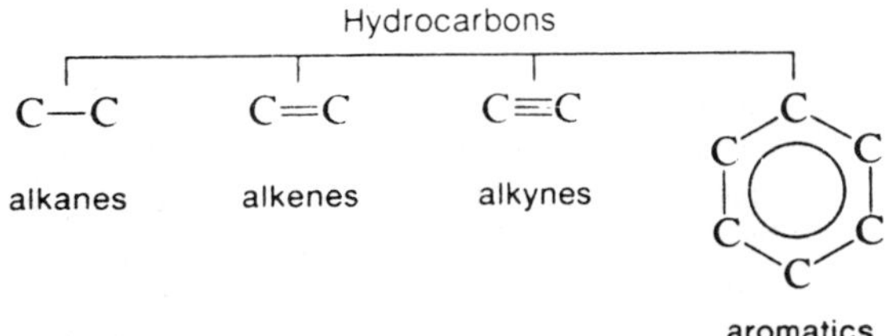

Fig. 11.1 : Characteristic grouping of the four subclasses of hydrocarbons.

Alkanes are characterized by having only single bonds between the carbon atoms. You are already familiar with the simplest alkane, methane (CH_4). The alkanes are also known as saturated hydrocarbons. The word *saturated* means the molecule contains only single bonds.

A general formula for alkanes is $C_nH_{(2n+2)}$.

Alkenes are the group of hydrocarbons containing a carbon-carbon double bond in the molecule. Occasionally, these compounds are also referred to as olefins.

Alkenes have the general formula C_nH_{2n}.

Alkynes, or acetylenes as they are also known, contain carbon-carbon triple bonds.

The general formula of alkynes is $C_nH_{(2n-2)}$.

The aromatic hydrocarbons have a six-membered carbon ring containing three double bonds.

Writing Structures and Naming the Alkanes

The names of alkanes always end in *ane*. This ending signifies the compound is hydrocarbon containing only single bonds.

The remaining portion of the name tells the number and the arrangement of the carbon atoms in the molecule. Table 11.1 gives the relationship between the name and the number of carbon atoms contained in a continuous chain in an alkane. This table should be memorized now.

Table 11.1 : Names and Formulas of Unbranched Alkanes

Numer of Carbon Atoms in a Continuous chain	Name	Molecular Formula $[C_nH_{(2n+2)}]$	Condensed Structural Formula
1	methane	CH_4	CH_4
2	ethane	C_2H_6	CH_3CH_3
3	propane	C_3H_8	$CH_3CH_2CH_3$
4	butane	C_4H_{10}	$CH_3CH_2CH_2CH_3$
5	pentane	C_5H_{12}	$CH_3CH_2CH_2CH_2CH_3$
6	hexane	C_6H_{14}	$CH_3CH_2CH_2CH_2CH_2CH_3$

Let us examine the first few members of the alkane family more closely. The structural formula for CH_4 and C_2H_6 are discussed in chapter 5.

The Lewis structure of propane is written

```
    H   H   H
    |   |   |
H — C — C — C — H
    |   |   |
    H   H   H
```

There are several important features about this structural representation for us to notice. One of these features is that the bonding of all the hydrogen atoms is identical. Two of the hydrogen atoms in propane are bonded to a carbon atom that is bonded to two other carbon atoms. The other two carbon atoms each bear three hydrogen atoms. The propane molecule is said to have two different kinds of hydrogen, meaning there are two different bonding environments for the hydrogen atoms in propane.

Another feature of the propane structure is that the three carbon atoms are *not* in a straight line. The four bonds around each carbon atom are directed toward the corners of a tetrahedron (see Fig 11.2).

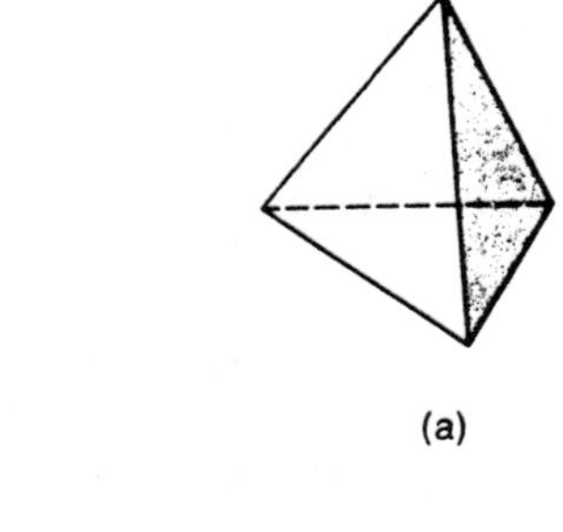

(a)

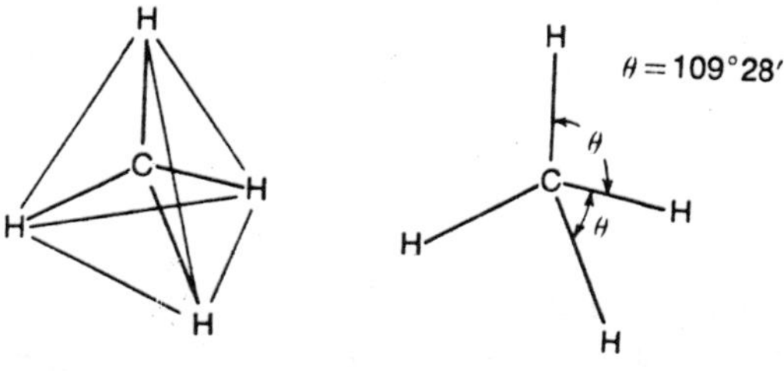

(b)

Fig. 11.2 : (a) A tetrahedron. (b) The tetrahedral shape of the methane molecule. Tetrahedral bonding angles equal 109°28′

This means the actual shape of the molecule is

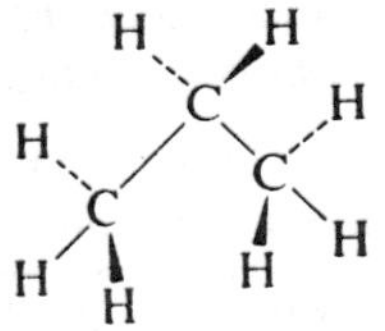

with bond angles of 109°28′. A line indicates that the bond is in the plane of the page. A wedge ( ) indicates that the bond is coming out of the plane of the page toward you. A dashed line indicates that the bond is going behind the plane of the page, away from you.

Yet another bonding feature we need to be aware of is that atoms (or groups of atoms) connected by a *single* bond can rotate about the bond without breaking or affecting the strength of the bond. Furthermore, the basic structure of the molecule is unchanged by rotation about the single bond. Although there is rotation about each single bond, the three carbon atoms in propane are still in continuous chain.

The four carbon atoms in butane form a continuous chain with the Lewis structure*

```
    H  H  H  H
    |  |  |  |
 H—C—C—C—C—H
    |  |  |  |
    H  H  H  H
```

Branched Alkanes

The atoms in C_4H_{10} can be arranged in more than one way. The four carbon atoms do not have to form a continuous chain as they do in butane. Another possible arrangement is

```
     H   H   H
     |   |   |
 H—C—C—C—H
     |   |   |
     H   |   H
      H—C—H
         |
         H
```

Although this compound has the same molecular formula as butane (C_4H_{10}), the structure is different. Since the longest continuous chain of carbon atoms is three, the compound is named as a propane (*prop* means three carbon atoms in a continuous chain). Clearly, the compound is not propane though. Propane has the formula C_3H_8. *Alkanes in which all the carbon atoms are not included in continuous chain are called branched alkanes.*

Branched alkanes are named according to the number of carbon atoms forming the *longest continuous* chain.

After we decide upon the longest continuous chain of carbon atoms and the corresponding name, the branches (also called side chains)

* These four carbons are not actually in a straight line.

are named, If these side chains contain only carbon and hydrogen atoms joined by single bonds, they are called *alkyl groups.* The names of alkyl groups are based on the number of carbon atoms contained in them. *An alkyl group has one less hydrogen atom than the corresponding alkane.* The simplest alkyl group is the methyl group (—CH_3). The name of an alkyl group is often obtained by changing the *ane* ending of the alkane to *yl* (e.g., CH_4, meth*ane*;—CH_3, meth*yl*). Thus,

```
      H     H     H
      |     |     |
   H—C-----C-----C—H
      |     |     |
      H  H—C—H  H
            |
            H
```

is named 2-methylpropane. *Methyl* and *propane* are written as one word. The 2 indicates the methyl group is joined to the second carbon of the propane chain. A condensed structural formula for 2- methylpropane can be written

```
CH3CHCH3
    |
   CH3
```

It is important to note that the name of the group is not obtained directly from the name of the alkane. *The names and structures of the alkyl groups are given in Table 11.2 should be memorized now.*

Table 11.2 : Alkyl groups

Condensed Structural Formula	Group	Parent Alkane
CH_3—	Methyl	Methane
CH_3CH_2—	Ethyl	Ethane
$CH_3CH_2CH_2$—	Propyl	Propane
$(CH_3)_2$CH— (two CH_3 joined to CH—)	Isopropyl	Propane

The alkyl groups are *not stable molecules.* One carbon atom of each alkyl group has one less than the desired number of bonds and does not possess a noble gas electron arrangement. This remaining bonding site on the alkyl group is used to attach it to the longer carbon chain of the branched alkane.

An alkyl group is formed by (hypothetical) removal of one hydrogen atom from an alkane, leaving an unused bonding site. For example, removal of a hydrogen from methane laves a methyl group.

$$\underset{\text{methane molecule}}{CH_4} \xrightarrow[\text{of hydrogen atom}]{\text{hypothetical removal}} \underset{\text{methyl group}}{-CH_3}$$

The removal of hydrogen atom from an end carbon of propane yields a propyl group; whereas, the (hypothetical) removal from the middle hydrogen of propane gives an isopropyl group.

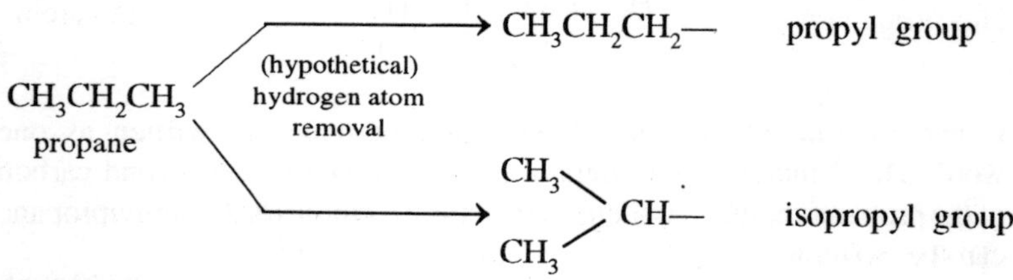

Naming Branched Alkanes

To name unbranched alkanes, recognize the longest continuous chain, and identify the side chins for branched alkanes. The rules for naming compounds are referred to as the IUPAC System of Nomenclature* and are summarized in Table 11.3

Table 11.3 : Nomenclature of Alkanes

1. Saturated hydrocarbons are classed as alkanes and always have the ending ane.
2. Branched alkanes are named on the basic longest continuous chain of carbon atoms in the molecule. (Unbranched alkanes are named as given in Table 11.1.
3. To indicate the location of the branches, the carbon atoms of the longest continuous chain are numbered consecutively. The chain is numbered such that the first branch has the lowest possible number.
4. The side chains (branches) are named as indicated in Table 11.2
5. When two or more identical side chains are contained in the same molecule, the number of identical groups is given by di for two, tri for three, etc. The location of each side chain is given by the number assigned to the carbon to which the branch alkyl group is attached.
6. In the name, hyphens are used to separate numbers from letters; commas, to separate numbers. Letters are not separated by spaces or punctuation marks.

* IUPAC is the abbreviation for International Union of Pure and Applied Chemistry. The IUPAC rules are internal ones and agreed upon by chemical societies all over the world.

ISOMERISM

There may be more than one way to bond the atoms of covalent compounds. For example, there are two ways to arrange to atoms in a compound with the formula C_2H_6O.

(i) By bonding the two carbon atoms together and then attach the oxygen atom:

```
  |   |
—C—C—O—
  |   |
```

This arrangement leaves six bonding sites, one for each hydrogen atom:

```
   H   H
   |   |
H—C—C—O—H
   |   |
   H   H
```

(ii) By attaching both carbon atoms to the oxygen:

```
  |       |
—C—O—C—
  |       |
```

This also leaves six bonding sites for the. hydrogen atoms:

```
   H       H
   |       |
H—C—O—C—H
   |       |
   H       H
```

Both of the structures obey all the bonding rules. Both structure are correct The first structure drawn is that of ethyl alcohol. The second is that of dimethyl ether.

The chemical and physiological properties of ethyl alcohol and dimethyl ether are vastly different, although the chemical formula for both is the same: C_2H_6O. *Compounds possessing the same molecular formula bud different structural arrangements of atoms are called structural isomers.* The structural arrangement of atoms in a molecule can only be determined from the way they are formed in a chemical reaction. There are several different kinds of isomerism that will be studied in this chapter. The greatest number of compounds displaying isomerism are those containing carbon. The compounds most intimately associated with life contain carbon.

Example **11.1**

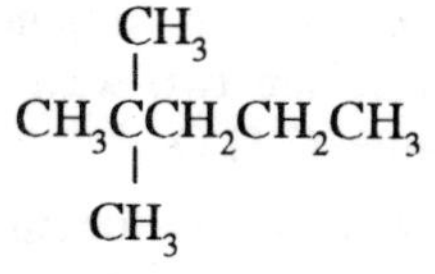

$$\begin{array}{c} CH_3 \\ | \\ CH_3CCH_2CH_2CH_3 \\ | \\ CH_3 \end{array}$$

The longest continuous chain is five carbons. Therefore, the compound is named as a pentane. Since there are two methyl groups present as side chains, the compound is a dimethylpentane. The location of each methyl group is carbon number 2. Hence, the compound's name is 2,2-dimethylpentane. (Why not 4,4-dimethylpentane?)

Example **11.2**

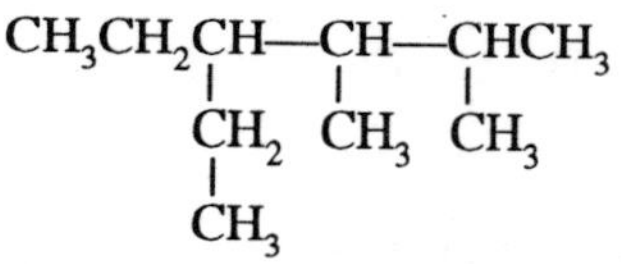

$$\begin{array}{l} CH_3CH_2CH—CH—CHCH_3 \\ \qquad\quad | \qquad\; | \qquad | \\ \qquad\; CH_2 \;\; CH_3 \;\; CH_3 \\ \qquad\quad | \\ \qquad\; CH_3 \end{array}$$

The longest continuous chain of carbon atoms is six. Two methyl and one ethyl side chains are present. Accordingly, the compound is named 4-ethyl-2,3-dimethylhexane. (Remember rule 3 dealing with the numbering of carbon chains.)

ALKENES

Alkenes, commonly called olefins, are the group of hydrocarbons containing carbon-carbon double bonds. The names of alkenes have the characteristic ending ene.

The rules governing the naming of alkenes are relatively simple and are given in Table 11.4

Table 11.4 Alkene Nomenclature

1. Characteristic ending of the name is ene.
2. Select the longest continuous chain containing the double bond. Name the chain by changing the *ane* ending of the corresponding alkane to *ene*.
3. Number the chain so as to give the first carbon atom of the double bond the lowest possible number.
4. Identify side chains in the same manner as for alkanes. Specify the location of side chains according to the numbering assigned in rule 3.

Example **11.3**

$CH_3CH_2CH_2CH{=}CH_2$

1-pentene

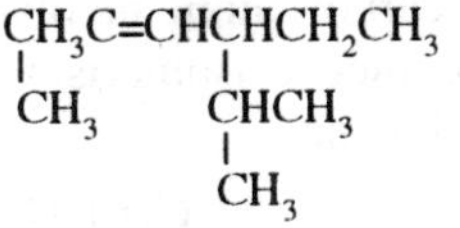

$$\begin{array}{l} CH_3C{=}CHCHCH_2CH_3 \\ \;|\qquad\quad\;| \\ CH_3\quad\;\; CHCH_3 \\ \qquad\qquad\;| \\ \qquad\qquad CH_3 \end{array}$$

4-isopropyl-2-methyl-2-hexene

or

4-ethyl-2,5-dimethyl-2-hexene

Sometimes more than one name is acceptable for a compound, as in this example. The first name is preferred.

The simplest alkene ($CH_2{=}CH_2$) is correctly named ethene; however, it is generally referred to by the older name ethylene. We shall often find examples of carbon compounds that are generally referred to by older or trivial names. Although inconvenient, it will be necessary to learn the trivial names for several compounds to be able to converse meaningfully with chemists, pharmacists, physicians, and nutritionists.

Cis and Tans Isomerism

The presence of a double bond in alkenes makes possible another type of isomerism. This new kind of isomerism is geometric or cis-trans isomerism. Geometric isomerism is the result of the fact that rotation around a double bond is not possible.

Consider 2-butene ($CH_3CH{=}CHCH_3$). Because there is no rotation around the double bond, there are two ways of arranging the carbon atoms (structures 1 and 2).

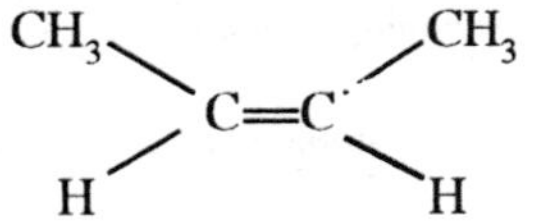

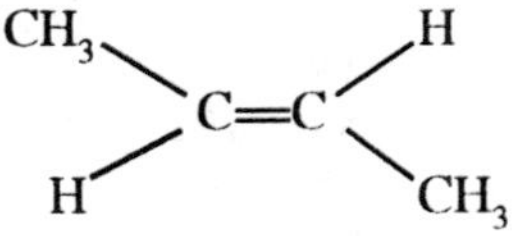

$$\begin{array}{ccc} CH_3 & & CH_3 \\ & C{=}C & \\ H & & H \end{array} \qquad\qquad \begin{array}{ccc} CH_3 & & H \\ & C{=}C & \\ H & & CH_3 \end{array}$$

The methyl group in structure 1 are on the same side of the double bond; whereas in structure 2 methyl groups are one opposite sides of the double bond. Structure 1 cannot be changed to structure 2 without breaking a bond, an event that does not happen under normal circumstances. Compounds corresponding to both structures have been isolated and studied.

When two identical groups (or atoms) are on the same side of a double bond, they are said to be cis to each other. If two identical groups are on opposite sides of the double bond, they are trans to each other. The compound whose molecules have structure 1 is cis-2 butene; structure 2 is that of *trans*-2-butene.

Notice that if there are two identical group (or atoms) on the same carbon atom involved in a double bond, there is no possibility for geometric isomerism. Thus, there is only one possible structure for 1-butene:

$$\begin{matrix} CH_3CH_2 & & H \\ & \diagdown\!C{=}C\!\diagup & \\ H & & H \end{matrix}$$

There are two requirements for geometric isomerism:

1. Restricted rotation about a bond must be present in a molecule. This restricted rotation may be caused by the presence of either a double bond or a ring formation.
2. No carbon atom whose rotation is restricted may bear two identical groups or atoms.

ALKYNES

Hydrocarbon containing a carbon-carbon triple bond are classified as alkynes. The characteristic suffix indicating the presence of a carbon-carbon triple bond is *yne*. See Table 11.5 for alkyne nomenclature. Compounds containing multiple (double or triple)bonds are said to be unsaturated. The first member of this series of compounds is ethyne

Table 11.5 : Alkyne Nomenclature

1. Characteristic ending of the name is yne.
2. Select the longest continuous carbon chain containing the triple bond. Name the chain by changing the *ane* ending of the corresponding alkane to *yne*.
3. Number the chain so as to give the first carbon atom of the triple bond the lowest possible number.
4. Identify side chains in the same manner as alkenes.

(H—C≡C—H). Ethyne is more generally known by its trivial name, acetylene. In fact, the whole class of compounds is more frequently referred to as acetylenes rather than alkynes. There are few alkynes of biological significance in mammals, although they are prevalent in plants and molds. A few alkynes are important drugs. For example, norethindrone, the principal component of the birth control pill Enovid contains a carbon-carbon triple bond. Several antibiotics (mycomycin, isomycomycin, and nemotin) also contain triple bonds between carbon atoms.

CYCLE COMPOUNDS

All the hydrocarbons considered so far are open-chain compounds. The molecule has two ends, like a piece of string has two ends. The ends of the string can be tied together to form a loop or "string ring". One might ask if the ends of a carbon atom chain can be bonded (tied) together to produce hydrocarbon rings. To join the terminal carbon atoms of an open chain into a ring requires one additional bonding site on each of these carbon atoms. For example, a four-carbon open-chain alkane has a combined total of 10 additional bonding sites on the carbon atoms; the corresponding carbon ring has only 8 bonding sites left after joining the carbon atoms together. Compare

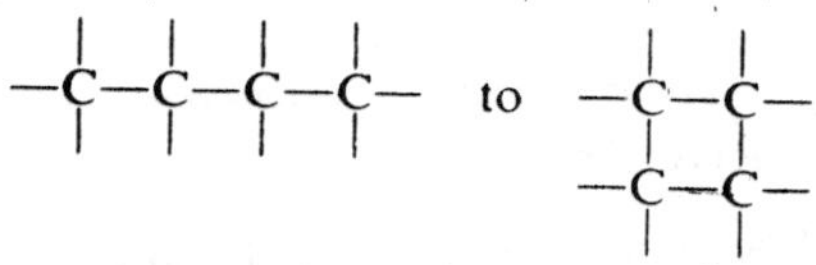

$$-\overset{|}{\underset{|}{C}}-\overset{|}{\underset{|}{C}}-\overset{|}{\underset{|}{C}}-\overset{|}{\underset{|}{C}}- \quad \text{to} \quad \begin{array}{c} -\overset{|}{C}-\overset{|}{C}- \\ \;\;| \quad\; | \\ -\underset{|}{C}-\underset{|}{C}- \end{array}$$

Compounds in which the carbon atoms form a ring are called *cyclic compounds.* Cyclic hydrocarbons containing only single bonds are cycloalkanes. They are named by using the prefix *cyclo*-before the name of the corresponding open-chain alkane. Thus,

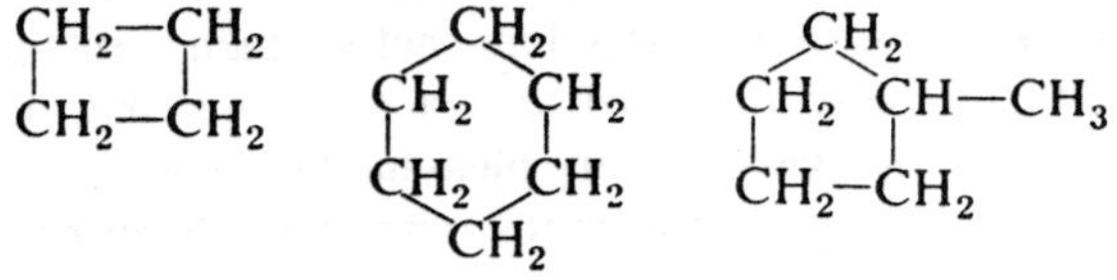

$$\begin{array}{c} CH_2-CH_2 \\ |\qquad\quad| \\ CH_2-CH_2 \end{array} \qquad \begin{array}{c} CH_2 \\ CH_2 \quad CH_2 \\ |\qquad\quad| \\ CH_2 \quad CH_2 \\ CH_2 \end{array} \qquad \begin{array}{c} CH_2 \\ CH_2 \quad CH-CH_3 \\ |\qquad\quad| \\ CH_2-CH_2 \end{array}$$

cyclobutane cyclohexane methylcyclopentane

When two or more groups are attached to carbon atoms in the ring, their relative positions must be given. Position of the groups are specified by numbering the atoms comprising the ring. Numbers are assigned the ring atoms so as to give the attached groups the lowest possible numbers.

Because carbon atoms always form four bonds, a shorthand notation for writing cyclic compounds is used. This illustrated below:

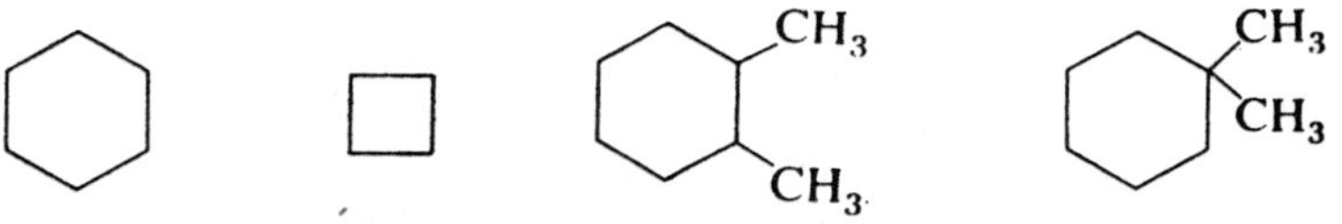

[hexagon] [square] [hexagon with CH_3 on adjacent carbons: CH_3, CH_3] [hexagon with two CH_3 on one carbon: CH_3, CH_3]

cyclohexane cyclobutane 1,2-dimethylcyclohexane 1,1-dimethylcyclohexane

Each point of the figure represents the position of a carbon atom, while each line represents a bond between two carbon atoms. It is *understood* that the remaining bonding sites on each carbon atom are occupied by hydrogen. Thus, carbons 1 and 2 of 1,2-dimethylcyclothexane each have one hydrogen atom bonded to them; the remaining carbon atoms in the ring are each bonded to two hydrogen atoms.

Many cyclic hydrocarbons contain carbon-carbon double bonds in the ring. These compounds are cycloalkenes. (Cycloalkenes are rare) Two examples of cycloalkenes are

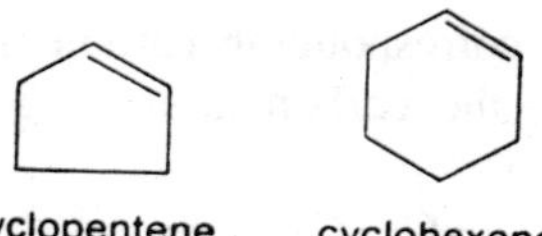

cyclopentene cyclohexene

AROMATIC HYDROCARBONS

All the hydrocarbons discussed so far are collectively known as *aliphatic hydrocarbons.* Aromatic hydrocarbons are a special class of hydrocarbons whose molecules have a six-membered carbon ring containing what appears to be alternating single and double carbon-carbon bonds. The term *aromatic* (a derivative of the word *aroma*) was applied to these compounds long before their structure became known. Thus, aromatic was used to denote their strong odor. While many aromatic compounds are very pleasant smelling compounds used in the perfume and flavorings industries, others have distinctly unpleasant odors.

The simplest aromatic compound is benzene, discovered by Faraday in 1824. Benzene has the molecular formula C_6H_6 and its structure can be written

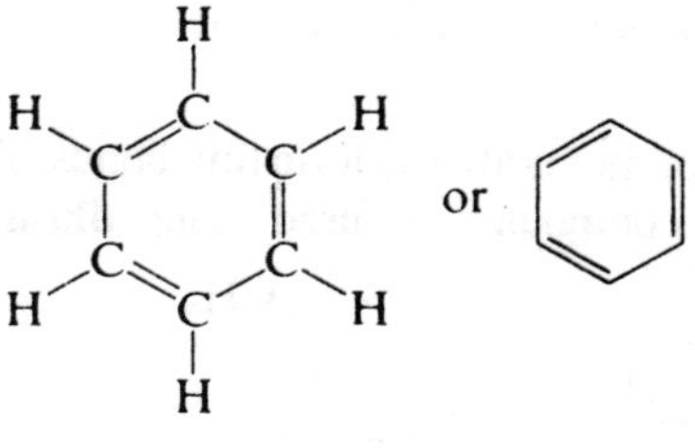

Chemists have experimentally determined that all the bonds in the ring are identical.

Our modern view of aromatic compounds describes the six electrons used for double-bond formation as being in a molecular electron shell. Thus, these six electrons are shared by all the carbon atoms in the ring. Chemists, in an effort to show the distribution of these six electrons in a molecular orbit, frequently depict the aromatic ring as

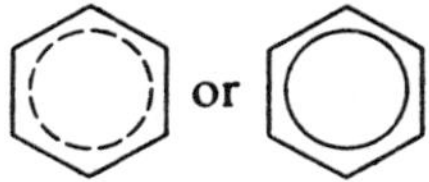

The (dotted) circle in the ring represents the six electrons in the molecular electron shell.

REACTIVITY OF HYDROCARBONS

All hydrocarbons can be burned to produce carbon dioxide and water. Combustion of hydrocarbons is a primary source of energy in our lives. We burn petrol, a mixture of hydrocarbons, to power our cars.

Balanced equations for the combustion of methane and for one component of gasoline are

$$CH_4 + 2O_2 \longrightarrow CO_2 + 2H_2O$$

$$2C_8H_{18} + 25O_2 \longrightarrow 16CO_2 + 18H_2O$$

Regardless of the class of hydrocarbons (alkane, alkene, or aromatic) or the isomer (structural or geometric), all hydrocarbons burn to ultimately produce carbon dioxide and water. In contrast to the combustion reaction, the vast majority of organic reactions are dependent on classification of the reactants.

Alkanes undergo very few reactions. Alkenes, however, undergo a large number of reactions. Typically, reactions of alkenes involve the double bond. During these reactions the carbon-carbon double bond is converted to a carbon-carbon single bond. Simultaneously, a new bond is formed by each of the carbon atoms of the original double bond. Such reactions are called *addition reactions.*

An example of an addition reaction is that between 1-butene and hydrogen in the presence of platinum.

$$CH_3CH_2CH{=}CH_2 + H_2 \xrightarrow{Pt} CH_3CH_2\underset{\displaystyle H}{\underset{|}{C}}H{-}\underset{\displaystyle H}{\underset{|}{C}}H_2$$

Notice that in this reaction only the double-bond portion of the alkene molecule reacts. The rest of the molecule remains unchanged. This observation suggests that other alkenes should react with hydrogen in the presence of platinum. Experimentation shows this to be true. For example, cis-2-butene and cyclothexene will both with hydrogen.

$$\underset{\text{H}}{\overset{\text{CH}_3}{>}}\text{C=C}\underset{\text{H}}{\overset{\text{CH}_3}{<}} + H_2 \xrightarrow{Pt} CH_3CH_2CH_2CH_3$$

and

$$\text{cyclohexene} + H_2 \xrightarrow{Pt} \text{cyclohexane (with H, H added)}$$

Under appropriate conditions, water will react with compounds containing double bonds. In the presence of sulphuric acid (H_2SO_4) both cis-2-butene and cyclohexane react with water to produce an alcohol.

$$\underset{\text{H}}{\overset{\text{CH}_3}{>}}\text{C=C}\underset{\text{H}}{\overset{\text{CH}_3}{<}} + H_2O \xrightarrow{H_2SO_4} CH_3\underset{|}{\underset{H}{C}}H\text{—}\underset{|}{\underset{OH}{C}}H\text{—}CH_3$$

$$\text{cyclohexene} + H_2O \xrightarrow{H_2SO_4} \text{cyclohexane (with H, OH added)}$$

Note that only the double-bond portion of the alkene takes part in the reaction. The rest of the molecule remains unchanged. We will find the hydrocarbon portions of the molecules containing only single bonds are unreactive. For this reason, chemists look for the reactive atom or group of atoms in organic molecules. This reactive atom or group of atoms is called a functional group.

Alkenes contain the functional group $>C=C<$. The presence of a carbon-carbon double bond is also the basis for classification of a particular hydrocarbon as an alkene. Thus, *a functional group both classifies a compound and tells us something about the chemical properties and reactivity of the compound.*

FUNCTIONAL GROUPS OF ORGANIC MOLECULES

We are already familiar with three functional groups: (1) carbon-carbon double bonds, (2) carbon-carbon triple bonds, and (3) aromatic rings. Some functional groups and illustrative compounds are given in Table 11.6

Table 11.6 : Some Functional Groups of Organic Compounds

Name of Compound Class	Functional Group	Example
alkane		$CH_3CH_2CH_2CH_3$
alkene	$>C=C<$	CH_3[$CH=C$]H_2
alkyne	$-C\equiv C-$	CH_3–[$C\equiv C$]–CH_2CH_3
aromatic	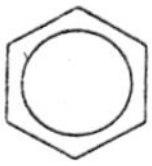 (benzene ring)	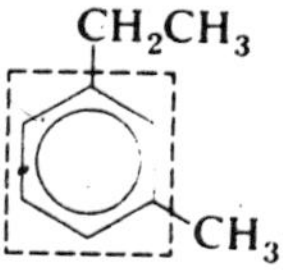 (benzene ring with CH_2CH_3 and CH_3)
alcohol	$-\overset{\mid}{\underset{\mid}{C}}-OH$	CH_3[CH_2OH]
ether	$-\overset{\mid}{\underset{\mid}{C}}-O-\overset{\mid}{\underset{\mid}{C}}-$	CH_3[CH_2-O-CH_2]CH_3
acid	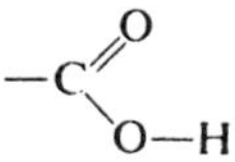 $-C(=O)-O-H$	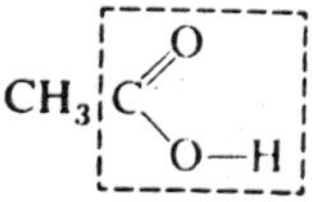 CH_3[$C(=O)-O-H$]
amine	$-\overset{\mid}{\underset{\mid}{C}}-NH_2$	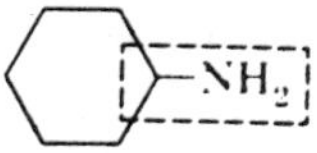 (cyclohexane ring)[$-NH_2$]
amide	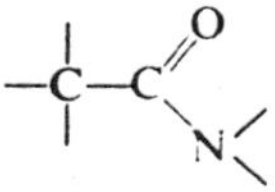 $-\overset{\mid}{\underset{\mid}{C}}-C(=O)-N<$	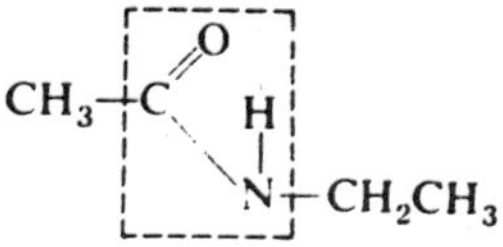 CH_3–[$C(=O)-\overset{H}{N}$]–CH_2CH_3
ester	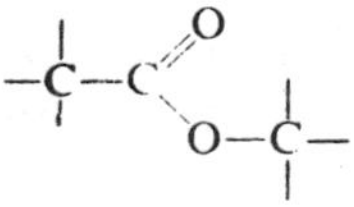 $-\overset{\mid}{\underset{\mid}{C}}-C(=O)-O-\overset{\mid}{\underset{\mid}{C}}-$	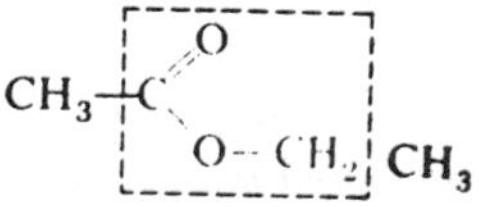 CH_3–[$C(=O)-O-CH_2$]CH_3

POLARITY OF ORGANIC COMPOUNDS

The electronegativity of carbon and hydrogen are very similar; consequently, there is very little polar character associated with carbon-hydrogen bonds. Accordingly, hydrocarbons are essentially nonpolar compounds.

Nitrogen and oxygen are more electronegative than carbon. Therefore C—N, C—O, N—H, and O—H bonds have a substantial polar character. The polar character of these bonds manifests itself in the behavior of alcohols, acids, and amines. All of these compounds are polar.

Let us consider alcohols first. The alcohols have the characteristic group $-\overset{|}{\underset{|}{C}}-O-H$. Both C—O bond and the O—H bond are polar. The Lewis structure of ethyl alcohol is

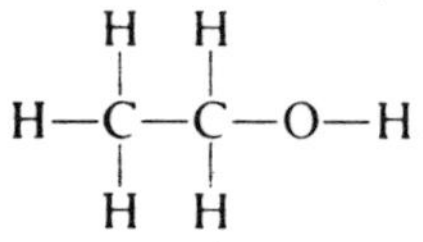

```
     H   H
     |   |
  H—C—C—O—H
     |   |
     H   H
```

The shape of the molecule around the oxygen atom is bent, or V-shaped. Thus its shape is similar to that of water.

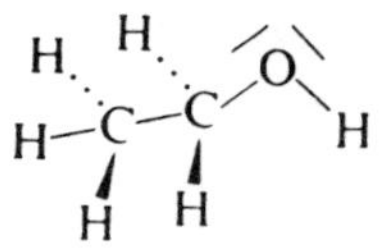

```
   H  H      O
    ·. ·.  /   \
  H—C—C        H
     ▲   ▲
     H   H
```

A result of the bond angles around oxygen and the polar character of O—H and C—O bonds is that the alcohol molecule is polar.

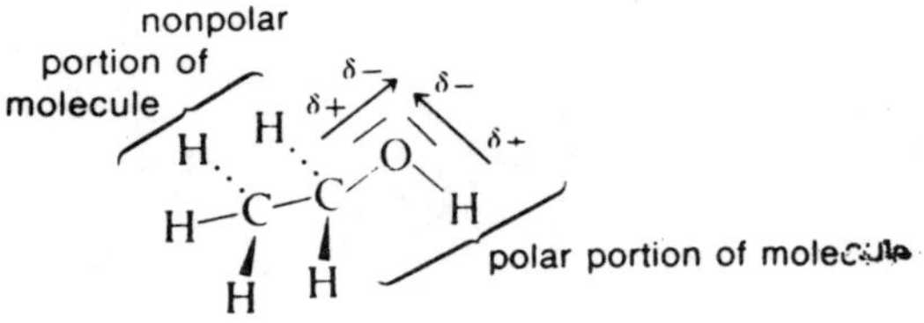

nonpolar portion of molecule

```
          δ−     δ−
      δ+ ↗     ↖   δ+
   H  H      O
    ·. ·.  /   \
  H—C—C        H
     ▲   ▲
     H   H
```

polar portion of molecule

In this respect the alcohol molecule is also similar to water. We might then expect to find some similarities in the behavior of alcohols and water. One such similarity is the observance of hydrogen bonding. Figure 11.3 shows hydrogen bonding in liquid water and Fig. 11.4 shows hydrogen bonding in liquid ethyl alcohol.

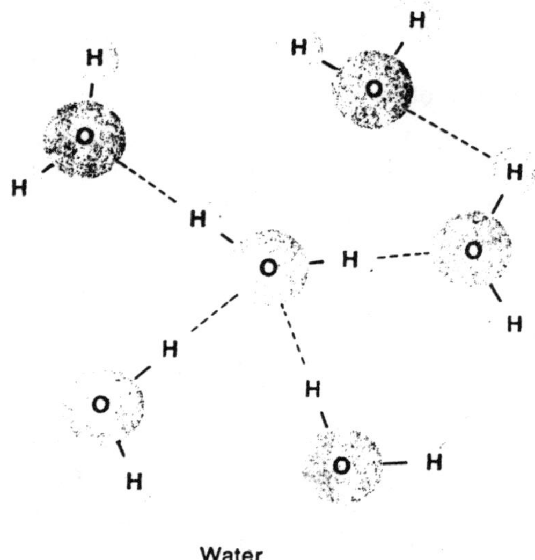

Fig. 11.3 : Hydrogen bonding in liquid water.

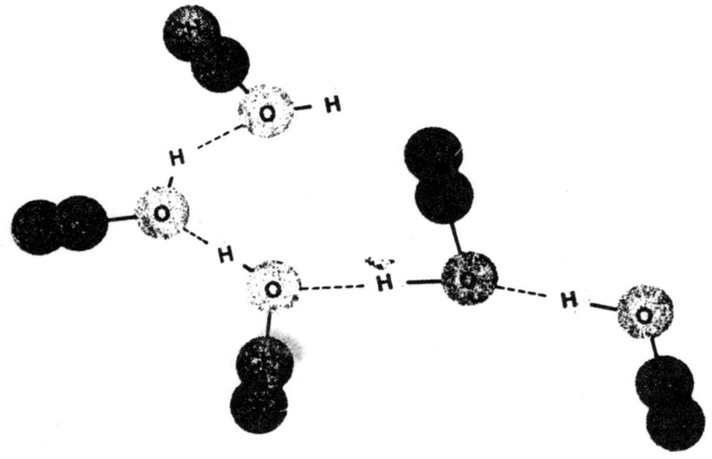

Fig. 11.4 : Hydrogen bonding in liquid ethyl alcohol.

The electronegative of nitrogen is less than that of oxygen. Thus, N—C and N—H bonds are less polar than C—O and O—H bonds. Nevertheless, amines are polar molecules and hydrogen bonding can be experimentally observed.

Organic acids are even more polar than alcohols or amines. The Lewis structure depicting the shape of acetic acid molecules shows why the molecule is so polar.

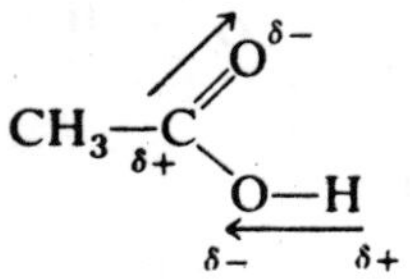

$$CH_3-\overset{\delta+}{C}(=O^{\delta-})-\overset{\delta-}{O}-\overset{\delta+}{H}$$

The particle positive charge on the carbon atom created by the double-bonded oxygen actually causes the bonding pair of electrons between the oxygen and hydrogen atoms of the O—H portion to move further away from the hydrogen than might be expected.

12

Alcohols and Their Derivatives

Classification and Nomenclature

An alcohol is characterized by a hydroxyl group (—O—H) bonded to a saturated carbon atom. Since the hydrocarbon group in a simple alcohol is an alkyl group, we may write ROH as a general symbol for simple alcohols. Alcohols vary in their water solubility. Low molecular weight alcohols, those containing up to five or six carbon atoms, are water-soluble by virtue of their ability to hydrogen bond with water. Monohydroxy alcohols containing more than six carbon atoms are water-insoluble as the behavior of the hydrocarbon portion of the molecule predominates.

The possibility of structural isomerism exists if the alcohol molecules contain three or more carbon atoms. Alcohols are divided into subclasses depending on the number of carbon atoms bonded to the carbon atom attached to the hydroxyl group. The four possible structures for alcohols with the molecular formula $C_4H_{10}O$ are shown in Fig. 12.1 (see page no. 250). A *primary* alcohol is one in which the hydroxyl group is attached to a carbon atom forming only one carbon-carbon bond-structures (a) and (b) in Fig. 12.1. *Secondary* alcohols—structure (c)—have two carbon-carbon bonds on the carbon with the hydroxyl group, while *tertiary* alcohols—structure (d) —have *three*. General symbols for each of the alcohol subclasses are

```
   H              R              R
   |              |              |
R—C—OH         R—C—OH         R—C—OH
   |              |              |
   H              H              R
```

primary alcohol   secondary alcohol   tertiary alcohol

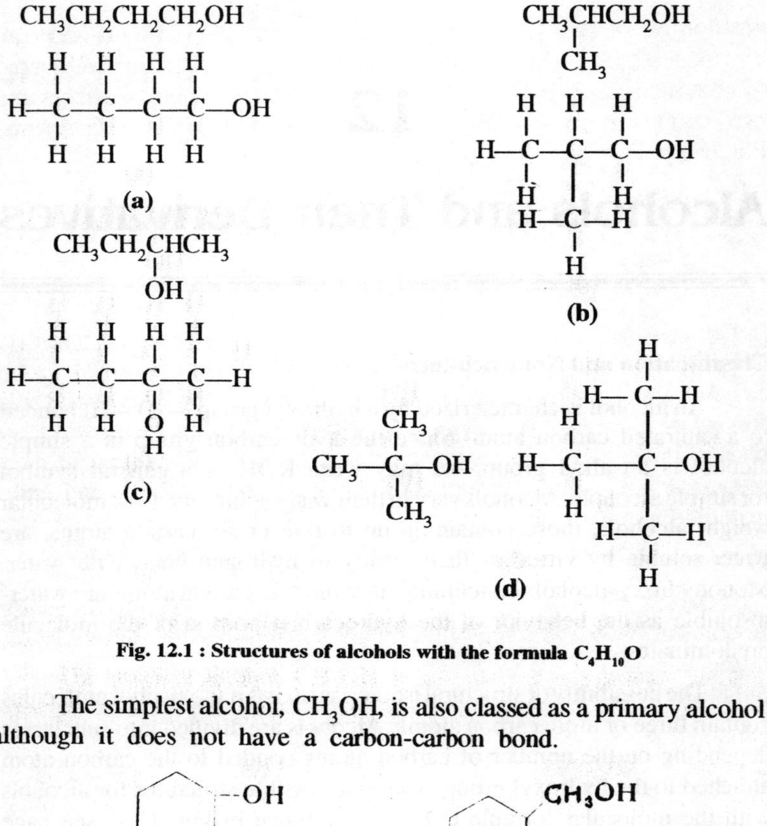

$CH_3CH_2CH_2CH_2OH$

```
   H   H   H   H
   |   |   |   |
H—C—C—C—C—OH
   |   |   |   |
   H   H   H   H
```

(a)

CH_3CHCH_2OH (with CH_3 on the second carbon)

```
    H   H   H
    |   |   |
H—C—C—C—OH
    |   |   |
    H   |   H
     H—C—H
        |
        H
```

(b)

$CH_3CH_2CHCH_3$ (with OH on the third carbon)

```
   H   H   H   H
   |   |   |   |
H—C—C—C—C—H
   |   |   |   |
   H   H   O   H
           H
```

(c)

```
        CH3
        |
CH3—C—OH
        |
        CH3
```

```
               H
               |
            H—C—H
     H         |
     |         |
  H—C————C—OH
     |         |
     H         |
            H—C—H
               |
               H
```

(d)

Fig. 12.1 : Structures of alcohols with the formula $C_4H_{10}O$

The simplest alcohol, CH_3OH, is also classed as a primary alcohol although it does not have a carbon-carbon bond.

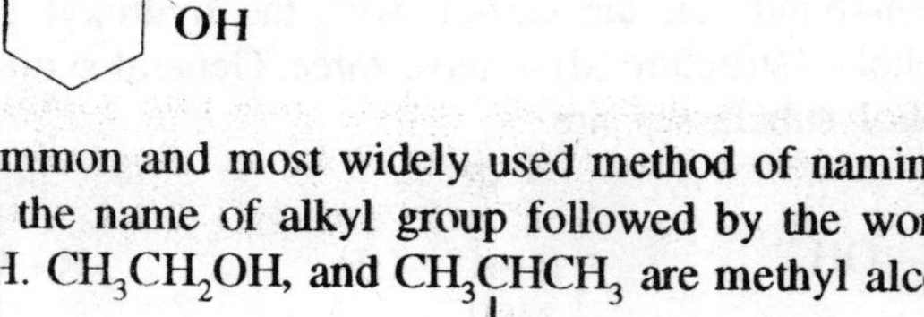

Cyclohexane ring –OH

Cyclohexane ring –CH_2OH

Cyclohexane ring with CH_3 and OH on the same carbon

The common and most widely used method of naming alcohols simply gives the name of alkyl group followed by the word alcohol. Thus, CH_3OH. CH_3CH_2OH, and CH_3CHCH_3 (with OH on the middle carbon) are methyl alcohol, ethyl alcohol, and isopropyl alcohol, respectively. In the IUPAC system of naming, the *e* ending of the alkane name is replaced by an *ol*. The position of the OH group on the carbon chain is given by a numeral. The longest continuous carbon chain containing the OH group is numbered to give the hydroxyl group the lowest value. Appropriate numbers and prefixes are used to indicate the presence and position of other substituents, For example,

$$\underset{\displaystyle CH_3}{CH_3\underset{|}{C}H}CH_2\underset{\displaystyle OH}{\underset{|}{C}H}CH_3$$

4-methyl-2-pentanol

Common and IUPAC names for several simple alcohols are given in Table 12.1. Phenol, the last entry in the table, has a hydroxyl group attached to an aromatic ring instead of an alkyl group. Aromatic alcohols are collectively known as *phenols*, with the simplest member called *phenol*. Phenol is commonly called carbolic acid. In addition to their properties as alcohols, phenols are also very weakly acidic and quite irritating tissue.

Preparation of Alcohols

Alcohols can be prepared in both the living cell and in the chemistry laboratory by the addition of water to a carbon-carbon double bond. In the laboratory, the reaction requires a strong acid catalyst-usually sulphuric acid.

$$\begin{matrix} H & & H \\ & C{=}C & \\ H & & H \end{matrix} + H{-}O{-}H \xrightleftharpoons{H_2SO_4} \underset{\displaystyle H}{\underset{|}{CH_2}}{-}CH_2{-}OH$$

The reaction is a reversible one, and maximum yields of alcohol are obtained by using a large excess of water to drive the reaction to completion.

DEHYDRATION OF ALCOHOLS

In the previous section it was pointed out that the addition of water to an alkene is a reversible reaction. Thus, for example, ethylene can be prepared by the dehydration of ethanol:

$$CH_3CH_2OH \xrightleftharpoons{H_2SO_4} CH_2{=}CH_2 + H_2O$$

To maximize the yield of olefin, an excess of sulphuric acid is used. The sulphuric acid acts as dehydrating agent (absorbing the water)

Table 12.1 Common and IUPAC names of some alcohols

Structure	*Common Name*	*IUPAC Name*
CH_3OH	Methyl alcohol	Methanol
CH_3CH_2OH	Ethyl alcohol	Ethanol
$CH_3CH_2CH_2OH$	*n*-Propyl alcohol	1-Propanol
$CH_3CH(OH)CH_3$	Isopropyl alcohol	2-Propanol
$CH_3CH_2CH_2CH_2OH$	*n*-Butyl alcohol	1-Butanol
$CH_3CH(CH_3)CH_2OH$	Isobutyl alcohol	2-Methyl-1-propanol
$CH_3CH_2CH(OH)CH_3$	Secondary butyl alcohol	2-Butanol
$CH_3-C(CH_3)_2-OH$	Tertiary butyl alcohol	2-Methyl-2-propanol
$CH_2(OH)-CH_2(OH)$	Ethylene Glycol	1,2-Ethanediol
$CH_2(OH)-CH(OH)-CH_2(OH)$	Glycerol or glycerine	1,2,3-Propanetriol
C_6H_5-OH	Phenol (carbolic acid)	Phenol

as it is formed. Removal of the water in this manner drives the reaction to completion as predicated by Le Chatelier's principle.

So far we have discussed only *intramolecular* elimination of water from alcohols. *Intra*molecular means within the molecule; the molecule of water eliminated comes from within each molecule of alcohol. Alcohols may also undergo an *intermolecular* elimination of water. *Inter*molecular means between two molecules. Thus, two molecules of alcohol come together to eliminate a molecule of water. The other product formed during this reaction is an *ether*. In the general case the reaction can be written

$$R—O—H + H—O—R \xrightleftharpoons{H_2SO_4} H_2O + R—O—R$$

an ether

or simply

$$2ROH \xrightleftharpoons{H_2SO_4} H_2O + R—O—R$$

A specific example is the formation of diethyl ether from ethyl alcohol.

$$\underset{\text{ethyl alcohol}}{2CH_3CH_2OH} \xrightleftharpoons{H_2SO_4} \underset{\text{diethyl ether}}{CH_3\ CH_2—O—CH_2CH_3} + H_2O$$

Notice that the catalyst for the formation of ethers from alcohols is the same as for the formation of alkenes from alcohols. A consequence of this is that in the chemistry laboratory a mixture of alkene and ether is usually obtained from the dehydration of alcohols. In most cases chemists can adjust the reaction temperature to obtain primarily alkene or primarily ether. You will not be asked to learn the conditions under which either olefin or ether primarily produced.

Whenever ethers are mentioned, most people think of diethyl ether, used as an anesthetic. In fact, even chemists often refer to diethyl ether (a liquid at room temperature: bp, 35° C,) is an important solvent, dissolving fats and other slightly polar compounds. Medically, diethyl ether has long been employed as a general anesthetic, its use having first been reported for this purpose in 1842 by Dr. Crawford W. Long. Drawbacks to ether are its highly flammable nature, irritation of respiratory tissues, and the tendency to cause nausea. Other ethers, principally methylpropyl ether, are also used as general anesthetics.

Ethylene oxide is a cyclic ether used as a fumigant and is one of few ethers that will readily undergo reaction with water. It reacts with water to produce ethylene glycol, the principal ingredient of permanent antifreeze.

$$\underset{\text{ethylene oxide}}{\overset{CH_2—CH_2}{\diagdown_{O}\diagup}} + H_2O \longrightarrow \underset{\text{ethylene glycol}}{\overset{CH_2CH_2}{\underset{OH\ \ OH}{|\ \ \ \ |}}}$$

ethylene oxide — ethylene glycol (1,2-ethanediol

DEHYDROGENATION OF ALCOHOLS—OXIDATION

Under appropriate conditions hydrogen can be removed from primary and secondary alcohols. Two hydrogen atoms (the equivalent of H_2) are removed from each alcohol molecule. *One of the hydrogen atoms removed is the hydroxyl group hydrogen atom and the other is from the carbon atom bonded to the hydroxyl group.* Removal of the hydrogen atoms is accompanied by formation of a double bond between the carbon and oxygen atoms. An *aldehyde* results from the removal of the hydrogen atoms from a *primary alcohol*:

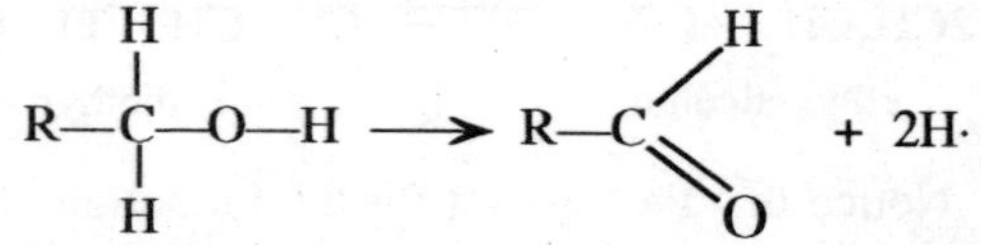

$$\underset{\text{primary alcohol}}{R-CH_2-O-H} \longrightarrow \underset{\text{aldeyde}}{R-C(H)=O} + 2H\cdot$$

Dehydrogenation of a *secondary alcohol* yields a *ketone*:

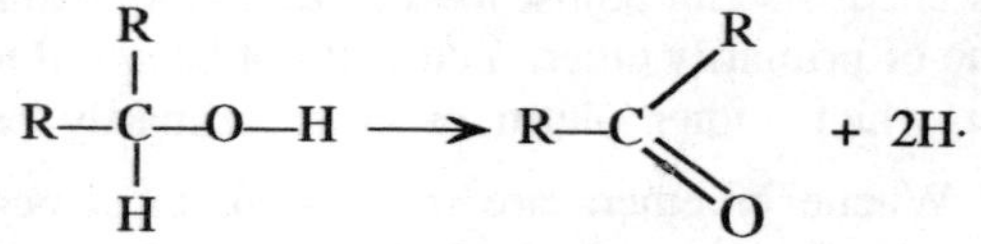

$$\underset{\text{secondary alcohol}}{R-CH(R)-O-H} \longrightarrow \underset{\text{ketone}}{R-C(R)=O} + 2H\cdot$$

Removal of hydrogen means electrons are removed from the alcohol molecule. Loss of electrons is *oxidation*; thus, alcohols may be oxidized to produce aldehydes and ketones. The loss of electrons by the alcohol is depicted in Fig. 12.2. Oxidation of a *primary alcohol* yield an *aldehyde*; oxidation of a *secondary* alcohol, a ketone.

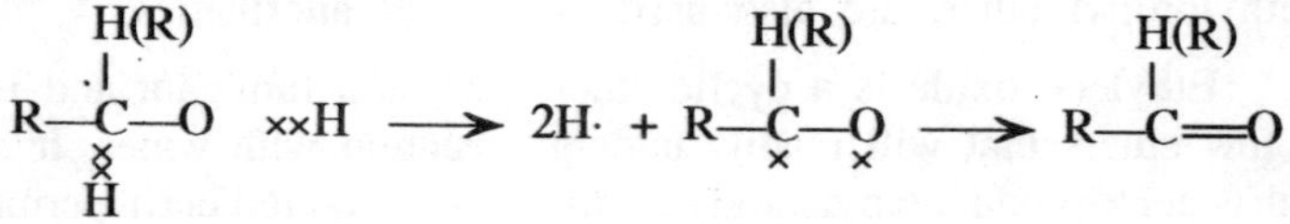

$$R-\overset{H(R)}{\underset{\overset{\times}{\cdot}H}{C}}-O\ {}^{\times\times}H \longrightarrow 2H\cdot + R-\overset{H(R)}{\underset{\times}{C}}-\underset{\times}{O} \longrightarrow R-\overset{H(R)}{C}=O$$

Fig. 12.2 Dehydrogenation of an alcohol is oxidation. The symbol — is two-electron covalent bond, × and · represent individual electrons.

Whenever oxidation occurs, reduction must simultaneously take place. Furthermore, an oxidizing agent is needed to cause oxidation. Suitable oxidizing agents to oxidize alcohols are potassium permanganates

—CHO is a short representation of an aldehyde group $-C(=O)H$

($KMnO_4$) and potassium dichromate ($K_2Cr_2O_7$). During reaction, manganese or chromium atoms gain electrons and thereby are reduced. The hydrogen atoms from the alcohol combine with oxygen from the oxidizing agent to produce water. Balanced equations for the general reactions in acidic reaction media are

$$5RCH_2OH + 2KMnO_4 + 3H_2SO_4 \longrightarrow 5RCHO + 2MnSO_4 + 8H_2O + K_2SO_4$$

$$3RCH_2OH + K_2Cr_2O_7 + 4H_2SO_4 \longrightarrow 3RCHO + Cr_2(SO_4)_3 + K_2SO_4 + 7H_2O$$

Balanced equations are written to convey quantitative information to the reader. Since organic reactions are frequently accompanied by side reactions, conversion of one organic compound to another often is not quantitative. For this reason, coupled with primary interest in the changes the organic compound undergoes, balanced equations are not usually written for reactions involving organic compounds. For instance, the oxidation of an alcohol by potassium dichromate is often simply written

$$R{-}CH_2OH + K_2Cr_2O_7 \xrightarrow{H^+} RCHO$$

or even as

$$\underset{\text{primary alcohol}}{R{-}CH_2OH} \xrightarrow[H^+]{K_2Cr_2O_7} \underset{\text{aldehyde}}{RCHO}$$

Tertiary alcohols cannot be oxidized because the carbon atom bonded to the hydroxyl group has no hydrogen atom bonded to it see Fig 12.3

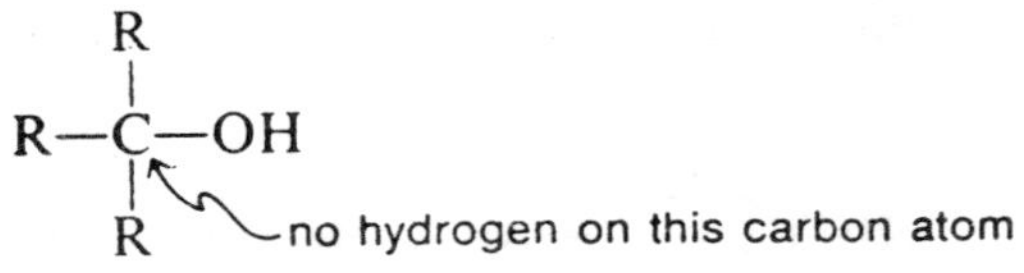

$$R_3C{-}OH$$

R—C(R)(R)—OH ← no hydrogen on this carbon atom

Fig. 12.3 : A tertiary alcohol cannot be oxidized.

OXIDATION OF ALDEHYDES

Aldehydes can be further oxidized to acids. This oxidation involves the incorporation of oxygen into the aldehyde molecule.

$$R{-}C(=O){-}H + (O) \longrightarrow R{-}C(=O){-}O{-}H$$

In order to oxidize a carbon-to-hydrogen bond, it is necessary that an activating group be bonded to the carbon atom. An oxygen atom may serve as an activating group. Carbon-carbon single bonds cannot normally be oxidized. Therefore, ketones are usually resistant to further chemical oxidation.

Both potassium permanganate and potassium dichromate can oxidize aldehydes to acids. In facts, aldehydes are more easily oxidized than are alcohols. This means that aldehydes are difficult to prepare. As some of the primary alcohol is oxidized to the aldehyde, the aldehyde produced will be further oxidized in preference to oxidizing the remaining alcohol. Fortunately, many aldehydes have lower boiling points than the alcohols from which they are prepared. Thus, low molecular weight aldehydes can be removed from the reaction mixture as they are formed. Removal of the aldehyde from the reaction mixture prevents further oxidation to the acid.

With one exception, acids are resistant to further oxidation. The exception is formic acid, which has a carbon atom bonded to both hydrogen and oxygen atoms.

$$H{-}C(=O){-}OH + (O) \longrightarrow H{-}O{-}C(=O){-}OH$$

formic acid

The product of formic acid oxidation is carbonic acid (H_2CO_3), which decomposes to carbon dioxide and water. Thus, the oxidation of formic acid can be written

$$H{-}C(=O){-}OH + (O) \longrightarrow H_2O + CO_2$$

Simple tests involving a visual change during reaction have been developed to identify the presence of aldehydes. These tests are based on the facts that aldehydes are more easily oxidized than alcohols. Strong oxidizing agents like $KMnO_4$ and $K_2Cr_2O_7$ are capable of oxidizing both aldehydes and alcohols. Weak oxidizing agents such as silver (1) and copper (II) ions are above to oxidize aldehydes but not alcohols.

An ammoniacal solution of silver ions known as *Tollen's reagent* oxidizes aldehydes to acids. The purpose of the ammonia is to form the complex ions $Ag(NH_3)_2^+$ preventing the precipitation of insoluble silver hydroxide. The silver ions are reduced to metallic silver, forming a silver mirror on the inside of the glass reaction vessel. The formation of the silver mirror gives visual confirmation of the presence of an aldehyde in the sample tested.

$$\underset{\text{Tollen's reagent}}{RCHO + 2Ag(NH_3)_2^+} + 2OH^- \rightarrow RCO_2^-NH_4^+ + \underset{\text{metallic silver}}{2Ag\downarrow} + 3NH_3 + H_2O$$

Notice that the ammonium slat of the organic acid is produced in the basic solution of Tollen's reagent.

The bright blue *Fehling's* and *Benedict's solutions* are basic solutions of copper (II) ions. Copper (II) sulfate is the usual source of copper ions. Fehling's solution contains sodium potassium tartrate and Benedict's solution contains sodium citrate as complexing agents. These complexing agent prevent the formation of insoluble copper (II) hydroxide. Both Fehling's and Benedict's solutions yield an orange to red precipitate of copper (I) oxide upon reaction with aldehydes. The overall reaction is

$$RCHO + \underset{\text{blue}}{2Cu^{2+}(aq)} \xrightarrow{NaOH} RCO_2^-Na^+ + \underset{\text{red}}{Cu_2O\downarrow}$$

Other easily oxidized compounds also react with Tollne's reagent and Fehling's or Benedict's solution. One such group of compounds are α-hydroxy ketones α-hydroxy ketone have an —OH group on the carbon next to the ketone group.

```
R—CH—C—R
   |  ||
   OH O
```

The alcohol group in these compounds is activated toward oxidation by the ketone group next to it. The reactions ol aldehydes and α-hydroxy ketones with these weak oxidizing agents are important in the study of sugars and other carbohydrates.

Addition Reactions of Aldehydes and Ketones

Just as carbon-carbon double bonds undergo addition reactions, carbon-oxygen double bonds also undergo addition reactions.

The addition of hydrogen to the carbon-oxygen double bonds converts aldehydes and ketones to alcohols. In the laboratory, a catalyst such as platinum metal is required for the hydrogenation of aldehydes or ketones:

$$R{-}C(=O){-}H + H_2 \xrightarrow{Pt} R{-}\underset{H}{\overset{H}{C}}{-}O{-}H$$

$$R{-}\overset{O}{\overset{\|}{C}}{-}R + H_2 \xrightarrow{Pt} R{-}\underset{OH}{\overset{H}{C}}{-}R$$

Aldehydes and ketones also undergo addition reactions with alcohols. *Addition of an alcohol to an aldehyde yield a hemiacetal:*

$$\underset{\text{aldehyde}}{R{-}C(=O){-}H} + \underset{\text{alcohol}}{R{-}OH} \rightleftharpoons \underset{\text{hemiacetal}}{R{-}\underset{H}{\overset{O-H}{C}}{-}O{-}R}$$

The hydroxyl group hydrogen of the alcohol bonds to the oxygen atom of the aldehyde, and the alkoxy portion (R—O—) of the alcohol bonds to the carbon atom of the original aldehyde group. The reaction does not require a catalyst and an equilibrium mixture of unreacted aldehyde, unreacted alcohol, an hemiacetal is formed spontaneously upon mixing of the aldehyde and alcohol. The position of the equilibrium depends on which aldehyde and alcohol are mixed.

Hemiacetals are said to react with Tollen's reagent and Fehling's or Benedict's solution since some free aldehyde is always present. As the aldehyde reacts, the equilibrium is upset, and some of the hemiacetal must decompose to free aldehyde and alcohol in order to restore equilibrium. Consequently, if sufficient oxidizing agent is present, all the hemiacetal will eventually decompose.

Reaction also occurs between alcohols and ketones to product hemiketals:

$$\underset{\text{ketone}}{R{-}\overset{\overset{O}{\|}}{C}{-}R} + \underset{\text{alcohol}}{R{-}O{-}H} \rightleftharpoons \underset{\text{hemiketal}}{R{-}\overset{HO\ \ OR}{C}{-}R}$$

Hemiketal formation is not as favorable as hemiacetal formation Since simple ketones cannot be oxidized further, their hemiketals do not react with Tollen's Benedict's and Fehling's Solutions.

Some molecules may contain both an alcohol group and an aldehyde group. In this case an intramolecular reaction may occur to form a *cyclic hemiacetal.* For example,

$$HO{-}CH_2CH_2CH_2CH_2C(=O)H \rightleftharpoons \text{(six-membered ring: } {-}O{-}\text{ in ring, ring carbon bearing } H \text{ and } O{-}H)$$

Cyclic hemiacetal (or cyclic hemiketal) formation is particularly favorable when a five-or six-membered ring can be formed. Under these conditions, almost all the material is present in the hemiacetal or hemiketal form.

Hemiacetals and hemiketals will react with additional alcohol in the presence of an acid catalyst to produce acetals and ketals, respectively. Water is also produced during reaction.

$$\underset{\text{hemiacetal}}{R{-}C(OH)(OR){-}H} + ROH \xrightleftharpoons{H^+} \underset{\text{acetal}}{R{-}C(OR)(OR){-}H} + HOH$$

$$\underset{\text{hemiketal}}{R{-}C(OH)(OR){-}R} + ROH \xrightleftharpoons{H^+} \underset{\text{ketal}}{R{-}C(OR)(OR){-}R} + HOH$$

Although the reaction is an equilibrium reaction, acetals and ketals can be isolated in the pure state if the catalyst is removed. After the catalyst is removed, neither the forward nor the reverse reactions occur at an appreciable rate. The easiest way to remove the acid catalyst is to add base to neutralize it. Thus, acetals, in contrast to hemiacetals, will *not* react with Tollen's reagent and Fehling's or Benedict's solutions.

13

Derivatives of Acids

Several classes of acid derivatives such as acid anhydrides, esters, and triglycerides are discussed in this chapter.

Anhydrides

Acid anhydrides make up a biologically important group of compounds. An anhydride group may be represented as

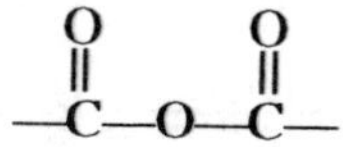

$$-\overset{\overset{\displaystyle O}{\|}}{C}-O-\overset{\overset{\displaystyle O}{\|}}{C}-$$

or

$$-\overset{\overset{\displaystyle O}{\|}}{P}-O-\overset{\overset{\displaystyle O}{\|}}{P}-$$

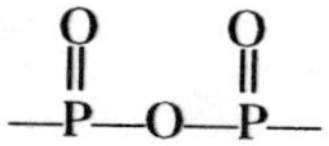

The term *anhydride* means *without water*. The anhydride is produced via intermolecular dehydration. Carboxylic acids form corresponding anhydrides as follows:

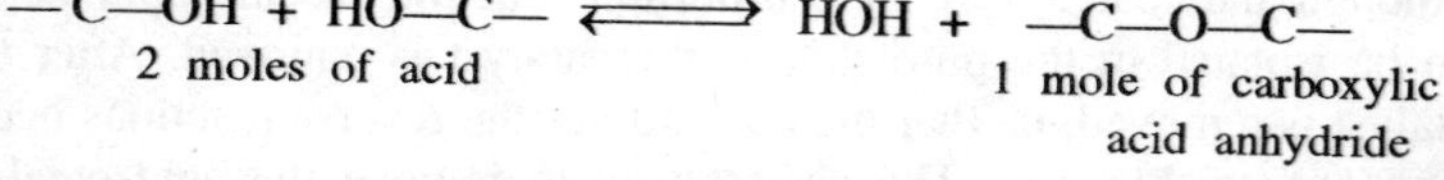

$$\underbrace{-\overset{\overset{\displaystyle O}{\|}}{C}-OH + HO-\overset{\overset{\displaystyle O}{\|}}{C}-}_{\text{2 moles of acid}} \rightleftharpoons HOH + \underbrace{-\overset{\overset{\displaystyle O}{\|}}{C}-O-\overset{\overset{\displaystyle O}{\|}}{C}-}_{\text{1 mole of carboxylic acid anhydride}}$$

Removal of water from 2 moles of acetic acid yields acetic anhydrides. For example,

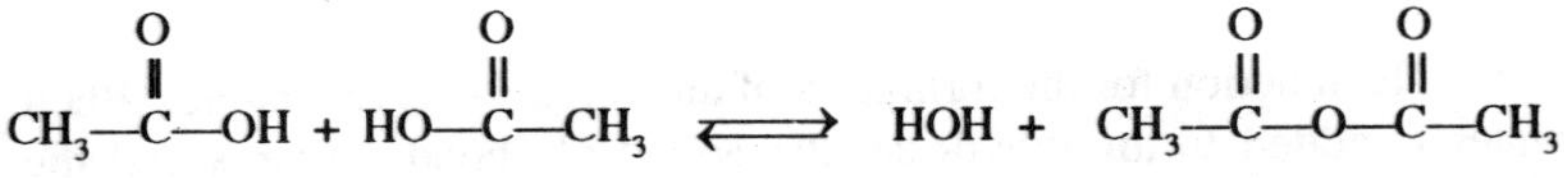

$$CH_3{-}\overset{\displaystyle O}{\overset{\|}{C}}{-}OH + HO{-}\overset{\displaystyle O}{\overset{\|}{C}}{-}CH_3 \rightleftharpoons HOH + CH_3{-}\overset{\displaystyle O}{\overset{\|}{C}}{-}O{-}\overset{\displaystyle O}{\overset{\|}{C}}{-}CH_3$$

acetic acid acetic acid acetic anhydride

The coupling of two phosphoric acid molecules by the elimination of a water molecule produces a phosphorus-containing anhydride:

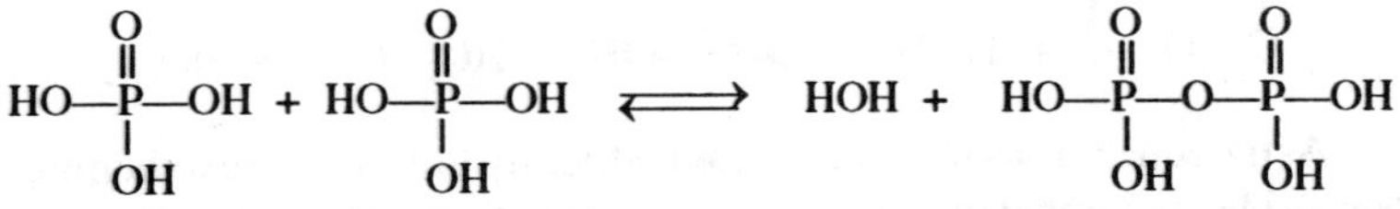

$$HO{-}\underset{\displaystyle OH}{\underset{|}{\overset{\displaystyle O}{\overset{\|}{P}}}}{-}OH + HO{-}\underset{\displaystyle OH}{\underset{|}{\overset{\displaystyle O}{\overset{\|}{P}}}}{-}OH \rightleftharpoons HOH + HO{-}\underset{\displaystyle OH}{\underset{|}{\overset{\displaystyle O}{\overset{\|}{P}}}}{-}O{-}\underset{\displaystyle OH}{\underset{|}{\overset{\displaystyle O}{\overset{\|}{P}}}}{-}OH$$

phosphoric acid a phosphoric acid anhydride

The coupling process can continue with the incorporation of a third molecule of phosphoric acid into its anhydride form.

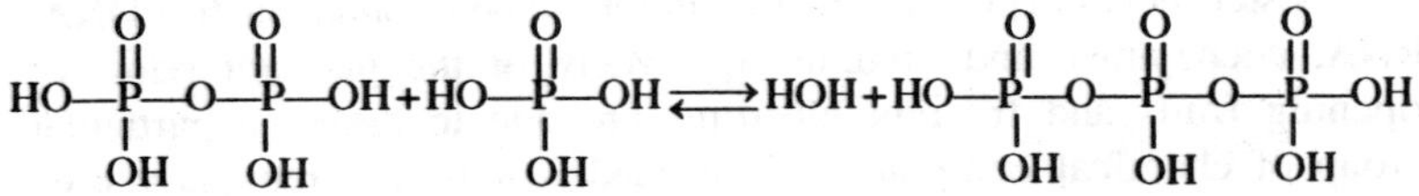

$$HO{-}\underset{\displaystyle OH}{\underset{|}{\overset{\displaystyle O}{\overset{\|}{P}}}}{-}O{-}\underset{\displaystyle OH}{\underset{|}{\overset{\displaystyle O}{\overset{\|}{P}}}}{-}OH + HO{-}\underset{\displaystyle OH}{\underset{|}{\overset{\displaystyle O}{\overset{\|}{P}}}}{-}OH \rightleftharpoons HOH + HO{-}\underset{\displaystyle OH}{\underset{|}{\overset{\displaystyle O}{\overset{\|}{P}}}}{-}O{-}\underset{\displaystyle OH}{\underset{|}{\overset{\displaystyle O}{\overset{\|}{P}}}}{-}O{-}\underset{\displaystyle OH}{\underset{|}{\overset{\displaystyle O}{\overset{\|}{P}}}}{-}OH$$

A third possible combination is the formation of a mixed anhydride by the coupling of two dissimilar acids. A carboxylic acid group and a phosphoric acid group can combine to form a mixed anhydride linkage.

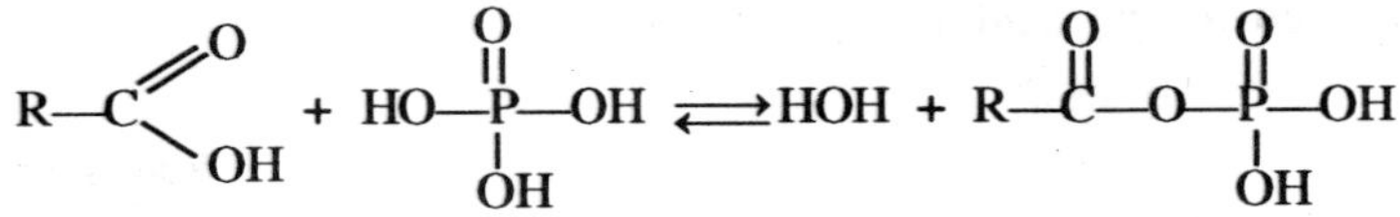

$$R{-}C{\lt}^{\displaystyle O}_{\displaystyle OH} + HO{-}\underset{\displaystyle OH}{\underset{|}{\overset{\displaystyle O}{\overset{\|}{P}}}}{-}OH \rightleftharpoons HOH + R{-}\overset{\displaystyle O}{\overset{\|}{C}}{-}O{-}\underset{\displaystyle OH}{\underset{|}{\overset{\displaystyle O}{\overset{\|}{P}}}}{-}OH$$

a carboxylic acid phosphoric acid a mixed carbon–phosphorus anhydride

Energy is stored during the formation of anhydrides

energy + acid + acid $\rightleftharpoons$ acid anhydride + water

The energy is stored in the anhydride bonding arrangement:

$$-\overset{\displaystyle O}{\overset{\|}{C}}{-}O{-}\overset{\displaystyle O}{\overset{\|}{C}}- \qquad -\overset{\displaystyle O}{\overset{\|}{P}}{-}O{-}\overset{\displaystyle O}{\overset{\|}{P}}-$$

Compounds such as adenosine triphosphate (ATP), adenosine diphosphate (ADP), and phosphoenolpyruvic acid (PEP) contain anhydride bonds.

The reaction for the formation of an anhydride is reversible. When water is added to an anhydride, the anhydride bond ruptures and the component acids are re-formed. Bond breakage due to the action of water is called *hydrolysis* (Greek: *hydro*, water, *lysis*, to cut).

$$-\overset{\overset{\displaystyle O}{\|}}{C}-O-\overset{\overset{\displaystyle O}{\|}}{C} + HOH \longrightarrow \overset{\overset{\displaystyle O}{\|}}{C}-OH + HO-\overset{\overset{\displaystyle O}{\|}}{C}- + \text{energy}$$

As the equation above indicates, anhydride hydrolysis is an exothermic process. The biological importance of anhydride formation and hydrolysis is that these two reactions provide a means by which an organism can store energy by anhydride formation and then release it upon hydrolysis.

ESTERS

Esters occur in physiologically important compounds such as DNA, RNA, coenzymes, and vitamin B_{12}, Many of the pleasant odors of ripening fruits and fragrant blossoms are due to esters, a particular group of chemical compounds. Man-made esters are frequently used as synthetic flavoring agents. Esters are produced by an intermolecular dehydration reaction between an alcohol and an acid. The alcohol and acid used determine the particular ester formed. The reaction may be catalyzed by enzymes or strong acids. Sulphuric acid (H_2SO_4) is used most often as the acid catalyst. In general,

$$\underset{\text{an alcohol}}{R-O-H} + \underset{\text{an acid}}{\overset{O}{\underset{HO}{\diagdown}}C-R'} \xrightarrow{\text{catalyst}} HOH + \underset{\text{an ester}}{\overbrace{R-O}^{\text{from alcohol}}-\overbrace{\overset{\overset{\displaystyle O}{\|}}{C}-R'}^{\text{from acid}}}$$

The H is removed from the alcohol and the —OH is removed from the acid. The remaining alcohol and acid fragments then combine to form the ester.

***Example* 13.1**

The esterification reaction between methyl alcohol and acetic acid can be represented as follows:

$$\underset{\text{methyl alcohol}}{CH_3—O—H} + \underset{\text{acetic acid}}{H—O—\overset{\overset{\displaystyle O}{\|}}{C}—CH_3} \xrightarrow{\text{catalyst}} \underset{\text{methyl acetate}}{\overbrace{CH_3—O}^{\text{from alcohol}}—\overbrace{\overset{\overset{\displaystyle O}{\|}}{C}—CH_3}^{\text{from acid}}} + H_2O$$

Note that the $\gt C{=}O$ and its alkyl group provide the clue that $CH_3\overset{\overset{\displaystyle O}{\|}}{C}—$ was part of the original acid

To relate this particular example to the general equation for ester formation, R is $—CH_3$ and R′ is also $—CH_3$. Recall the characteristic $R—O—\underset{\underset{\displaystyle O}{\|}}{C}—R'$ grouping of an ester.

ESTERS OF PHOSPHORIC ACID

Esterification is not limited to carboxylic acids. Phosphoric acid

$$HO{=}\overset{\overset{\displaystyle O}{\|}}{\underset{\underset{\displaystyle OH}{|}}{P}}—OH$$

can be esterified. We can see that phosphoric acid has a central nonmetal atom double-bonded to an oxygen and single-bonded to an —OH group, a feature common to carboxylic acids.

$$\underset{\text{an alcohol}}{R—O—H} + \underset{\text{phosphoric acid}}{HO—\overset{\overset{\displaystyle O}{\|}}{\underset{\underset{\displaystyle OH}{|}}{P}}—OH} \longrightarrow \underset{\text{a phosphate ester}}{R—O—\overset{\overset{\displaystyle O}{\|}}{\underset{\underset{\displaystyle OH}{|}}{P}}—OH} + HOH$$

ESTERS OF THIOLS

Thiols (thio alcohols) are sulphur-containing compounds analogous to alcohols, with sulphur substituting for oxygen in the alcohol grouping: R—S—H, a thiol; R—O—H, an alcohol. Thiols can react with acids to form the corresponding thioester:

$$\underset{\text{a thiol}}{R—S—H} + \underset{\text{an acid}}{H—O—\overset{\overset{\displaystyle O}{\|}}{C}—R'} \longrightarrow HOH + \underset{\text{a thio ester}}{R—S—\overset{\overset{\displaystyle O}{\|}}{C}—R'}$$

Coenzyme A (CoA—S—H) is a metabolically significant thiol that is esterified to produce acetyl coenzyme A. Another compound important to body metabolism. The abbreviation CoA represents a structurally complex portion of the coenzyme A molecule (see structure on page 265).

TRIGLYCERIDES

Glycerol is polyhydroxy alcohol containing three —OH groups per molecule:

```
     H
     |
H—C—OH
     |
H—C—OH
     |
H—C—OH
     |
     H
```

glycerol

It is an extremely important compound in metabolism because of the esterification reactions it undergoes. Its most important esterification reaction involves combination with those carboxylic acids called *fatty acids.* Fatty acids have the following general properties:

1. They usually contain one carboxylic acid group per molecule, i.e., monocarboxylic, and an alkyl group.
2. The alkyl group is usually an unbranched carbon chain.
3. The fatty acid molecule contains an even number of carbon atoms, usually from 12 to 26 carbons.
4. Due to the long hydrocarbon alkyl chain, fatty acids are water-insoluble.
5. The hydrocarbon portion of the chain may be saturated or unsaturated. If unsaturated, it may contain one, two, three, or four carbon-to-carbon double bonds. Those containing two to four double bonds are called polyunsaturated.
6. Usually, the greater the degree of unsaturation in a fatty acid, the lower its melting point. Table 13.1 lists several of the more important fatty acids.

Very few free (uncombined) fatty acids are found in a cell. Rather, they are chemically combined by the esterification reaction with glycerol

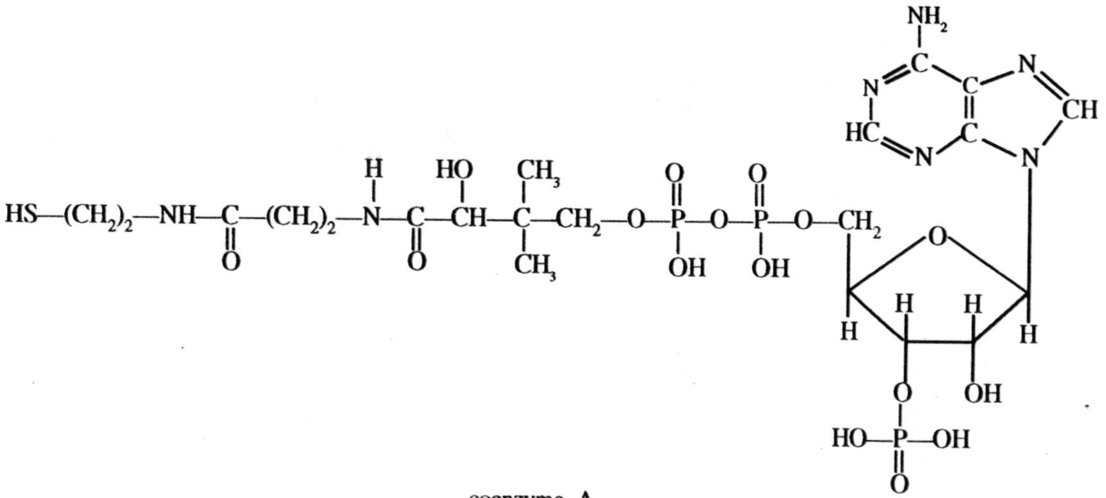

$$HS{-}(CH_2)_2{-}NH{-}\underset{\underset{O}{\|}}{C}{-}(CH_2)_2{-}\overset{\overset{H}{|}}{N}{-}\underset{\underset{O}{\|}}{C}{-}\overset{\overset{HO}{|}}{CH}{-}\overset{\overset{CH_3}{|}}{\underset{\underset{CH_3}{|}}{C}}{-}CH_2{-}O{-}\overset{\overset{O}{\|}}{\underset{\underset{OH}{|}}{P}}{-}O{-}\overset{\overset{O}{\|}}{\underset{\underset{OH}{|}}{P}}{-}O{-}CH_2{-}\text{(ribose–adenine; 3'-}O{-}PO(OH)_2\text{)}$$

coenzyme A

$CH_3{-}\underset{\underset{O}{\|}}{C}{-}$ is known as an acetyl group.

$$\underset{\text{coenzyme A}}{CoA{-}S{-}H} + \underset{\text{acetic acid}}{H{-}O{-}\overset{\overset{O}{\|}}{C}{-}CH_3} \longrightarrow \underset{\text{acetyl coenzyme A}}{CoA{-}S{-}\overset{\overset{O}{\|}}{C}{-}CH_3} + HOH$$

The prefix thio means an oxygen atom has been replaced by a sulphur atom.

Table 13.1 : Important fatty acids

Name	Formula	Type	Melting Point (°C)
Lauric acid	$CH_3(CH_2)_{10}\text{—}C(=O)OH$	saturated	44
Myristic acid	$CH_3(CH_2)_{12}\text{—}C(=O)OH$	saturated	58
Palmitic acid	$CH_3(CH_2)_{14}C(=O)OH$	saturated	63
Stearic acid	$CH_3(CH_2)_{16}C(=O)OH$	saturated	70
Oleic acid	$CH_3(CH_2)_7CH{=}CH(CH_2)_7C(=O)OH$	unsaturated	4
Linoleic acid*	$CH_3(CH_2)_4CH{=}CHCH_2CH{=}CH(CH_2)_7C(=O)OH$	polyun-saturated	—5
Linolenic acid*	$CH_3CH_2CH{=}CHCH_2CH{=}CHCH_2CH{=}CH(CH_2)_7C(=O)OH$	polyun-saturated	—11

* The body cannot synthesize these fatty acids and therefore they must be included in the diet.

to form a class of compounds called triglycerides. If all three fatty acids are the same, the product is a simple triglyceride. More commonly, a variety of both saturated an unsaturated fatty acids react with glycerol to produce a mixed triglyceride. This process can be represented as

$$\underbrace{R_1\text{—}C(=O)OH + R_2\text{—}C(=O)OH + R_3\text{—}C(=O)OH}_{\text{fatty acids}} + \underbrace{\begin{matrix} H \\ HO\text{—}C\text{—}H \\ HO\text{—}C\text{—}H \\ HO\text{—}C\text{—}H \\ H \end{matrix}}_{\text{glycerol}} \xrightarrow{\text{enzyme}} 3HOH + \underbrace{\begin{matrix} H \\ R_3\text{—}C(=O)\text{—}O\text{—}C\text{—}H \\ R_2\text{—}C(=O)\text{—}O\text{—}C\text{—}H \\ R_1\text{—}C(=O)\text{—}O\text{—}C\text{—}H \\ H \end{matrix}}_{\text{a mixed triglyceride}}$$

Note that 3 moles of water are formed during esterification of 1 mole of glycerol with 3 moles of fatty acids. The triglyceride is a solid or liquid at normal temperatures, depending on the nature of the component fatty acids. Triglycerides containing a large portion of unsaturated fatty acids are normal liquids and are described as oils (not the automobile-lubricating variety). Fat is the collective term for triglycerides containing a high portion of saturated fatty acid components and usually solids at room temperature. Triglycerides provide the greatest source of chemical energy during body metabolism.

Triglycerides make up the majority of a diverse group of compounds called *lipids*. Lipids are described as compounds of biological origin that are soluble in ether or other organic solvents. The diversity of these compounds ranges from salad oils, Butter, sex hormones, certain vitamins, and cholesterol to important components of nerve and brain tissue. Waxes and phospholipids are two other subclasses of lipids that will be discussed at this point.

A wax is the ester product of the reaction between a long-chain fatty acid and a monohydroxy alcohol. Thus its general formula is $R{-}\overset{\overset{\displaystyle O}{\|}}{C}{-}O{-}R'$. R is the fatty acid component containing a saturated hydrocarbon side chain ranging from 24 to 36 carbons; R' is a high molecular weight alcohol containing 16 to 36 carbon atoms. Waxes are insoluble in water and are chemically unreactive. Carbauba wax (used in automobile polishes), earwax, and beeswax are examples of naturally occurring waxes.

Phospholipids (phosphatides) are complex lipids containing phosphate, glycerol, fatty acids, and a nitrogen-containing base. A typical example is phosphatidyl choline, an important component of nerve tissue membranes. Its common name is lecithin:

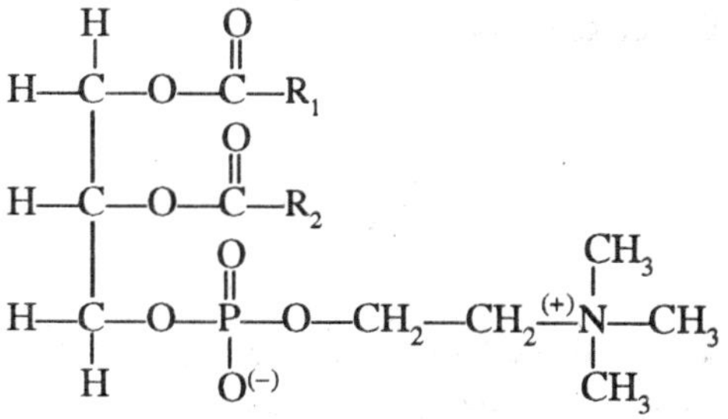

$$\begin{array}{l}
\quad\;\; H \qquad\;\; O \\
\quad\;\; | \qquad\;\;\; \| \\
H{-}C{-}O{-}C{-}R_1 \\
\quad\;\; | \qquad\;\; O \\
\quad\;\; | \qquad\;\;\; \| \\
H{-}C{-}O{-}C{-}R_2 \\
\quad\;\; | \qquad\;\; O \qquad\qquad\qquad\qquad\quad CH_3 \\
\quad\;\; | \qquad\;\;\; \| \qquad\qquad\qquad\qquad\qquad | \\
H{-}C{-}O{-}P{-}O{-}CH_2{-}CH_2{\overset{(+)}{-}}N{-}CH_3 \\
\quad\;\; | \qquad\;\;\; | \qquad\qquad\qquad\qquad\qquad | \\
\quad\;\; H \qquad O^{(-)} \qquad\qquad\qquad\quad\;\; CH_3
\end{array}$$

phosphatidyl choline (lecithin)

Note that the phosphatide above may be considered the glycerol ester of two fatty acids

$$R_1{-}C{\lesssim}^{O}_{OH} \qquad \text{and} \qquad R_2{-}C{\lesssim}^{O}_{OH}$$

plus a choline derivative of phosphoric acid

$$HO{-}\overset{\overset{\large O}{\|}}{\underset{\underset{\large OH}{|}}{P}}{-}OH$$

ESTER HYDROLYSIS

Esterification is a reversible process Ester bonds are susceptible to hydrolysis. The presence of an appropriate enzyme or dilute acid (Dilute acid is used so that there is a large excess of water present, driving the reaction to completion) catalyzes the hydrolysis reaction. Acid or enzymatic hydrolysis of an ester causes the ester bond to break with the re-formation of the original acid and alcohol components. Generalizing the ester hydrolysis,

$$\underset{\text{ester}}{R{-}C{\lesssim}^{O}_{OH}} + HOH \xrightarrow{H^+ \text{ or enzyme}} \underset{\text{acid}}{R{-}C{\lesssim}^{O}_{OH}} + \underset{\text{alcohol}}{R'{-}OH}$$

The acid or enzymatic hydrolysis of an ester is the reverse of the esterification process. During esterification, the —OH from the acid the H from the alcohol joining to form water and the ester forming by the union of the residual alcohol and acid fragments. During hydrolysis, the ester bond breaks at this point

$$-C{\lesssim}^{O}_{\times O-}$$

and it may be imagined that the OH from water adds to the *carbonyl* carbon and the H from water adds to the oxygen:

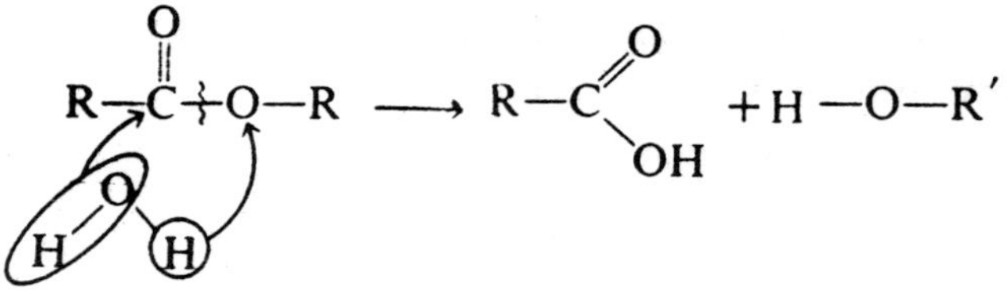

$$R-\overset{O}{\overset{\|}{C}}-O-R + H-O-H \longrightarrow R-C(=O)-OH + H-O-R'$$

Thus, the component acid and alcohol are liberated. Note that water is shown as a reactant in the balanced chemical equation.

***Example* 13.2**

What are the structural formula of the acid and alcohol formed upon acid hydrolysis of

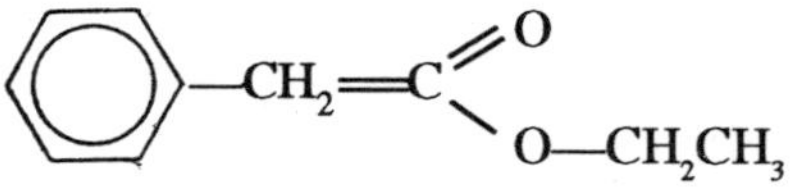

$$C_6H_5-CH_2=C(=O)-O-CH_2CH_3 \quad ?$$

First, set up the question in the form of a chemical equation;

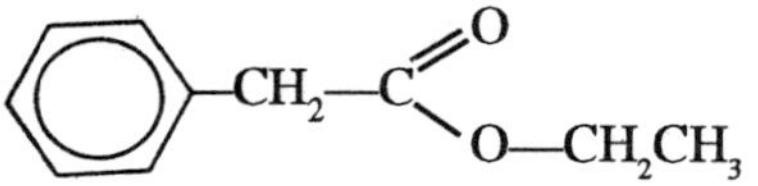

$$C_6H_5-CH_2-C(=O)-O-CH_2CH_3 + HOH \xrightarrow{H^+} \text{acid} + \text{alcohol}$$

Secondly, we know that

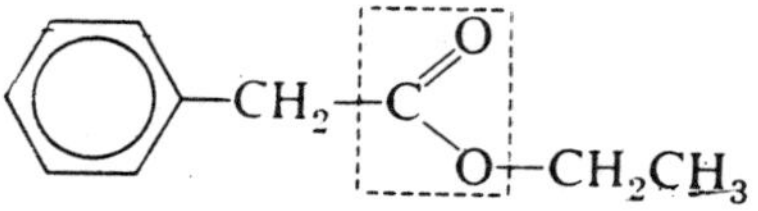

$$C_6H_5-CH_2-C(=O)-O-CH_2CH_3$$

contains an ester grouping outlined by the dotted lines:

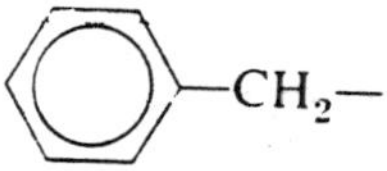

$$C_6H_5-CH_2-$$

is analogous to R in the general equation and $-CH_2CH_3$ is equivalent to R′.

Hydrolysis of the ester linkage occurs by rupture of the carbon-to-oxygen single bond marked with a wavy line:

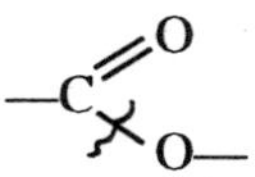

$$-C(=O)\wr O-$$

The carbon-oxygen *double* bond remains intact and will eventually become part of the acid liberated.

Next, imagine that the rupture of the ester bond generates two fragments:

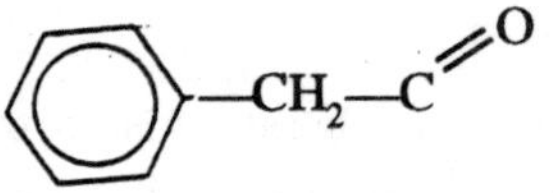

$$C_6H_5—CH_2—C{=}O \qquad \text{and} \qquad —O—CH_2CH_3$$

The OH form water adds to the carbonyl carbon; the H from water adds to the $—O—CH_2CH_3$ groups.

The net result is formation of the acid

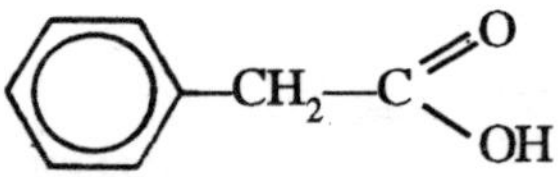

$$C_6H_5—CH_2—C(=O)—OH$$

and the alcohol $H—O—CH_2CH_3$.

We have previously seen that esterification with glycerol is a convenient way for the body is store fatty acids: however, esterification is a reversible process. The component fatty acids and glycerol can be obtained by hydrolysis.

Since 1 mole of a triglyceride is formed from the reaction of 3 moles of fatty acids with 1 mole of glycerol, acid or enzymatic hydrolysis of a triglyceride regenerates 3 moles of fatty acid and 1 mole of glycerol. For hydrolysis, 3 moles of water per 1 mole of triglyceride is required:

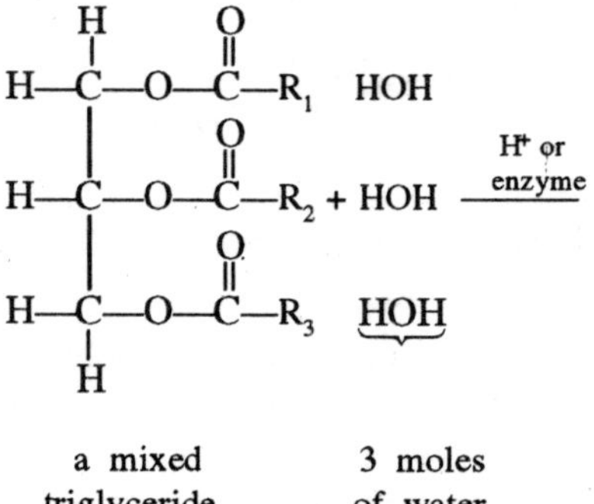

$$\begin{array}{l} H_2C—O—C(=O)—R_1 \\ \quad | \\ HC—O—C(=O)—R_2 \\ \quad | \\ H_2C—O—C(=O)—R_3 \end{array} + \begin{array}{c} HOH \\ HOH \\ HOH \end{array} \xrightarrow{H^+ \text{ or enzyme}}$$

a mixed triglyceride

3 moles of water

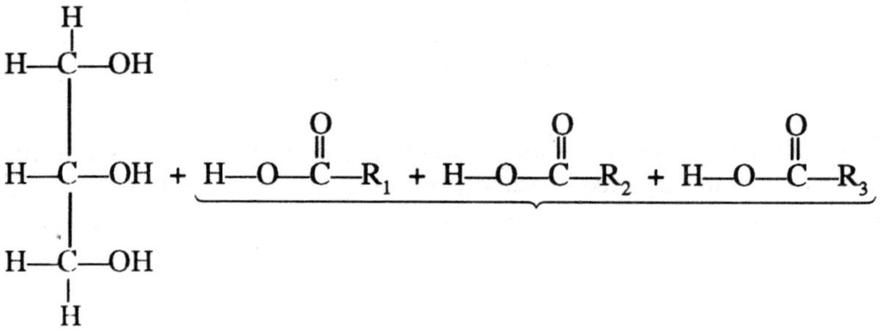

$$\underset{\text{glycerol}}{\begin{array}{c} H \\ | \\ H-C-OH \\ | \\ H-C-OH \\ | \\ H-C-OH \\ | \\ H \end{array}} + \underbrace{H-O-\overset{O}{\overset{\|}{C}}-R_1 + H-O-\overset{O}{\overset{\|}{C}}-R_2 + H-O-\overset{O}{\overset{\|}{C}}-R_3}_{\text{3 moles of fatty acids}}$$

SAPONIFICATION

Ester bonds can also be broken by the action of aqueous solutions of strong bases like sodium hydroxide. This process is called *saponification.*

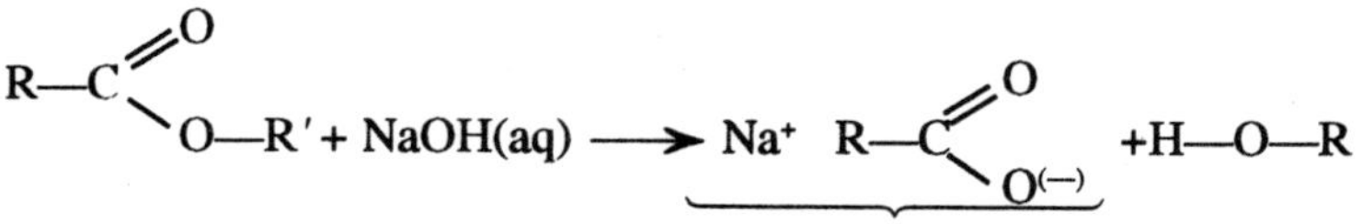

$$\underset{\text{ester}}{R-C(=O)-O-R'} + NaOH(aq) \longrightarrow \underbrace{Na^{+}\ R-C(=O)-O^{(-)}}_{\text{salt}} + \underset{\text{alcohol}}{H-O-R}$$

Saponification of an ester thus liberates the original alcohol as in acid or enzymatic hydrolysis. The original acid cannot be isolated, however, because it reacts with excess metal hydroxide present to form the corresponding metal salt.

If the saponification of a triglyceride is carried out in an alkali, metal hydroxide solution (usually NaOH or KOH,), the salts of the component fatty acids are formed. These salts are collectively described as *soaps.*

DECARBOXYLATION

One of the reactions carboxylic acids undergo is decarboxylation, the loss of carbon dioxide upon heating. The CO_2 loss is a way in which the number of carbon (and oxygen) atoms in a molecule can be decreased. Loss of carbon dioxide occurs via the breakdown of a carboxylic acid group:

$$R{-}C(=O){-}OH \longrightarrow CO_2 + R{-}H$$

Decarboxylation occurs during many physiologically important processes. In carbohydrate metabolism, oxalosuccinic acid is converted in this way to α-ketoglutaric acid in the presence of the appropriate decarboxylase enzyme.

$$HO_2C{-}CH_2{-}CH(C(=O)OH){-}C(=O){-}CO_2H \longrightarrow HO_2C{-}CH_2{-}CH_2{-}C(=O){-}CO_2H + CO_2$$

oxalosuccinic acid → α-ketoglutaric acid

14

Carbohydrates

INTRODUCTION

Carbohydrates (or saccharides) are composed of carbon, hydrogen, and oxygen. Their name was originally derived from the general formula $C_x(H_2O)_y$. Structurally, these compounds are *polyhydroxyaldehydes* or *polyhydroxyketones* and their derivatives. Those saccharides containing an aldehyde group are called *aldoses*, while those with a ketone group are *ketoses*.

Lower molecular weight carbohydrates are known as sugars and their names generally have the characteristic ending *-ose*. Some examples are sucrose (cane sugar), lactose (milk sugar), glucose (also known as dextrose), and fructose (levulose). *Monosaccharides* (simple sugars) are further classified according to the number of carbon atoms in a continuous chain. A three-carbon monosaccharide is a triose; four carbons, a tetrose; five carbons, a pentose; and six carbons, a hexose. A disaccharide results from the joining of two monosaccharides. Carbohydrates containing from three up to about eight monosaccharide units are *oligosaccharides*. Those containing more than about eight monosaccharide units are polysaccharides. The classifications of carbohydrates are given in Table 14.1.

Monosaccharides

The D-family of aldoses up to the hexoses is shown in Fig 14.1. The D in the name of sugars indicates the position of the —OH *group*

Table 14.1

The classification of carbohydrates

I. Monosaccharides
   <u>Trioses</u>
   1. Aldose example; *glyceraldehyde*
   2. Ketose example; *dihydroxyacetone*
   <u>Tetroses</u>
   <u>Pentoses</u>
   1. Aldose example: *arabinose, ribose*
   <u>Hexoses</u>
   1. Aldose example: *glucose, mannose*
   2. Ketose example: *fructose*

II. Disaccharides example: *maltose, lactose, sucrose*

III. Trisaccharides

IV. Oligosaccharides

V. Polysaccharides
   <u>Starch</u>
   1. Amylose
   2. Amylopectin
   <u>Glycogen</u>
   <u>Cellulose</u>
   <u>Dextrin</u>

on the carbon atom next to the CH_2OH group. When this —OH group is written on the right, a D precedes the name of the sugar; on the left, an L precedes the name. Only a few of these sugars are widespread in nature. The pentose p-ribose is common to all living organisms as a constituent of the ribonucleic acids (RNA). D-Arbionose and D-xylose are primarily found in plants. Gum arabic, which is used as a thickening agent in foods, is primarily a polymer of arbionose. D-Xylose can be obtained by hydrolysis of corncobs, straw, or wood. The hexoses are the most common monosaccharides. Almost all the carbohydrates in our diet are hexoses or their derivatives. Among the eight aldohexoses of the D-family, only three are accepted as naturally occurring: glucose, mannose, and galactose. Glucose is by far the most prevalent of the three.

Glucose (also called dextrose) is of great importance to human metabolism,. Glucose is the normal sugar occurring in blood and is one of the primary sources of energy to the cells. Galactose is found combined with glucose to form the disaccharide lactose or milk sugar. Additionally, galactose is found in nervous and brain tissues. Relatively little is known about the importance of mannose in human metabolism.

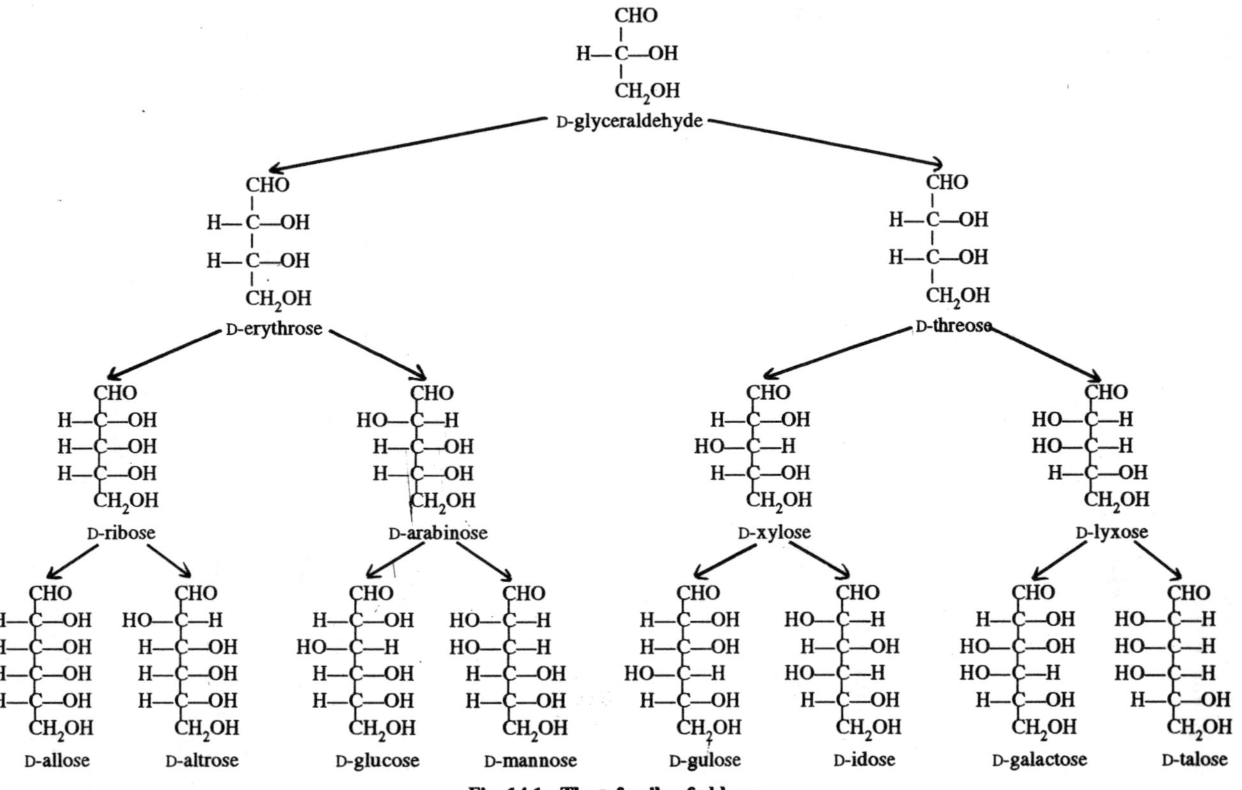

Fig. 14.1 : The D family of aldoses

In addition to the aldoses, There is one ketose of importance in human metabolism, D-fructose, whose structure is

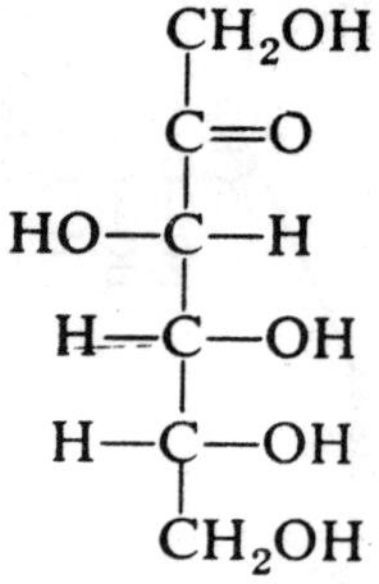

$$\begin{array}{rcl} & CH_2OH & \\ & | & \\ & C{=}O & \\ & | & \\ HO- & C & -H \\ & | & \\ H- & C & -OH \\ & | & \\ H- & C & -OH \\ & | & \\ & CH_2OH & \end{array}$$

Fructose is a member of the D-family of carbohydrates, as shown by the position of the —Oh group on the fifth carbon atom. Actually, the arrangement about four of the carbon atoms in fructose is identical to the arrangement in D-glucose.

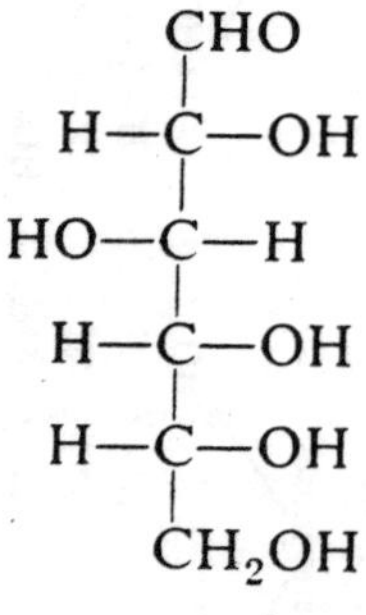

$$\begin{array}{rcl} & CHO & \\ & | & \\ H- & C & -OH \\ & | & \\ HO- & C & -H \\ & | & \\ H- & C & -OH \\ & | & \\ H- & C & -OH \\ & | & \\ & CH_2OH & \end{array} \qquad\qquad \begin{array}{rcl} & CH_2OH & \\ & | & \\ & C{=}O & \\ & | & \\ HO- & C & -H \\ & | & \\ H- & C & -OH \\ & | & \\ H- & C & -OH \\ & | & \\ & CH_2OH & \end{array}$$

D-glucose 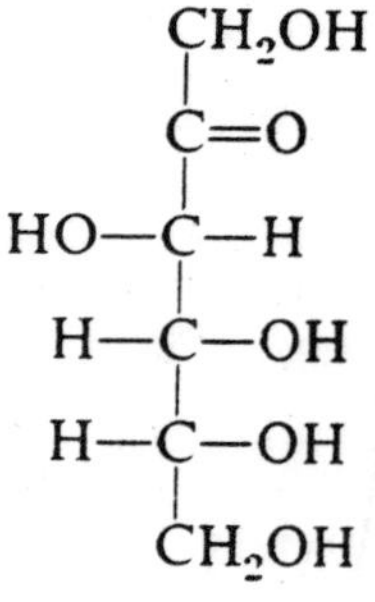 D-fructose

Fructose is the sweetest of all sugars. A standard sweetness scale assigns sucrose (cane sugar) a value of 100. On this scale fructose has a sweetness value of 135; glucose, 75. Greater than 50 percent of the sugar in honey is fructose, giving honey its extremely sweet taste. It is very difficult to make fructose crystallize from solution. Furthermore, the presence of fructose prevents glucose and sucrose from crystallizing. This peculiar characteristic of fructose keeps honey from crystallizing over long periods of time. Candy and jelly makers take advantage of this fact by adding invert sugar containing approximately 30 percent fructose to their products to prevent an undesirable grainy texture.

REACTIONS OF MONOSACCHARIDES

The reactions of monosaccharides are those to be expected from the functional groups present. For example, the aldehyde or ketone group can be reduced to produce a polyhydroxy alcohol. Reduction of D-glucose with H_2 gas in the presence of a catalyst produces the polyhydroxy alcohol sorbitol.

```
  H   O
   \ //
    C                              CH2OH
    |                               |
 H—C—OH                         H—C—OH
    |          H2/Pt                |
HO—C—H      ———————→           HO—C—H
    |                               |
 H—C—OH                         H—C—OH
    |                               |
 H—C—OH                         H—C—OH
    |                               |
   CH2OH                           CH2OH

 D-glucose                        sorbitol
```

In the presence of the appropriate enzymes, glucose can be reduced to sorbitol by living organisms. Biological reduction of glucose is important to reproduction of mammals. In the seminal vesicles, glucose is reduced to sorbitol. This reduction is followed by oxidation of sorbitol to fructose. Fructose is abundant in seminal fluid and is the primary energy source for sperm cells. This sequence of reactions can be represented as follows:

```
  H   O
   \ //
    C                      CH2OH                    CH2OH
    |                       |                        |
 H—C—OH                 H—C—OH                     C=OH
    |        NADH2          |          NAD           |
HO—C—H     ————→       HO—C—H       ————→       HO—C—H
    |                       |                        |
 H—C—OH                 H—C—OH                   H—C—OH
    |                       |                        |
 H—C—OH                 H—C—OH                   H—C—OH
    |                       |                        |
   CH2OH                   CH2OH                    CH2OH

 D-glucose                sorbitol                D-fructose
```

The reducing and oxidizing agents are $NADH_2$ and NAD, respectively.

Just as simple aldehydes and ketones undergo reaction with alcohols to produce hemiacetals and hemiketals, so will monosaccharides. A major difference is, however, that in monosaccharides the alcohol group and the aldehyde or ketone group are contained in the same molecule. The result of this is that monosaccharides undergo *intramolecular* reaction to produce hemiacetals or hemiketals.

The formation of five-and six-membered ring hemiacetals or hemiketals is particularly favorable. Aldohexoses form six-membered ring hemiacetals by reaction of the aldehyde group with the alcohol group of the No. 5 carbon atom.

```
 H     O                      H     O
  \   //                       \   /
   ¹C                           ¹C ─────────┐
    |                            |          |
H ─²C─ OH                    H ─²C─ OH      |
    |                            |          |
HO─³C─ H                     HO─³C─ H       |
    |                            |          |
H ─⁴C─ OH                    H ─⁴C─ OH      |
    |                            |          |
H ─⁵C─ OH                    H ─⁵C ─────────O
    |                            |
   ⁶CH2OH                       ⁶CH2OH
```

D-glucose (open-chain form) — α-D-glucose (hemiacetal form)

When the hemiacetal is formed, two isomers are possible. Since reaction involves many molecules rather than a single molecule, the result is a mixture of both isomers. For glucose the formation of both isomers can be represented as follows:

```
 H    O                     H    O                    H    O
  \  /                       \  //                     \  /
   C ───────┐                 C                         C ───────┐
   |        |                 |                         |        |
H──C──OH    |              H──C──OH                  H──C──OH    |
   |        |                 |                         |        |
HO─C──H     |    ⇌        HO──C──H        ⇌         HO──C──H     |
   |        |                 |                         |        |
H──C──OH    |              H──C──OH                  H──C──OH    |
   |        |                 |                         |        |
H──C ───────O              H──C──OH                  H──C ───────O
   |                          |                         |
  CH2OH                      CH2OH                     CH2OH
```

β-D-glucose (open-hemiacetal form) — D-glucose (open-chain form) — α-D-glucose (one hemiacetal form)

Very little glucose exists in the open chain form. The hemiacetal form of glucose in which the —OH group is written on the right-hand side of the structural formula is the alpha form: α--D-glucose. The beta form (β-D-glucose) has the —Oh group on the left-hand side. An equilibrium mixture of glucose consists of 0.02 percent of the open-chain form. 37 percent of the α-isomer, and 63 percent of the β-isomer.

Haworth Formulas for Sugars

The structural formulas as written in the preceding section for the cyclic hemiacetal forms of sugars are awkward looking and do not portray the structure very well. While we might write straight chains of carbon atoms, the carbon atoms are not in a straight line. A better, although still not entirely accurate, way of depicting the hemiacetal forms of sugars are *Haworth formulas.* The Haworth formula emphasizes the ring structure of the hemiacetal form of monosaccharides. The conventional means of writing the Haworth formula for a six-membered ring is to draw a hexagon. The oxygen atom of the ring is shown in the back right-hand corner of the hexagon; the remaining corners of the hexagon are understood to be occupied by carbon toms. Thus, to write the Haworth formula for any six-membered ring hemiacetal of a monosaccharide, you initially write

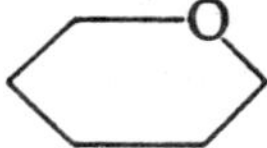

Since the hemiacetal is formed between the aldehyde group of the No.1 carbon and the —OH group on the No.5 carbon, the carbon atoms of the ring are numbered

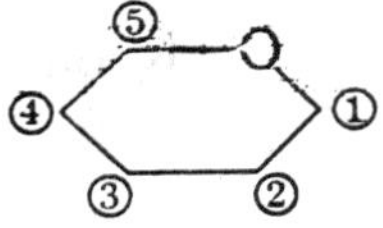

If the hexose is a member of the L-family, the No. 6 C atom is written downward. For all members of the D-family of hexoses, the No. 6 carbon is written attached to the No. Carbon in an upward position.

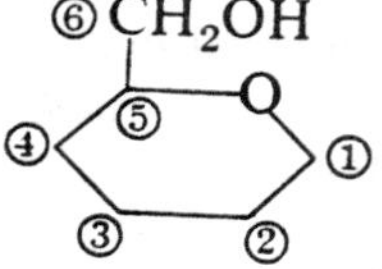

To complete the writing of the Haworth formula, all groups written to right in the straight chain formula are written downward; all groups on the left are written upward in the Haworth formula.

Both ways of writing structural formulas for β-D-glucose and α-D-glucose are below.

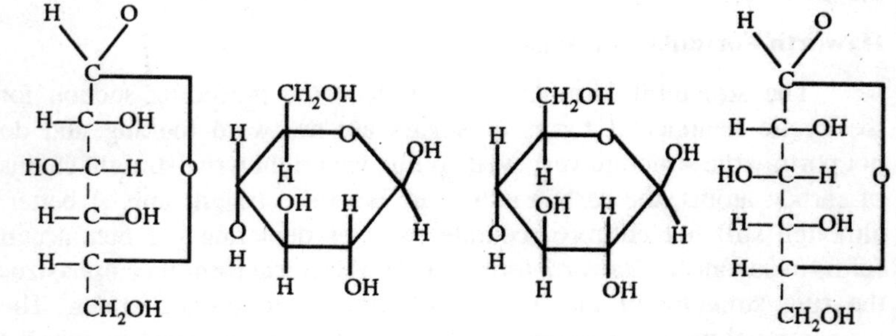

Haworth formula

Haworth formula

β-D-glucose

α-D-glucose

In keeping with the writing of other ring compounds, the hydrogen atoms are not shown but are understood to be present. Thus, the haworth formulas for β-D-glucose and α-D-glucose become

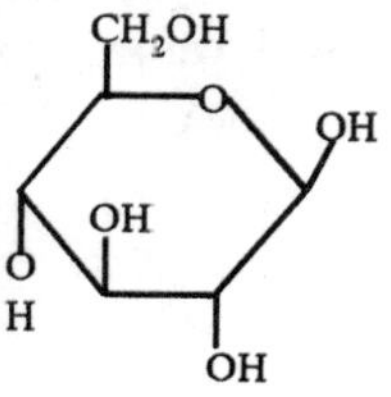

β-D-glucose

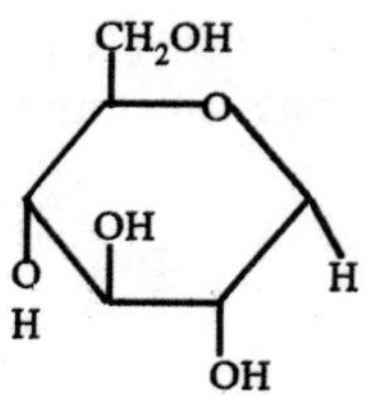

α-D-glucose

Fructose forms a cyclic hemiketal in equilibrium with the open-chain ketone form. The cyclic form of fructose is a five-membered ring. The formation of the hemiketal gives rise to a mixture of α-D-fructose and β-D-fructose. The open-chain and Haworth formulas for the cyclic forms of fructose are :

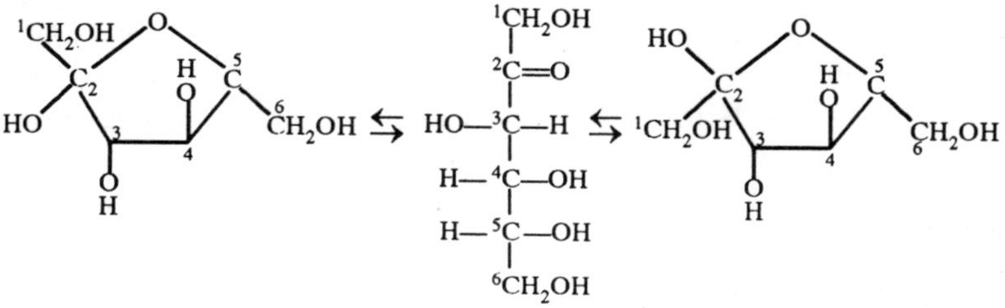

$^{1}CH_2OH$, HO, C_2, O, H, ^5C, $^6CH_2OH$, 3, 4 ⇆ $^{1}CH_2OH$ — $^{2}C{=}O$ — $HO{-}^{3}C{-}H$ — $H{-}^{4}C{-}OH$ — $H{-}^{5}C{-}OH$ — $^{6}CH_2OH$ ⇆ HO, C_2, $^{1}CH_2OH$, O, H, ^5C, CH_2OH (6), 3, 4

β-D-fructose fructose (open chains) α-D-fructose

The procedure for writing Haworth formulas for five-membered rings is different from that for six-membered rings

REDUCING PROPERTIES OF SUGARS

Aldehydes and α-hydroxy ketones are easily oxidized compounds, reacting readily with Tollen's reagent and Fehling's or Benedict's solutions. Since monosaccharides are polyhydroxy aldehydes and polyhydroxy ketones, we would expect them to react readily with mile oxidizing agents. Although monosaccharides exist primarily in the cyclic hemiacetal or hemiketal form at equilibrium, there is always present a small amount of the open-chain free aldehyde or α-hydroxy ketone form. As the open-chain form of the monosaccharide reacts with the oxidizing agent, some of the hemiacetal or hemiketal form reverts to the open-chain species in an attempt to maintain equilibrium. Eventually, if enough oxidizing agent is present, all the monosaccharide is converted to the open-chain species in an attempt to maintain equilibrium. Eventually, if enough oxidizing agent is present, all the monosaccharide is converted to the open-chain form and is oxidizing agent is present, all the monosaccharide is converted to the open-chain form and is oxidized. Recall that oxidation cannot occur without reduction and since the monosaccharides are oxidized, they must be reducing agents. *All sugars that will react with Tollen's reagent or either Fehling's or Benedict's solutions are known as reducing sugars. All monosaccharides are reducing sugars.*

Monosaccharides may appear as an abnormal constituent of urine. The most common cause of a monosaccharide in urine is diabetes mellitus resulting in the presence of glucose in the urine. The condition is known as glucosuria, and the glucose concentration in the urine of a diabetic individual may range from 0.5 to 12.0 percent.

Benedict's solution has been widely used to test urine for the presence of glucose. The amount and color of the precipitate formed presumably gives an indication of the amount of glucose in the urine sample. Unfortunately, the Benedict's test is not specific for glucose. Although glucose is the most likely cause of reaction when urine is treated with Benedict's solution, it is not the only possible cause. A reaction will occur due to presence of many different possible reducing substances.

Other possible substances in urine capable of reducing the copper (II) ions are the five-carbon sugars, arabinose and xylulose; the seven-carbon sugar, mannoheptulose; galactose; the disaccharide, lactose; creatinine; and glucuronic acid or its derivatives. Creatinine and glucuronic acid are normal components of urine. Arabinose in urine may result from ingestion of large amounts of fruit of fruit juices. Similarly, mannoheptulose is found in the urine of individuals who have eaten large amounts of avocado. Lactose is frequently found in the urine of lactating women. The presence of either xylulose or galactose in urine results from hereditary defects. The excretion of xylulose is apparently harmless. Galactose in urine is a symptom of galactosemia. Fortunately, galactosemia is relatively rare, ordinarily easily detectable in early infancy, and can be controlled by excluding galactose from the diet. Galactosemia results from the congenital absence of an enzyme that catalyzes the conversion of galactose to glucose. The absence of the enzyme allows galactose to accumulate in the body, eventually causing damage to the brain, liver, and eyes.

To lessen the possibility of mistaking other reducing substances for the presence of glucose in urine, more selective tests have recently been developed. One of these tests involves the use of the enzyme glucose oxidase. The enzyme is specific for reaction with glucose; consequently, no reaction is observed in the absence of glucose.

DISACCHARIDES

Hemiacetals will react with alcohols in the presence of an acid catalyst. Since aldohexoses exist primarily in the hemiacetal form, we would expect them to react with alcohols. Glucose, for example, would be expected to react with methyl alcohol to produce a methyl acetal (see structure on page 283). (The hemiacetal carbon of α-D-glucose and the acetal carbon of methyl-α-D-glucoside are shown in color.) The acetals of aldoses are called *glycosides*; those of glucose are specifically known as *glucosides*.

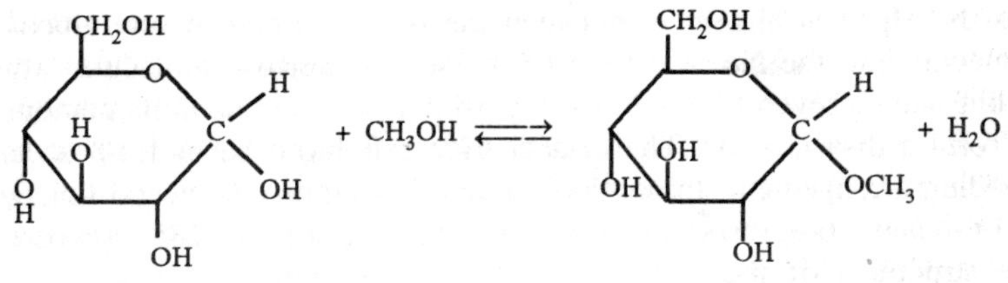

α-D-glucose + CH_3OH ⇌ methyl-α-D-glucoside + H_2O

A disaccharide can be formed by the reaction of two monosaccharide molecules. One serves as the hemiacetal; the other, as the alcohol. For example, two glucose molecules could combine to form the disaccharide maltose. If the hemiacetal group of one glucose molecule reacts with the —OH group on carbon 4 of the second glucose molecule, the disaccharide maltose is produced. The two glucose molecules are joined by an acetal group. Notice that maltose contains both an acetal group and a hemiacetal group. The hemiacetal groups of both glucose and maltose and the acetal group of maltose are shown in color. The presence of a hemiacetal group in maltose means that it is a reducing sugar.

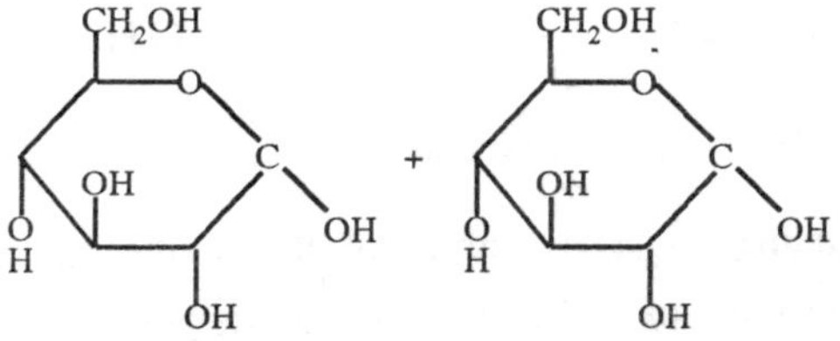

α-D-glucose + α-D-glucose

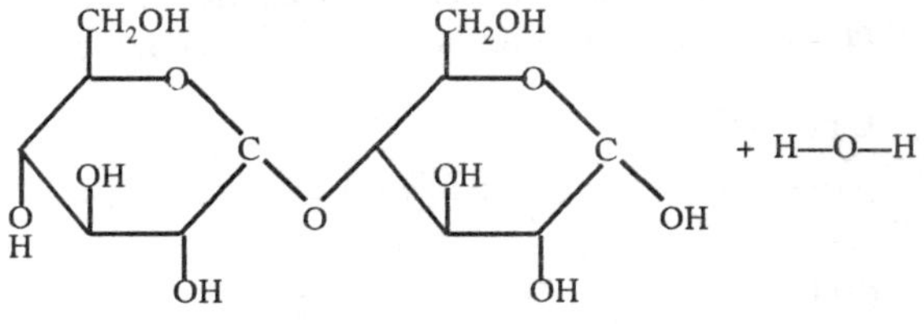

maltose + H—O—H

Maltose is not abundant in nature; however, it is easily prepared from starch by the action of either enzymes or dilute solutions of a strong acid. The primary natural source of maltose is sprouting grain seeds. For this reason maltose is often referred to as *malt sugar.*

Notice the acetal linkage between the glucose units in maltose has the alpha arrangement between the No. 1 carbon of one glucose molecule and the No. 4 carbon of the second glucose molecule. This is but one of several known ways in which two glucose units combine to form a disaccharide. The disaccharide cellobiose has a 1,4-β acetal bonding arrangement. In gentiobiose and isomaltose the acetal linkage is 1,6-β and 1,6-α, respectively. Figure 14.2 (see at p. no. 285) compares the structures of these four disaccharides formed from glucose.

Two additional disaccharides of biological importance are lactose and sucrose. Lactose is found solely in the milk of mammals; hence, its common name is milk sugar. The acetal bonding arrangement is between the data hemiacetal form of galactose and the —OH group on the No. 4 carbon of glucose.

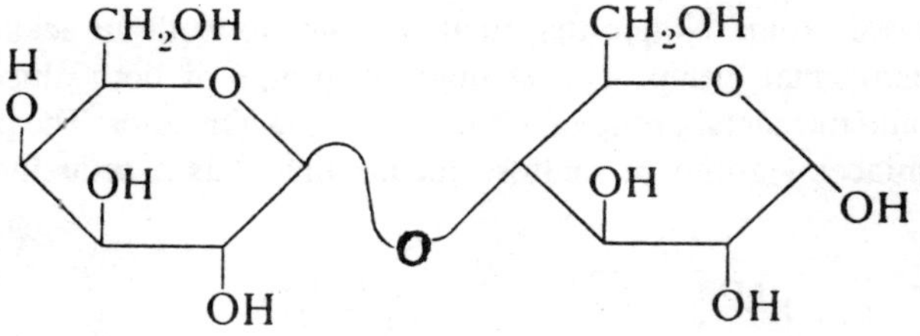

α-lactose

Lactose is a reducing sugar by virtue of the hemiacetal group.

Sucrose, composed of a molecule of glucose and one of fructose, is the most common and familiar disaccharide. Commercially, sucrose is obtained from sugar cane and sugar beets. It is also found in varying concentrations in most fruits, flowers, seeds, and roots. In contrast to most other disaccharides, sucrose is not a reducing sugar.

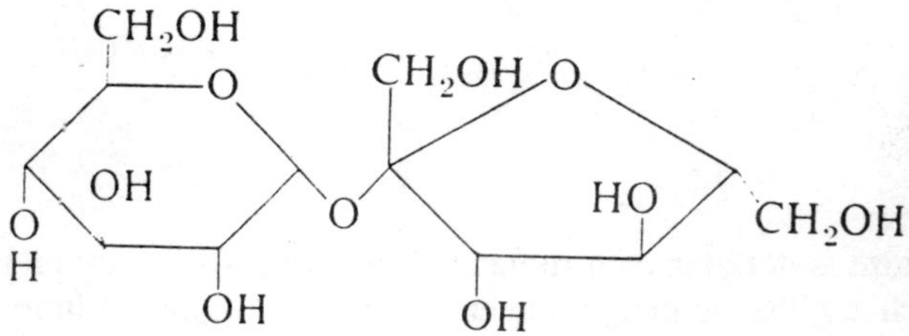

sucrose

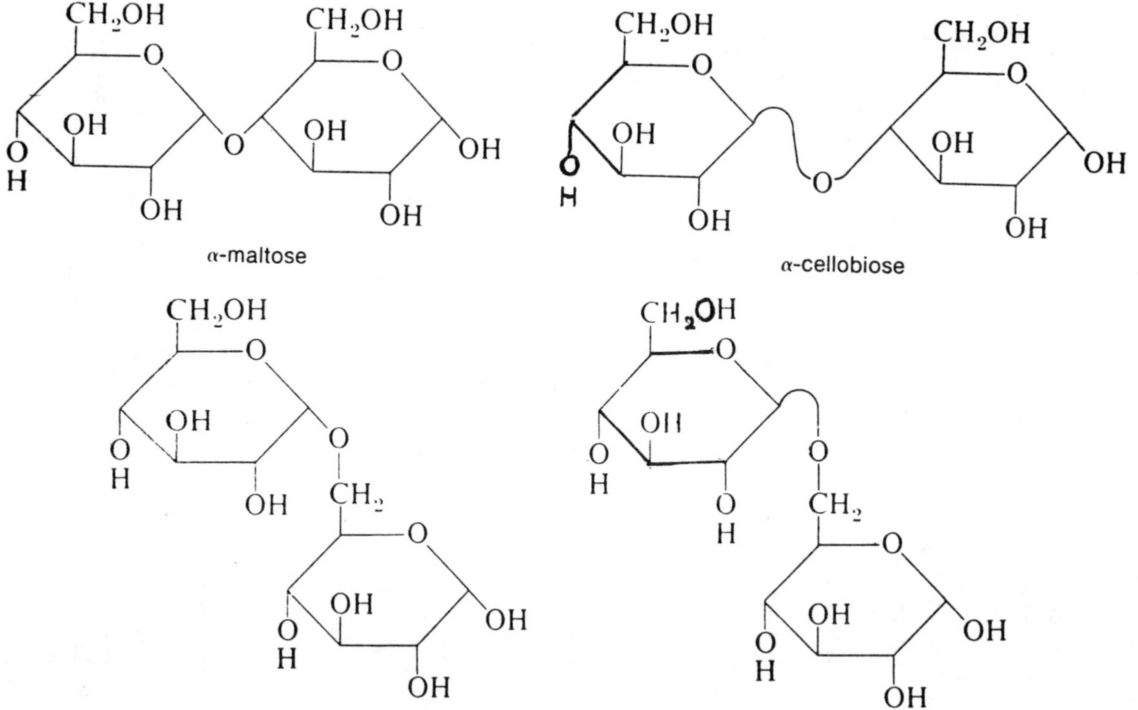

Fig. 14.2 : The structures of four disacharides formed from glucose.

POLYSACCHARIDES

Polysaccharides are polymers of monosaccharides and are very complex molecules. The great majority of all carbohydrates in nature exist as polysaccharides. These molecules are generally composed of several hundred to several thousand monosaccharide units. As a result, their properties are considerably different than those of monosaccharides and disaccharides. High molecular weight compounds (greater than 100,000) are generally insoluble. Most are insoluble in water, do not have a sweet taste, and fail to react with mild oxidizing agents.

Although there are naturally occurring polysaccharides of both pentoses and hexoses, the most important are those of glucose. The polysaccharides of glucose can be divided into two basic classes, depending on whether the glucose units are joined by 1,4-α or 1,4-β acetal linkages. Those joined by 1,4-β linkages primarily fulfill the structural needs of plant cells; whereas those in which the glucose molecules are joined by 1,4-α linkages serve nutritional needs in plants and animals.

Cellulose is the most abundant organic compound in the word. It is made up of long chains of glucose joined by 1,4-β acetal linkages.

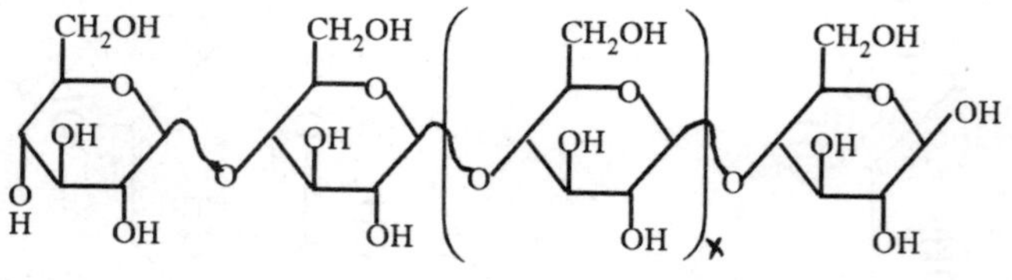

cellulose
(x = 300 to 3000)

The purest natural source of cellulose is cotton, which is greater than 90 percent cellulose. In additional to the use of natural cotton to weave fabrics, large quantities of cotton are used to prepare cellulose derivatives. Cellulose acetate, for example, is used to manufacture film; plastics; yarns, and fabrics; and, when dissolved in an organic solvent, fingernail polish. Rayon and cellophane are also produced from cellulose.

Starch and glycogen are the glucose polymers in which the glucose molecules are joined by 1,4-α acetal linkages. Starch is a plant polysaccharide, while glycogen is produced by animals. These substances are a form of carbohydrate storage.

Starch is divided into two types, Amylose and amylopectin; 10 to 20 percent of starch is amylose. Amylose is thought to consist of long chains of glucose molecules held together by 1,4-α linkages. Amylopectin, the major component of starch, is known to be a branched polysaccharide. Although the glucose units are primarily joined by 1,4-α linkages, in about every 24 to 30 glucose units there is also a 1,6-α linkage. This 1,6-α linkage allows the formation of a branch chain. At each branch point there is a glucose molecule that is bonded to another glucose molecule via the No. 6 carbon in addition to those bonded to it on the No. 1 and No.4 carbons. A portion of an amylopectin molecule showing the branching is illustrated in Fig 14.3

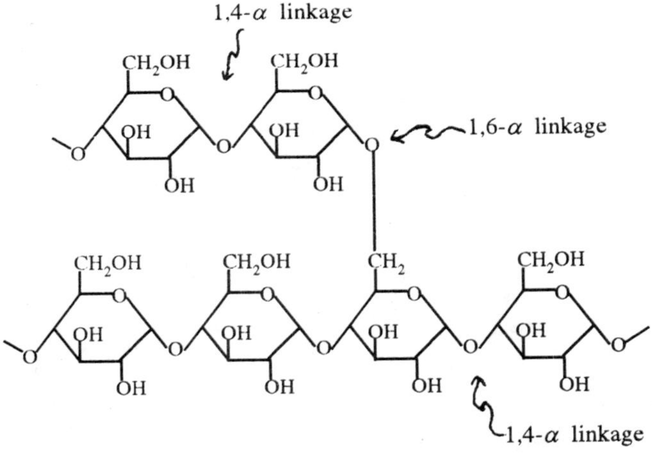

Fig. 14.3 : A portion of an amylopectin molecule showing how branching occurs through 1,6-α linkages.

Starch is easily hydrolyzed by dilute solutions of strong acids, ultimately leading to the formation of glucose. This breakdown occurs by the random hydrolysis along the chain of the acetal bonds linking the glucose units together. Thus, amylopectin is first broken down into smaller polysaccharide fragments. These smaller fragments are subsequently hydrolyzed into progressively smaller fragments until maltose and finally glucose is produced. The intermediate-sized fragments between maltose and starch are collectively known as *dextrins* (see Fig. 14.4)

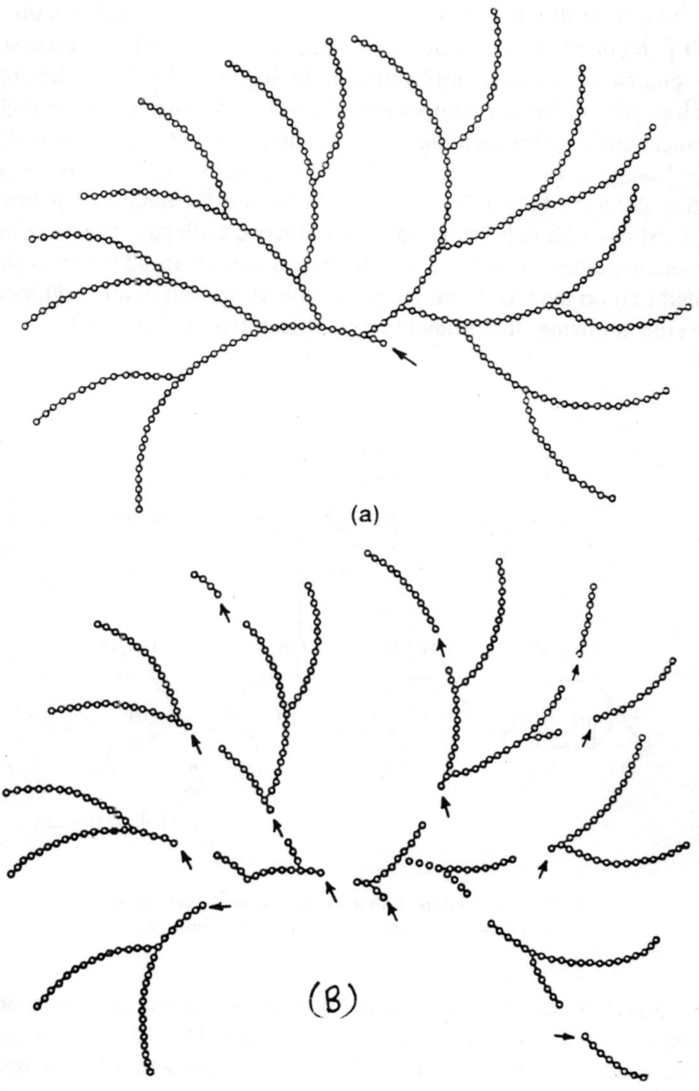

Fig. 14.4 : Action of a-amylase on amylopectin. (a) Amylopectin model. (b) Dextrins of medium molecular weight giving violet, purple, or red iodine color produced by splitting of 4 percent of the glucosidic linkages of amylopectin.

A characteristic reaction of starch is with iodine to produce an intense blue color. Dextrins react with iodine to produce various shades and color intensities of red through orange and yellow to colorless solutions. The shade and intensity of color is an indication of the chain length of the fragments produced by starch hydrolysis. These color changes are used to follow the progress of starch hydrolysis reactions by removing small samples of the reaction mixture at various times and treating them with an iodine solution.

Glycogen, found in the liver and muscle tissue of animals, is the animal counterpart of starch. It is a very highly branched polymer of glucose with a structure resembling that of amylopectin. Branching of the chains apparently occurs every 8 to 12 glucose units.

The ability to make glycogen form glucose reversibly is extremely important to the body, which uses glucose as its primary source of energy. We ordinarily take in carbohydrates in large quantities at widely spaced intervals. These carbohydrates are rapidly broken down in the digestive process and the resultant glucose is quickly absorbed. This absorption of large quantities of low molecular weight molecules could cause fatal disturbances to the osmotic pressure relationships in the body. Thus, it becomes a matter of great importance that the organism be able to store large quantities of material required as an energy source in a manner that will have little effect on osmotic pressure. Recall that osmotic pressure depends mainly on the number of dissolved particles present and not their kind or size. The reversible formation of the extremely large slightly soluble molecules of glycogen is a means of storing glucose in a readily accessible form that has little effect on osmotic pressure.

15

Carbohydrate Metabolism

Introduction

This chapter deals with the fate of carbohydrates following ingestion; how they are broken down and absorbed, how they react in the cells to produce energy, and how they can be stored in the body. In addition of study their interaction with lipid metabolism and comparison of the energy available to the body from carbohydrate metabolism to that available from lipids is also included. Finally, the chemical basis and consequences of abnormalities in carbohydrate metabolism, such as galactosemia and diabetes mellitus, will be studied.

Digestion and Absorption of Carbohydrates

When we eat a meal, we take in a vast assortment of compounds. These are usually classified as belonging to one of the three major groups of foodstuffs; proteins, fats, and carbohydrates. Among the compounds in our diet classed as carbohydrates are starch, cellulose, sucrose, glucose, fructose, and lactose, plus minor amounts of maltose, sorbitol, and arabinose. The first three compounds account for the majority of the carbohydrates ingested, probability greater than 95 percent of them.

Prior to absorption into the body, all carbohydrates are broken down to monosaccharides in a process called *digestion.* In order for the nutrients to be absorbed into the bloodstream they must first pass through the mucosal cells of the small intestine. Remember that cell membranes are only permeable to small molecules. Thus, it is necessary that the large food molecules be broken down to their simpler components before absorption can occur. Any foodstuffs that are not broken down are simply excreted. The breakdown of some carbohydrates begins in the mouth, although most occurs in the upper portion of the small intestine known as the duodenum.

Saliva contains the enzyme *sahvary α-amylase*, which catalyzes the hydrolysis of starch (both amylose and amylopectin) to dextrins and maltose. The contribution of salivary amylase to the overall digestion of starch is uncertain because of the variability of the length of time of contact between the enzyme and the starch. The pH required for the optimum activity of salivary amylase is between 6 to 8. This means that after the food is swallowed, the enzyme is deactivated by the low pH of the stomach juices. A small amount of starch hydrolysis may occur in the stomach due to the acid-catalyzed hydrolysis of the acetal bonds linking the glucose units together.

Most carbohydrate digestion takes place in the slightly basic media of the small intestine. The pancreas releases *pancreatic amylase,* which catalyzes the hydrolysis of starch of maltose and a mixture of oligosaccharides (sugars containing three to eight glucose units). We can see the specificity of enzymes and the biological importance of the stereochemistry of molecules in the digestive process. Both salivary and pancreatic amylases are enzymes specific for the hydrolysis of 1,4-α acetal linkages between glucose units. Neither of these enzymes is capable of catalyzing hydrolysis of 1,4-*β* acetal linkages between glucose molecules. Consequently, we cannot digest cellulose.

In humans and most other animals, cellulose is of negligible nutritional value. There are certain bacteria and other microorganisms that have enzymes capable of catalyzing the hydrolysis of 1,4-*β* linkages. Thus, in the overall scheme of nature, a means is provided for the breakdown and utilization of the bulk of dead vegetation. A few mammals (such as cows, horses, and other ruminants) and certain insects (such as termites) harbor in their gastrointestinal tract bacteria whose enzymes will hydrolyze cellulose.

The intestinal digestive juices also contain the enzymes *maltase, lactase,* and *sucrase*, which catalyze the hydrolysis of maltose, lactose, and sucrose, respectively. Also contained in the intestinal juices is *oligo-1,6-glucosidase*, which completes the hydrolysis of the oligosaccharides containing 1,6-*α* acetal bonds and the isomaltose produced during the breakdown of starch. Thus, the digestion of the main carbon hydrates contained in the diet leads to formation of the three monosaccharides: glucose, galactose, and fructøse, Glucose is overwhelmingly the predominant monosaccharide.

The mixture of monosaccharides if transported across the intestinal barrier into the blood. Although subjected to intensive study, relatively little is yet known about the manner in which these sugars penetrate the intestinal barrier. Diffusion across membranes due to concentration differences may play a part in sugar transport. It has long been known, however, that transport involves more than simple diffusion. In recognition of this the term active transport is applied to the movement of monosaccharides from the intestines into the blood. The idea of *active transport* implies the existence of a carrier of some type that combines with the sugar and enables the sugar molecule to move from one side of the membrane to the other.

Glucose Metabolism

Glucose entering the cells of an organism carrier with it energy that was stored in its chemical bonds during photosynthesis. This energy is released to the cells during a process called *metabolism*. The overall reaction for the metabolism of glucose by respiration is

$$C_6H_{12}O_6 + 6O_2 \longrightarrow 6H_2O + 6CO_2 + \text{energy}$$

The reaction is not carried out in a single step but is accomplished by a series of reactions, some of which release a small amount of the total available energy. By release of the energy in small amounts, a large portion of the total energy can be used for endergonic reactions and thus stored by the organism.

The first step in the use of glucose by the body is its conversion to *glucose-6-phosphate*. This step is endergonic since an ATP is used.

$$\text{glucose} + ATP \xrightarrow{\text{hexokinase}} \text{glucose-6-phosphate} + ADP$$

CH₂OH / OH / OH / OH / OH (glucose ring) + ATP —hexokinase→ CH₂—O—P(=O)(OH)—OH / OH / OH / OH / OH (glucose-6-phosphate ring) + ADP

glucose | glucose-6-phosphate

The enzyme *hexokinase* catalyzes the phosphorylation of glucose by ATP to form the ester glucose -6- phosphate.

After the formation of glucose-6-phosphate, there are several paths for its utilization. One path leads to glycogen and is called *glycogenesis*—the formation of glycogen. The other path leads to the breakdown of glucose to pyruvic acid and then eventually to carbon dioxide and water. The conversion of glucose to pyruvic acid is called *glycolysis.*

Glycolysis

In glycolysis, glucose, a six-carbon compound, is broken down to two three-carbon fragments of pyruvic acid. The initial steps of glycolysis are the same for most organisms. Some of the later steps may differ, however. Consequently biochemists refer to *fermentation, aerobic glycolysis,* and *anaerobic glycolysis,* depending on the final products produced during glycolysis. Fermentation leads to formation of ethanol: aerobic glycolysis, to carbon dioxide and water; and anaerobic glycolysis, to lactic acid. In each of these processes, pyruvic acid is produced as an intermediate. Aerobic means that oxygen is required, while *anaerobic* glycolysis occurs in the absence of oxygen. These processes are summarized in Fig 15.1.

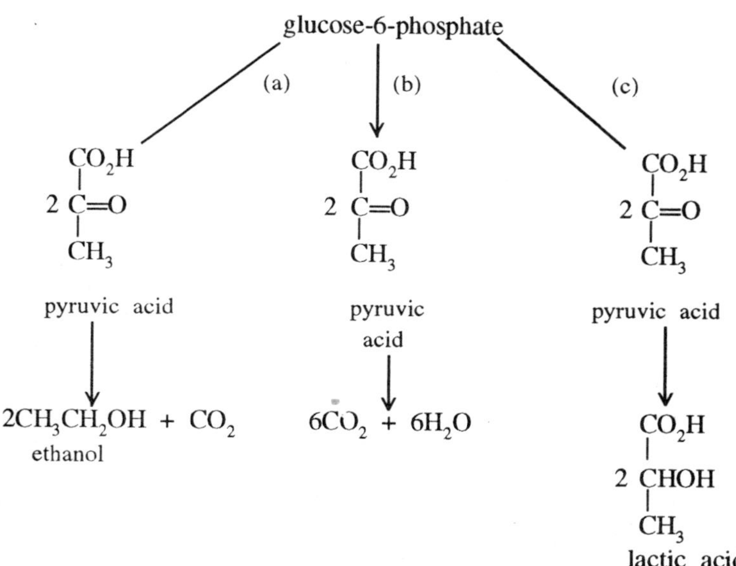

Fig. 15.1 : The three possible fates of glucose during glycolysis. Path (a) is called fermentation, path (b) is aerobic glycolysis, and path (c) is anaerobic glycolysis.

Fermentation is essentially restricted to microorganisms, primarily yeast. Some microorganisms are capable of aerobic glycolysis, while others are restricted to anaerobic glycolysis. Mammals are capable of both anaerobic and aerobic glycolysis, although the latter is the normal reaction pathway.

The overall sequence for the anaerobic conversion of glucose to lactic acid is shown in Fig. 15.2 (see page no. 295).

Pyruvic Acid And ATP Production

The initial step in the breakdown of pyruvic acid is its conversion to acetyl coenzyme A. The reaction involves both decarboxylation and oxidation as well as ester formation. The oxidation is accomplished by NAD.

$$\text{coenzyme A} + \underset{\text{pyruvic acid}}{\overset{CO_2H}{\underset{CH_3}{C=O}}} \xrightarrow{NAD \quad NADH_2} \underset{\text{acetyl coenzyme A}}{CH_3C(=O)CoA} + CO_2$$

Since each glucose molecule yields two pyruvic acid molecules, two acetyl coenzyme A molecules are produced from each glucose. Additionally, two $NADH_2$ molecules are formed from metabolism of each glucose. Upon interaction with the respiratory chain, each of these $NADH_2$ molecules leads to the formation of three additional ATP molecules. Thus, the cell obtains six more ATP molecules from the original glucose molecule. To this point, the cell has a net gain of 14 ATP molecules from a single glucose molecule.

Each of the acetyl coenzyme A molecules produced from glucose are converted to CO_2 and water via the Krebs cycle. For acetyl coenzyme A reacting via the Krebs cycle, 12 more ATP molecules are produced. This allows the formation of 24 more ATP molecules from the initial glucose molecule, *for a grand total of 38 ATP molecules gained by the cell during the conversion of a single molecule of glucose to carbon dioxide and water.* The accounting for these 38 ATP molecules is summarized in Fig. 15.3 (see page no. 296).

Mechanism of Glycolysis

Step 1. The first step in glycolysis is the conversion of glucose-

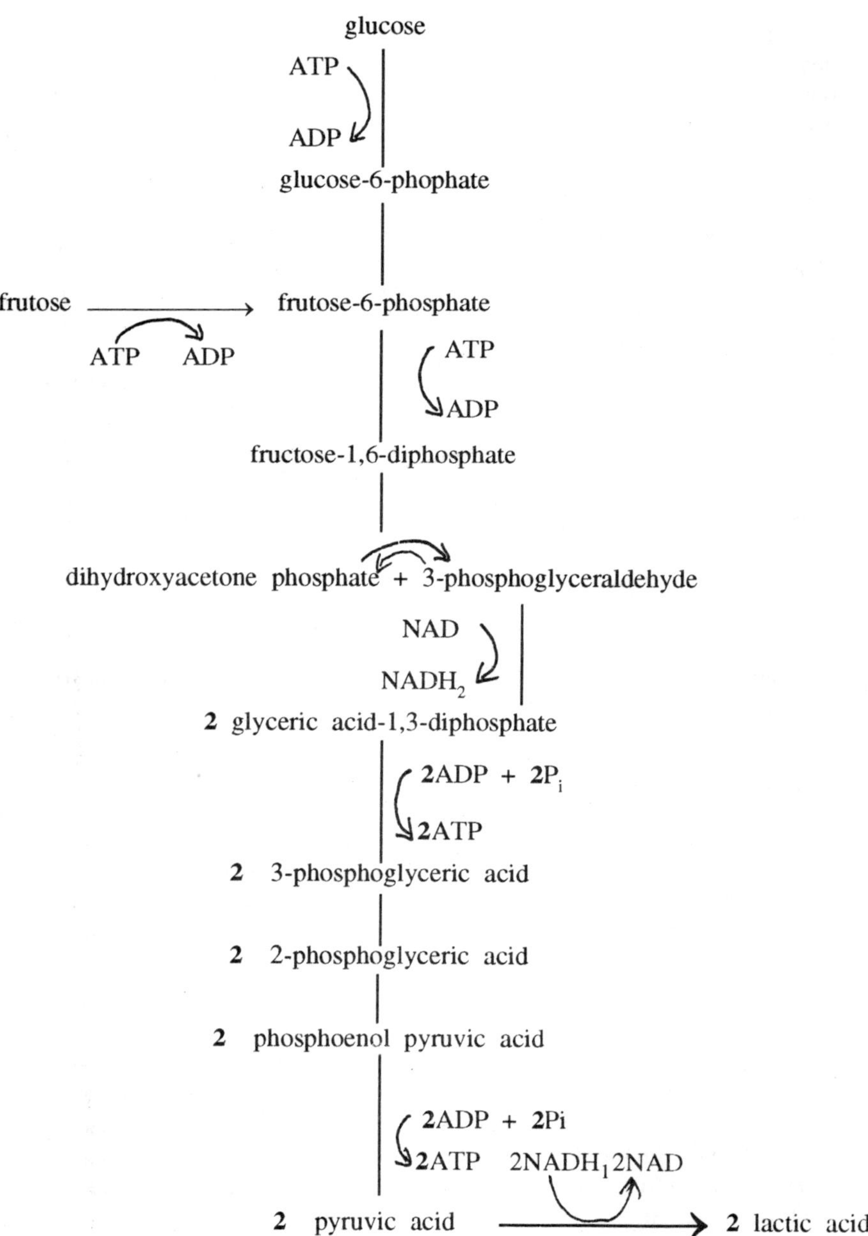

Fig. 15.2 : Summary of glycolysis leading to lactic acid formation.

$$\text{glucose} \xrightarrow{\text{glycolysis}} 2\ \text{pyruvic acid} + 2ATP + 2NADH_2$$

$$\downarrow$$

$$2CO_2 + 2\ \text{acetyl coenzyme A} \quad + \quad 2NADH_2$$

$$\downarrow \text{Krebs cycle and respiratory chain}$$

$$4CO_2 + 6H_2O + 24ATP$$

Subtotal : $6CO_2 + 6H_2O + 26ATP + 4NADH_2$

$4NADH_2 \xrightarrow{\text{respiratory chain}} 12ATP$

Grand total: $6CO_2 + 6H_2O + 38ATP$

Fig. 15.3 : Summary of ATP production from glucose.

6-phosphate to fructose-6-phosphate, catalyzed by the enzyme *phosphohexose isomerase.*

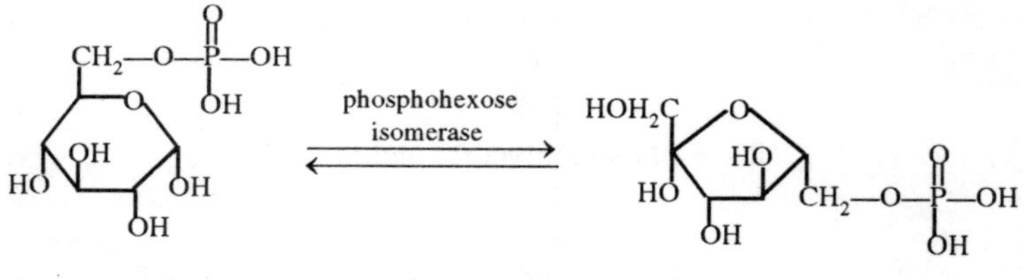

glucose-6-phosphate $\underset{}{\overset{\text{phosphohexose isomerase}}{\rightleftharpoons}}$ fructose-6-phosphate

As in most biochemical reactions, this isomerism is a reversible reaction. At equilibrium there would be 70 percent glucose-6-phosphate and 30 percent fructose-6-phosphate. As the fructose-6-phosphate is used in subsequent steps, however, more glucose-6-phosphate will be isomerized to fructose-6- phosphate in an attempt to maintain equilibrium Thus, the reaction is eventually driven to completion and the organism can utilize all the available glucose.

Glucose-6-phosphate is not the sole source of fructose-6-phosphate. Any fructose present from a dietary source can be phosphorylated directly by ATP under the action of fructokinase. Thus, a cell can use fructose as an energy source as well as glucose. These relationships are shown in Fig. 15.4.

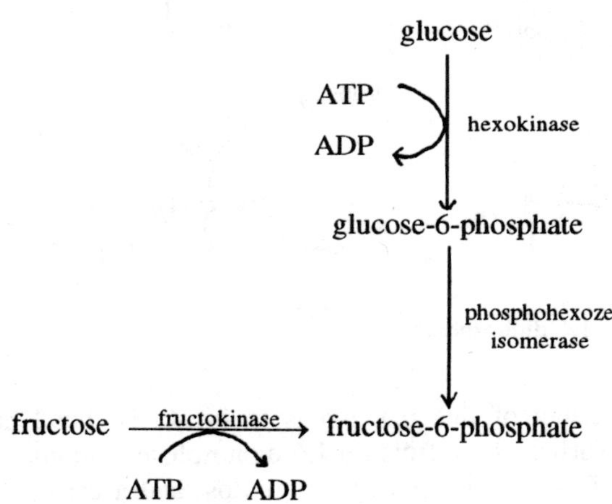

Fig. 15.4 : The incorporation of fructose into glycolysis.

Step 2. The next step in glycolysis is phosphorylation of fructose-6-phosphate to fructose-1,6-diphosphate. This requires ATP and is catalyzed by the enzyme phosphofructokinase.

CH_2OH O HO HO OH CH_2—O—P(=O)(OH)—OH

phosphofructo - kinase →

ATP ADP

O—P(=O)(OH)—OH CH_2 O HO HO OH CH_2—O—P(=O)(OH)—OH

fructose-6-phosphate fructose-1,6-diphosphate

To this point in glycolysis, the cell has invested two ATP molecules in each monosaccharide molecule and, as yet, has not received back. Thus, so far the process has been *endoergic* rather than exoergic.

Step 3. The fructose-1,6-diphosphate molecule is now split into two three carbon fragments.

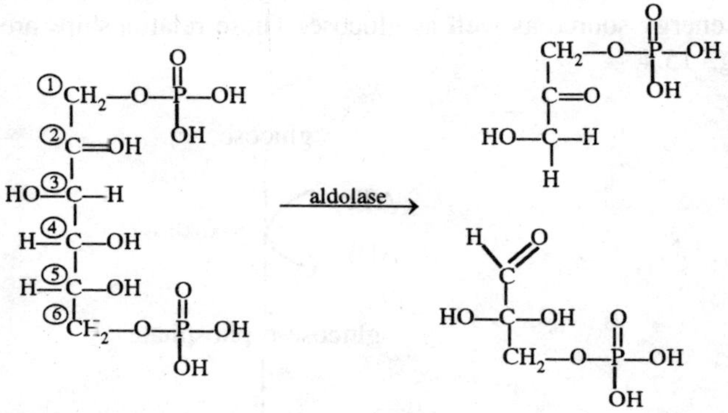

```
                 O
                 ||
(1)CH2—O—P—OH
   |             |
(2)C=OH          OH
   |
HO(3)C—H                 aldolase
   |              ———————————→
H(4)C—OH
   |
H(5)C—OH          O
   |              ||
(6)CH2—O—P—OH
                  |
                  OH
```

fructose- 1,6-diphosphate

```
              O
              ||
CH2—O—P—OH
|             |
C=O           OH
|
HO—C—H
   |
   H

H     O
 \\ //
  C
  |
HO—C—OH     O
   |        ||
   CH2—O—P—OH
            |
            OH
```

The net effect of this reaction is to break the bond between carbon 3 and carbon 4 of fructose-1,6-diphosphate. Simultaneously, the hydrogen of the —OH on carbon 4 is lost and a carbon oxygen double bond is formed producing 3-phosphoglyceraldehyde. Carbon of the fructose-1,6-diphosphate gains a hydrogen to yield dihydroxy-acetone phosphate.

Step 4. Cells can only utilize the 3-phosphoglyceraldehyde directly. This doesn't mean that the dihydroxyacetone phosphate is wasted. An enzyme capable of isomerizing dihydroxyacetone phosphate to 3-phos-phoglyceraldehyde is present in cells.

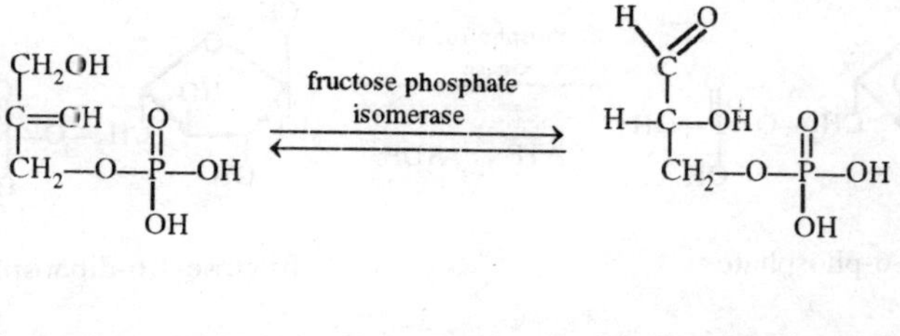

```
                                                     H    O
                                                      \\ //
CH2OH                    fructose phosphate            C
|                            isomerase                 |
C=OH      O             ———————————→              H—C—OH     O
|         ||            ←———————————                   |          ||
CH2—O—P—OH                                             CH2—O—P—OH
          |                                                       |
          OH                                                      OH
```

dihydroxyacetone phosphate   3-phosphoglyceraldehyde

The enzyme effects the movement of a hydrogen atom from carbon 1 to 2. Also, a hydrogen from the —OH group of carbon 1 is moved into the oxygen of carbon 2. In essence, the primary alcohol of carbon 1 is moved onto the oxygen of carbon 2 · is reduced to an alcohol.

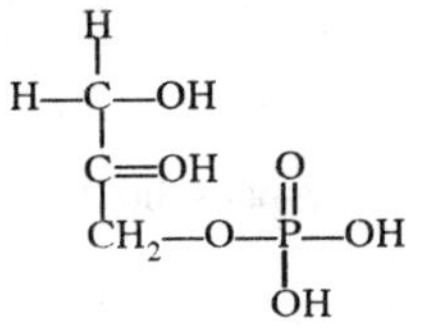

```
      H                                  H   O
      |                                   \\//
 H—C—OH                                    C
      |                                    |
     C=OH   O                         H—C—OH   O
      |     ||                             |    ||
     CH2—O—P—OH                           CH2—O—P—OH
            |                                   |
            OH                                  OH
```

dihydroxyacetone phosphate 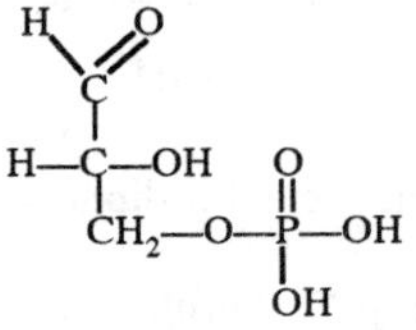   3-phosphoglyceraldehyde

Step 5. The 3-phosphoglyceraldehyde undergoes oxidation and phosphorylation in the next step. The oxidizing agent for this step is NAD.

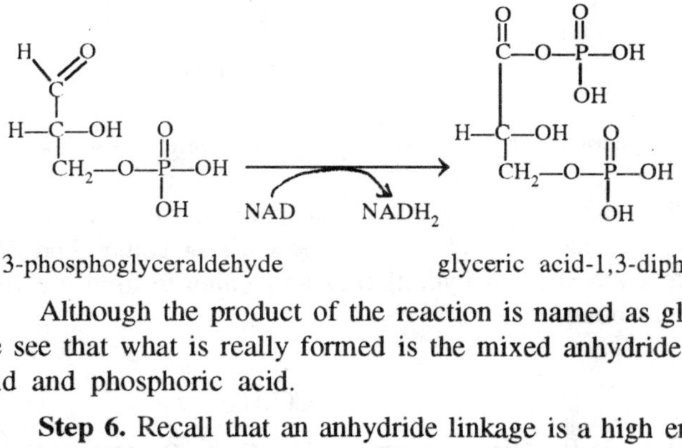

```
                                                  O     O
                                                  ||    ||
  H   O                                           C—O—P—OH
   \\//                                           |     |
    C                                             |     OH
    |                                             |
H—C—OH   O                                   H—C—OH   O
   |     ||                                       |    ||
  CH2—O—P—OH  ---------------------------->      CH2—O—P—OH
         |                                              |
         OH       NAD          NADH2                    OH
```

3-phosphoglyceraldehyde   glyceric acid-1,3-diphosphate

Although the product of the reaction is named as glyceric acid, we see that what is really formed is the mixed anhydride of glyceric acid and phosphoric acid.

Step 6. Recall that an anhydride linkage is a high energy group and hydrolysis of the linkage releases energy. The hydrolysis of the mixed anhydride of glyceric acid-1,3 diphosphate releases sufficient energy for the formation of ATP (see structure formula on page no. 300).

Each glucose produces tow glyceric acid-1,3-diphosphate molecules. From each glyceric acid-1,3-diphosphate, one ATP is formed. Thus, we see the formation of two ATP molecules form each glucose molecule. *At this point in the breakdown of glucose there has been no net change in the number of ATP molecules available to the cell.* Remember two ATP molecules have been used in glycolysis and now two ATP molecules have been regenerated. More energy is available from 3-phosphoglyceric acid, as we shall see.

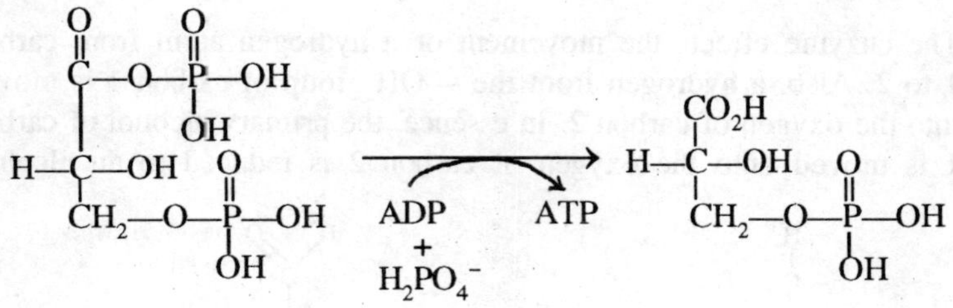

$$\begin{array}{l} \overset{O}{\overset{\|}{C}}-O-\overset{O}{\overset{\|}{P}}-OH \\ \quad\quad\quad\ \ | \\ \quad\quad\quad\ OH \\ H-C-OH \quad \overset{O}{\|} \\ \quad CH_2-O-P-OH \\ \quad\quad\quad\quad\quad | \\ \quad\quad\quad\quad\ OH \end{array} \xrightarrow[\text{ADP} + H_2PO_4^- \quad \text{ATP}]{} \begin{array}{l} \quad CO_2H \\ H-C-OH \quad \overset{O}{\|} \\ \quad CH_2-O-P-OH \\ \quad\quad\quad\quad\quad | \\ \quad\quad\quad\quad\ OH \end{array}$$

glyceric acid-1,3-diphosphate 3-phosphoglyceric acid

Step 7. Another isomerization occurs next, in which 3-phosphoglyceric acid is converted to 2-phosphoglyceric acid.

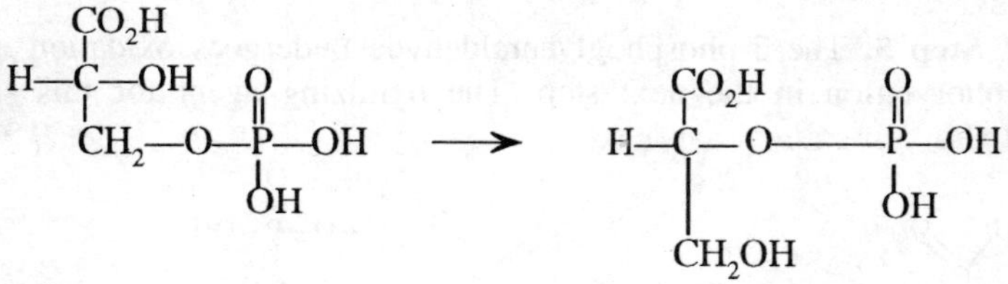

$$\begin{array}{l} \quad CO_2H \\ H-C-OH \quad \overset{O}{\|} \\ \quad CH_2-O-P-OH \\ \quad\quad\quad\quad\quad | \\ \quad\quad\quad\quad\ OH \end{array} \longrightarrow \begin{array}{l} \quad CO_2H \quad\quad\quad \overset{O}{\|} \\ H-C-O-P-OH \\ \quad | \quad\quad\quad\quad | \\ \quad | \quad\quad\quad\ OH \\ \quad CH_2OH \end{array}$$

3-phosphoglyceric acid 2-phosphoglyceric acid

Step 8. Water is removed from 2-phosphoglyceric acid to produce phosphoenolpyruvic acid. Phosphoenolpyruvic acid is sometimes abbreviated PEP. This reaction is the dehydration of an alcohol to form a carbon-carbon double bond.

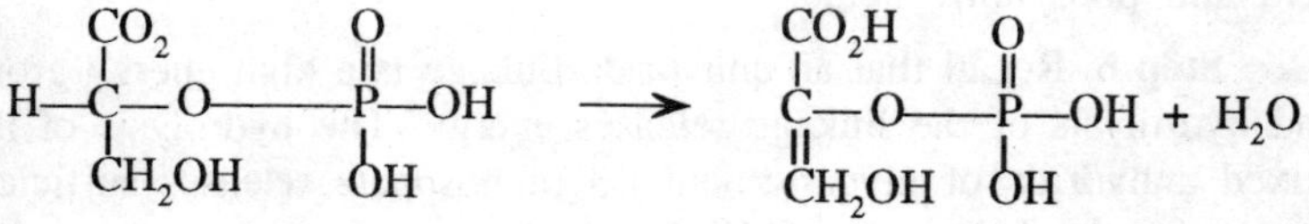

$$\begin{array}{l} \quad CO_2 \quad\quad\quad \overset{O}{\|} \\ H-C-O-P-OH \\ \quad CH_2OH \quad\ OH \end{array} \longrightarrow \begin{array}{l} CO_2H \quad \overset{O}{\|} \\ C-O-P-OH \\ \| \quad\quad\quad | \\ CH_2OH \ \ OH \end{array} + H_2O$$

2-phosphoglyceric acid phosphoenolpyruvic acid

Step 9. The structure of phosphoenolpyruvic acid is rather interesting if we compare it to the structure of a mixed. Carbon-phosphate anhydride. Pay particular attention to the colored portion of phosphoenolpyruvic acid in the structure below in comparison to the structure of an anhydride.

```
CH2      O                 O     O
||       ||                ||    ||
C—O——P—OH                  C—O—P—OH
|        |                 |     |
CO2H     OH                R     OH
```

phosphoenolpyruvic acid

a mixed carbon-phosphate anhydride

Except for the fact that phosphoenolpyruvic acid has a carbon-to-carbon double bond instead of a carbon-to-oxygen double bond, the structures look very similar. (The —CO_2H group is merely the R group of the mixed anhydride.) This similarity in structure is apparently sufficient to make the phosphate linkage in phosphoenolpyruvic acid a high-energy one. Chemists find that the hydrolysis of phosphoenolpyruvic acid releases enough energy to allow the cell to produce ATP from ADP and inorganic phosphate.

```
CO2H     O                              CO2H
|        ||                             |
C—O——P—OH     ——————————————→           C=O
||       |                              |
CH2      OH      ADP         ATP        CH3
                 +Pi
```

phosphoenolpyruvic acid

pyruvic acid

Since two phosphoenolpyruvic acid molecules result from each molecule of glucose undergoing glycolysis, there has been a net gain to the cell of two ATP molecules from each glucose. Thus, the cell has been able to obtain and store for its own purposes some of the energy looked into the glucose molecule during photosynthesis.

Step 10. The fate of the pyruvic acid depends both on the organism and the circumstances in which it finds itself. Remember that NAD is reduced to $NADH_2$ during the oxidation of 3-phosphoglyceric acid to glyceric acid 1,3-diphosphate. Thus, from one molecule of glucose, the cell has had a net gain of two ATP molecules and is left with two $NADH_2$ molecules.

The cell has only a limited amount of NAD; thus the $NADH_2$ must be reoxidized by some means. A mammal operating under normal body conditions reoxidizes the $NADH_2$ to NAD via the respiratory chain. Recall that three ATP molecules are produced for each $NADH_2$ interacting with the respiratory chain. Thus, six more ATP molecules are available to a mammal from each glucose molecule. *This leads to a net formation of eight ATP molecules* (six from the two $NADH_2$ molecules plus the two produced directly during glycolysis) *from each glucose molecule undergoing glycolysis.*

Anaerobic microorganisms reoxidize the $NADH_2$ to NAD by reducing pyruvic acid to lactic acid.

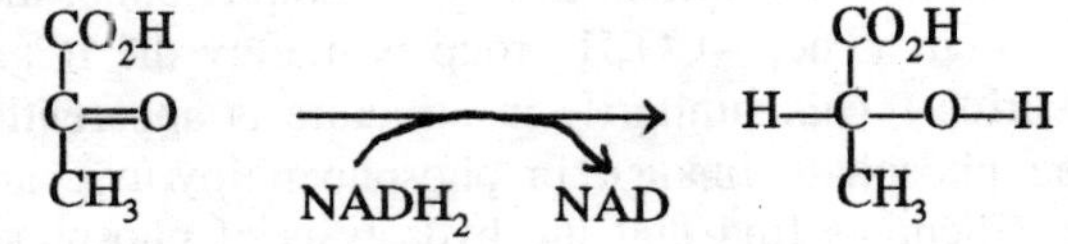

$$\underset{\text{pyruvic acid}}{\begin{array}{c} CO_2H \\ | \\ C{=}O \\ | \\ CH_3 \end{array}} \xrightarrow[NADH_2 \;\curvearrowright\; NAD]{} \underset{\text{lactic acid}}{\begin{array}{c} CO_2H \\ | \\ H{-}C{-}O{-}H \\ | \\ CH_3 \end{array}}$$

The conversion of glucose to lactic acid is often referred to as anaerobic glycolysis. Notice that anaerobic glycolysis yields only a net gain of two ATP. Not all anaerobic organisms reduce pyruvic acid to lactic acid. Yeasts, for example, decarboxylate the pyruvic acid to acetaldehyde and then reduce the acetaldehyde to ethyl alcohol. We call the breakdown of glucose to ethanol and carbon dioxide *fermentation*

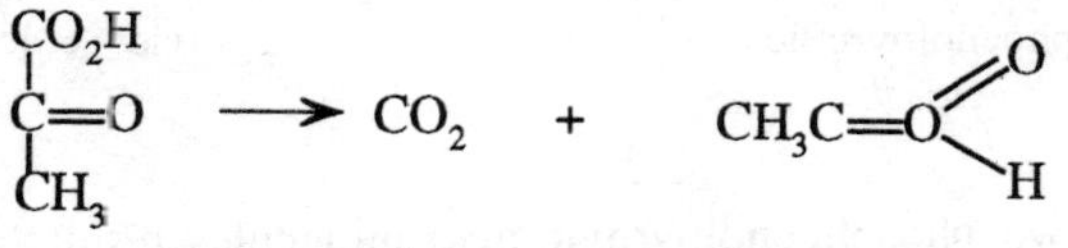

$$\begin{array}{c} CO_2H \\ | \\ C{=}O \\ | \\ CH_3 \end{array} \longrightarrow \underset{\text{carbon dioxide}}{CO_2} + \underset{\text{acetaldehyde}}{CH_3C{=}O\!\!\begin{array}{l} \nearrow O \\ \searrow H \end{array}}$$

$$CH_3C{=}O\!\!\begin{array}{l} \nearrow O \\ \searrow H \end{array} \xrightarrow[NADH_2 \;\curvearrowright\; NAD]{} \underset{\text{ethyl alcohol}}{\begin{array}{c} CH_3CH{-}OH \\ | \\ H \end{array}}$$

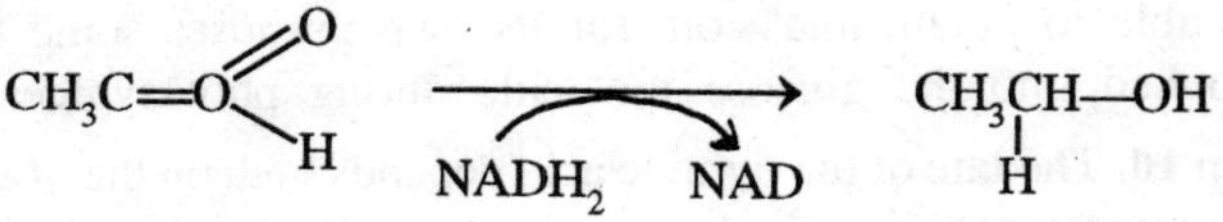

The reactions involved in glycolysis are summarized in Fig. 11.5 (see page No. 303).

The Fate of Pyruvic acid under Aerobic Conditions

We have discussed the fate of pyruvic acid by anaerobic organisms and have seen that it is converted either to lactic acid or to ethanol.

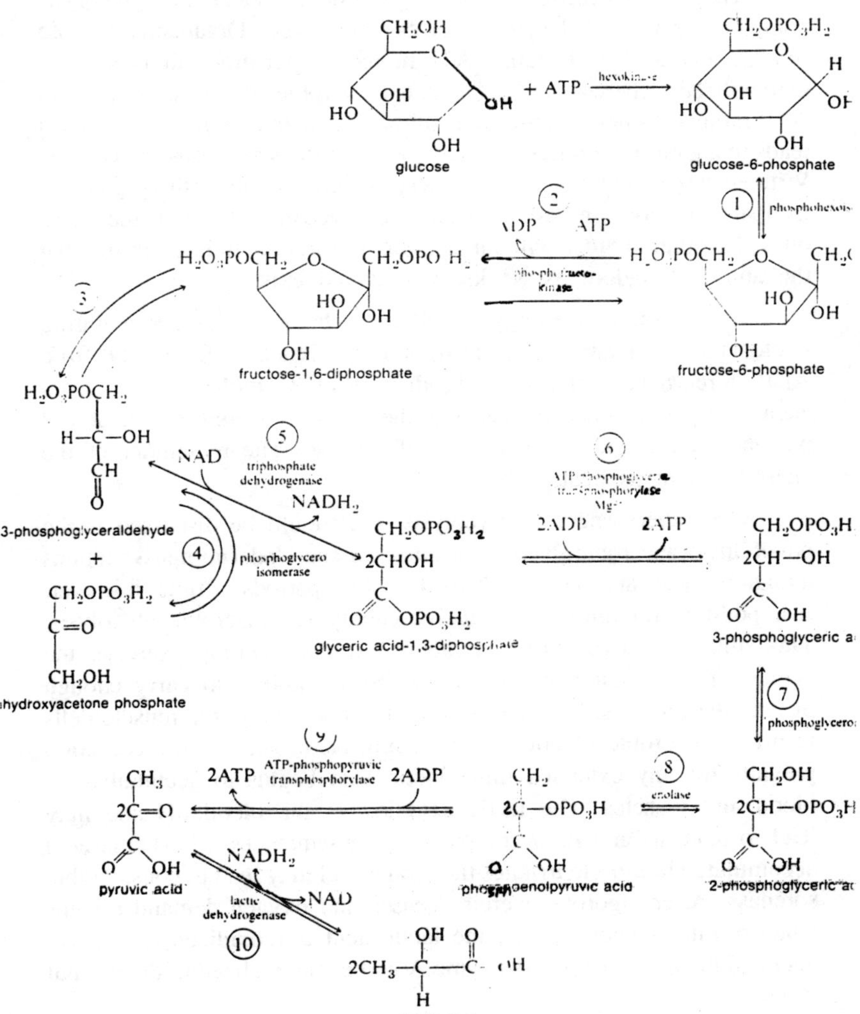

Fig. 15.5 : The anaerobic conversion of glucose to lactic acid.

In mammals and other organisms capable of respiration, the pyruvic acid undergoes further reactions to produce eventually CO_2, water, and ATP via the Krebs cycle. During aerobic oxidation, 38 molecules of ATP are produced per molecule of glucose.

Anaerobic Versus Aerobic Glycolysis

There is a great difference in energy produced by anaerobic glycolysis compared to that produced by aerobic glycolysis. Organisms capable of anaerobic glycolysis gain 2 ATP molecules per molecule of glucose metabolized; whereas, those capable of aerobic glycolysis obtain 38 ATP molecules per glucose molecule metabolized. Aerobic glycolysis leads to a gain of 19 times as much energy as does anaerobic glycolysis. Without the development of aerobic respiration it is doubtful that organisms incapable of photosynthesis would have evolved much beyond unicellular ones. In other words, without aerobic respiration it is doubtful that the animal kingdom as we know it would exist.

The amount of energy stored in each mole of glucose during photosynthesis is 686 kcal. The formation of each mole of ATP from ADP stores 8 kcal of energy. Production of 38 moles of ATP from each mole of glucose means that the cell has stored 304 kcal (44 percent) of the energy in a mole of glucose. The remainder of the energy is lost as heat.

One other aspect of glycolysis that should be discussed is the possibility of anaerobic glycolysis in muscle tissue. Under normal conditions a mammal uses aerobic glycolysis. For short periods of time, however, it is possible for muscles to receive energy via anaerobic glycolysis. This situation arises during strenuous exercise. During exercise, the demand for oxygen may exceed the body's ability to carry enough of it to the muscles. In order to continue functioning, the muscle cells resort to anaerobic glycolysis, producing lactic acid. If the condition prevails for any extended time, lactic acid begins to accumulate in blood and muscles. The lactic acid lowers the blood pH and may lead to acidosis and its accompanying consequences. The lactic acid accumulated in muscle irritates the tissues and may lead to considerable soreness. After vigorous exercise ceases and oxygen demand regains balance with oxygen supply, the lactic acid is reoxidized to pyruvic acid and thence to acetyl coenzyme A and is metabolized in the normal fashion.

The Formation of Glycogen Glycogenesis

Glycogenesis is the making glycogen from glucose or from pyruvic and lactic acids. Earlier it was mentioned that glucose could be stored in the body as glycogen. This storage can occur in both liver and muscles. Glycogen formation is a form of short-term storage of carbohydrates and in normal individuals it occurs when the glucose level in the blood exceeds approximately 100 mg of glucose per 100 ml of blood. The first step in *glycogenesis* is the isomerization of glucose-6-phosphate to glucose-1-phosphate. Following the formation of glucose-1-phosphate the glucose molecules are linked together by 1,4-α acetal linkages and 1,6-α acetal linkages to form glycogen. The individual steps by which this is accomplished are very complex and will not be studied in this text. We should note however, that glycogenesis is an energy-consuming process. The amount of glycogen stored is controlled by several hormones.

Glycogen can also be formed from lactic acid. The lactic acid produced in muscles during vigorous exercise must be removed from the muscle tissue. Seldom is this lactic acid immediately oxidized to carbon dioxide and water. The oxidation occurs only if ATP is needed at the time. If the demand for ATP is small, the body reconverts the majority of the lactic acid back to glucose and then to glycogen. This process occurs primarily in the liver. Muscle cells do not contain a sufficient concentration of the enzymes required to catalyze the conversion of lactic acid to glucose. The overall relationships among blood glucose, liver glycogen, muscle glycogen, and lactic acid are shown in Fig. 15.6.

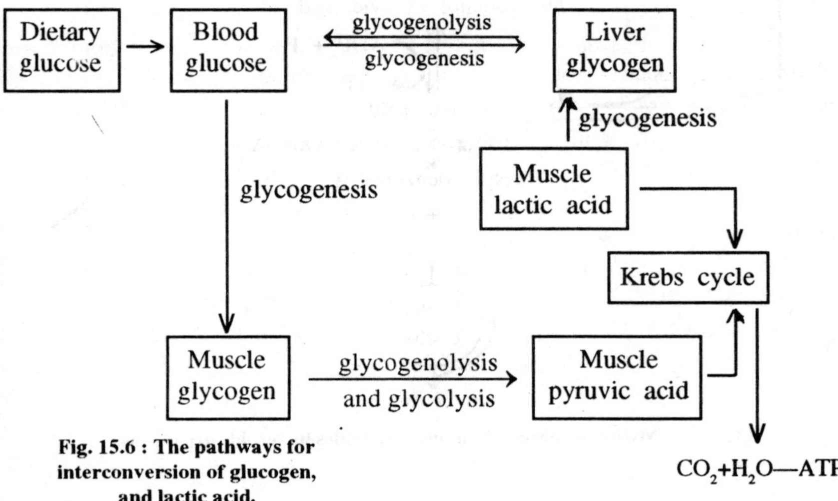

Fig. 15.6 : The pathways for interconversion of glucogen, and lactic acid.

The relationships between glycogenesis and glycolysis are shown in Fig 15.7.

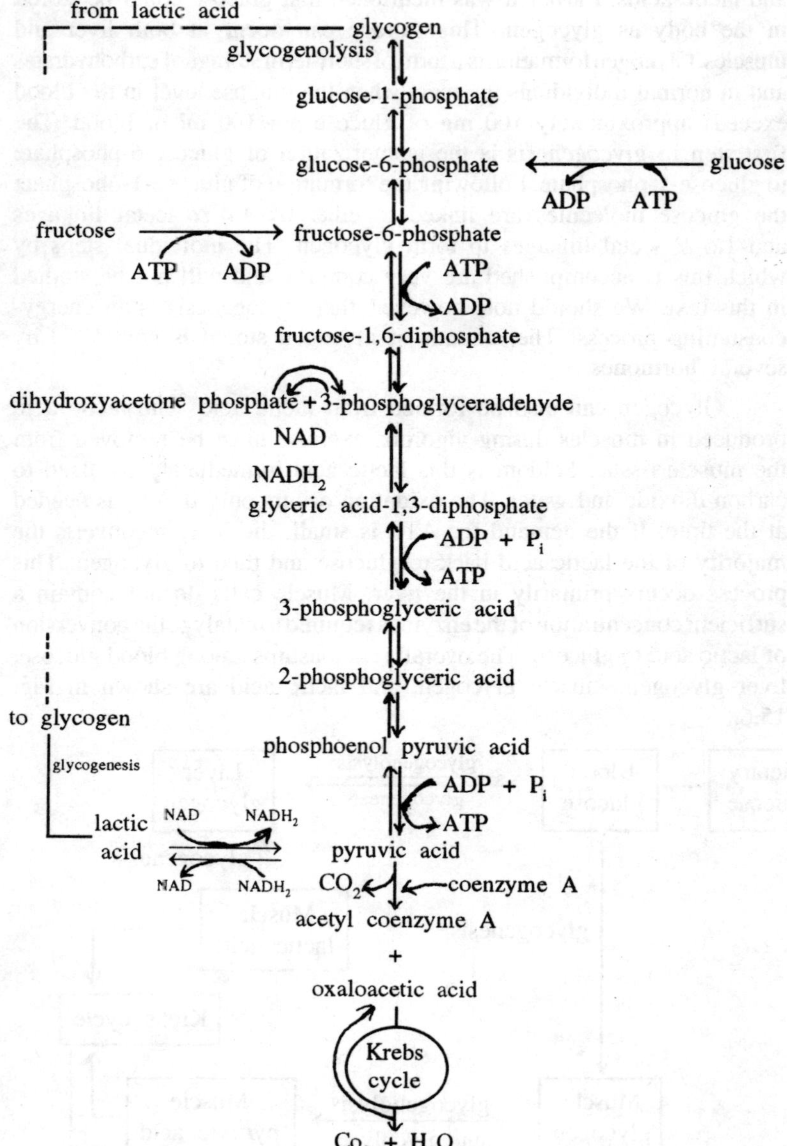

Fig. 15.7 : Metabolic paths for monosaccharides in aerobic organisms.

Interaction of Carbohydrate and Lipid Metabolism Long-term Storage

If we eat more than necessary, we gain weight. To say we are gaining weight is merely a slightly less offensive way of saying we are getting fat. This fat is manufactured by the body from excess nutrients included in the diet. The production of fat is *lipogenesis.*

Lipogenesis is the body's means of long-term storage of excess nutrients. The lipids can be made from both fats and the carbohydrates in our diet. Lipids are deposited in regions widely distributed over various parts of the body known as adipose tissue or fat depots. When we speak of lipid storage, it is not meant that the fat is totally inactive. The cells within the adipose tissue are constantly producing new fat molecules and consuming old fat molecules. Thus, there is a continuing turnover of the stored lipids.

The energy reserves present in the fat depots of the normal person are very large compared to other energy reserves within the body. At least 10 percent of body weight is lipid and is sufficient to fulfil the energy requirements of an individual for over a month. In contrast, the body has only enough glycogen to take care of energy requirements for about 24 hr. An advantage to the body of storing energy reserves as fat rather than as glycogen is the high-energy density of fats. There are over twice as many calories per gram of fat as there are per gram of carbohydrate. Fat contains about 9 kcal/g versus 4 kcal/g for carbohydrates.

Correspondingly, when the body metabolizes fats, it obtains a little more than twice as much ATP per gram of fat as it does per gram of carbohydrate metabolized. The body obtains 0.211 mole of ATP per gram of glucose while a gram of fat yields 0.50 mole of ATP.

Fat Synthesis

The synthesis of fats within the body involves the synthesis of both glycerol and fatty acids. We shall examine the reactions of fatty acid synthesis first. The beginning point is acetyl coenzyme A. Thus, we can see immediately how carbohydrates in the diet can be converted to fat since sugars can be converted to acetal coenzyme a by glycolysis.

The initial reaction is the combination of acetyl coenzyme A with carbon dioxide to yield malonyl coenzyme A.

$$CH_3C(=O)S\text{—}CoA + CO_2 \xrightarrow{ATP \rightarrow ADP} HO_2C\text{—}CH_2\text{—}C(=O)S\text{—}CoA$$

acetyl coenzyme A — malonyl coenzyme A

The second step is the combination of another acetyl coenzyme A with malonyl coenzyme A. This step is accompanied by the loss of carbon dioxide. It is interesting to note that although CO_2 is required to start the process, it is lost almost immediately.

$$HO_2C\text{—}CH_2\text{—}C(=O)\text{—}S\text{—}CoA + CH_3C(=O)S\text{—}CoA \xrightarrow{\text{coenzyme A},\ CO_2} CH_3C(=O)\text{—}CH_2C(=O)S\text{—}CoA$$

malonyl coenzyme A — acetyl coenzyme A — acetoacetyl coenzyme A

Other than the gain and loss of carbon dioxide, the synthesis of fatty acids is essentially the reverse of beta oxidation of fatty acids. These reactions are shown in Fig 15.8 (see page no. 309).

The final step of fat synthesis is the combination of the fatty acids with glycerol. The source of glycerol is the $NADH_2$ reduction of dihydroxyacetone phosphate, produced during glycolysis to glycerol phosphate.

$$CH_2\text{—}O\text{—}P(=O)(OH)\text{—}OH,\ C{=}O,\ CH_2OH \xrightarrow{NADH_2 \rightarrow NAD} CH_2\text{—}O\text{—}P(=O)(OH)\text{—}OH,\ H\text{—}C{=}OH,\ CH_2OH$$

dihydroxyacetone phosphate — glycerol phosphate

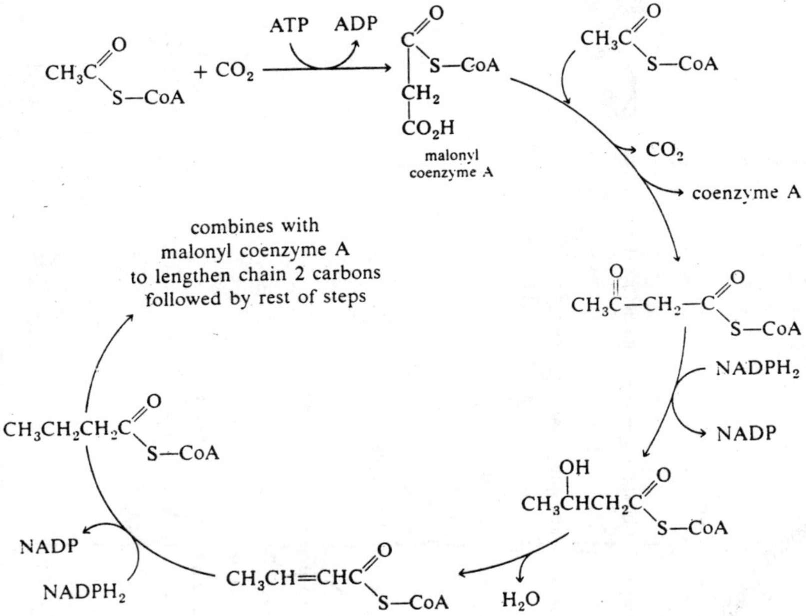

$CH_3C(=O)S{-}CoA + CO_2 \xrightarrow{ATP \to ADP}$ $HO_2C{-}CH_2{-}C(=O)S{-}CoA$

malonyl coenzyme A

$CH_3C(=O)S{-}CoA$

CO_2

coenzyme A

$CH_3C(=O){-}CH_2{-}C(=O)S{-}CoA$

$NADPH_2$

NADP

$CH_3CH(OH)CH_2C(=O)S{-}CoA$

H_2O

$CH_3CH{=}CHC(=O)S{-}CoA$

$NADPH_2$

NADP

$CH_3CH_2CH_2C(=O)S{-}CoA$

combines with malonyl coenzyme A to lengthen chain 2 carbons followed by rest of steps

Fig. 15.8 : Reactions involved in fatty acid synthesis.

Three moles of fatty acids react with 1 mole of glycerol phosphate to yield triglyceride plus inorganic phosphate.

The reduction of dihydroxyacetone phosphate to glycerol phosphate is a reversible reaction. In the catabolism of lipids, glycerol is released. This glycerol is phosphorylated by ATP to produce glycerol phosphate (see structure on page 312).

The glycerol phosphate is oxidized to dihydroxyacetone by NAD. The dihydroxyacetone enters into the glycolysis scheme just as if its source were glucose (see Fig. 15.9 on page 310 & 311).

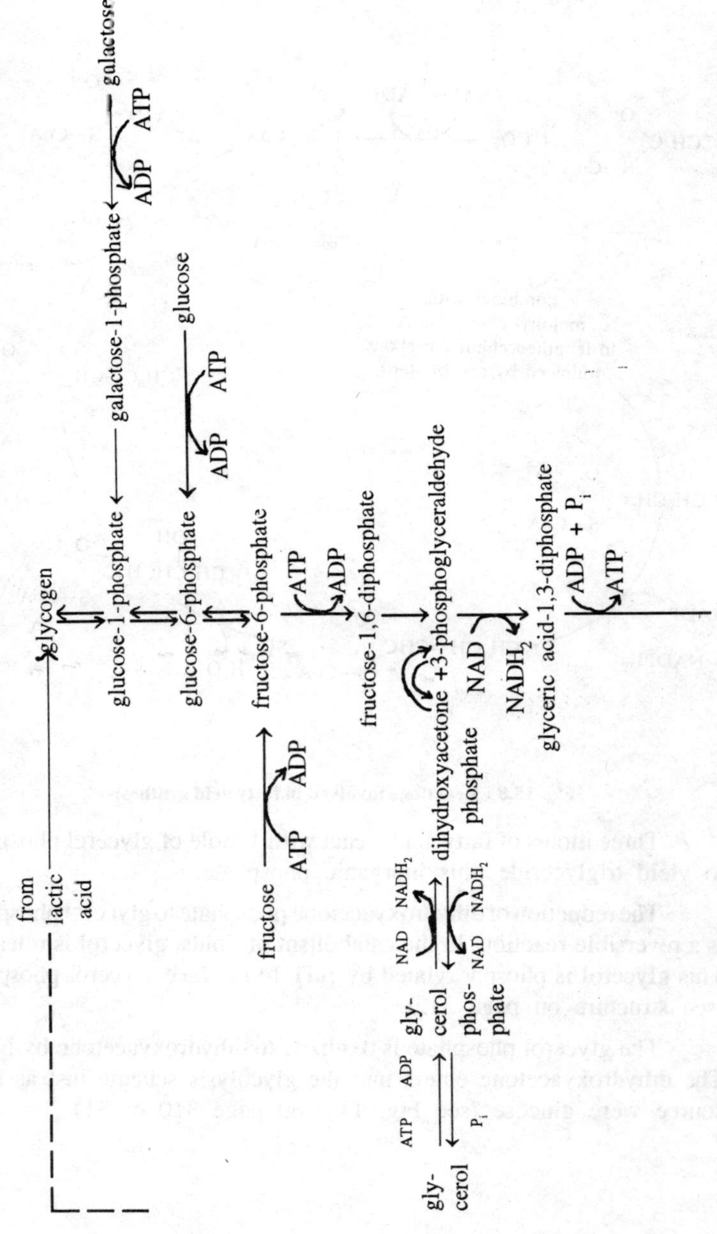
galactose
ATP
ADP
galactose-1-phosphate
glucose
ATP
ADP
from lactic acid
glycogen
glucose-1-phosphate
glucose-6-phosphate
fructose-6-phosphate
ATP
ADP
fructose-1,6-diphosphate
fructose
ATP
ADP
dihydroxyacetone phosphate
+3-phosphoglyceraldehyde
NAD
$NADH_2$
glyceric acid-1,3-diphosphate
ADP + P_i
ATP
glycerol
ATP
ADP
P_i
glycerol phosphate
NAD
$NADH_2$

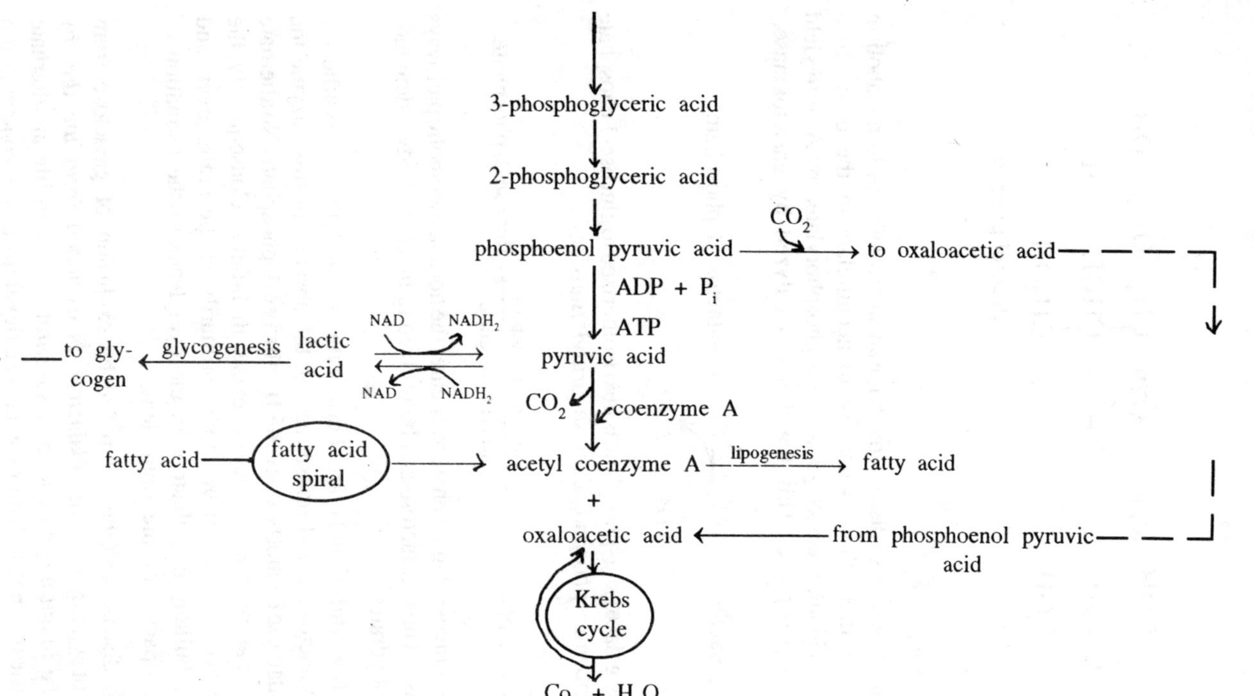

Fig. 15.9 : Overall summary of carbohydrate and lipid metabolism in a normal individual.

$$\begin{matrix} CH_2OH \\ | \\ CHOH \\ | \\ CH_2OH \end{matrix} \xrightarrow{ATP \quad ADP} \begin{matrix} CH_2-O-\overset{\displaystyle O}{\overset{\|}{P}}-OH \\ | \qquad\qquad | \\ CHOH \qquad OH \\ | \\ CH_2OH \end{matrix}$$

glycerol → glycerol phosphate

Galactosemia

Galactosemia is a disease involving an inborn defect in the metabolism of galactose that allows galactose to accumulate in the body. In a normal individual, ingested galactose is phosphorylated by ATP to yield galactose-1-phosphate. This reaction is catalyzed by galactokinase.

$$\text{galactose} \xrightarrow[\text{ATP} \quad \text{ADP}]{\text{galactokinase}} \text{galactose-1-phosphate}$$

The galactose-1-phosphate is next isomerized to glucose-1 phosphate by the enzyme *phosphogalactose uridyl transferase.*

$$\text{galactose-1-phosphate} \xrightarrow[\text{aridyl transferase}]{\text{phosphogalactose}} \text{glucose-1-phosphate}$$

The glucose-1-phosphate may enter the normal metabolic pathways for glucose. Thus, galactose can be utilized to meet the body's demands for carbohydrates.

A few individuals are born with an inability to synthesize phosphogalactose uridyl transferase. In the absence of this enzyme the dietary galactose cannot be converted to glucose-1-phosphate. No alternate metabolic pathway for galactose exists in infants. Consequently the galactose accumulates in the body, primarily in the brain, liver, and eyes. The buildup of galactose in the eyes leads to the formation of *cataracts*, opacity of the optic lens.

The disease can be controlled by exclusion of galactose from the diet of young humans. *Galactose is excluded from the diet by feeding the infant a synthetic milk substitute.* Later in life an alternate pathway for the metabolism of galactose develops as the concentration of the enzyme required for the alternate pathway increases. The presence of this enzyme in the liver allows an adult lacking the transferase to metabolize normal amounts of galactose.

DIABETES

Diabetes mellitus is more widespread abnormality of carbohydrate metabolism than is galactosemia. Diabetes is the result of the inability of the pancreas to provide adequate insulin, which regulates glucose concentration in blood by promoting glycogenesis in the liver and muscles and by controlling cellular oxidation of glucose by influencing glucose transport across cell membranes. In the absence of insulin, blood glucose concentration rises and if it exceeds the renal threshold, glucose appears in the urine (see Fig 15.10 on page no. 314). The presence of glucose in urine is called *glucosuria.* The appearance of glucose in the urine is indicative that other adverse conditions may exist in the body as a consequence of the inability of the person to metabolize glucose properly. Possible conditions are ketosis, acidosis, electrolyte imbalance, and fatty liver.

Since glucose cannot enter the cells without insulin, it cannot be oxidized to provide energy to the cells. The body responds to the cellular energy demand by resorting to its alternate energy source, the oxidation of fats. The depot fat is mobilized and brought to the liver by the bloodstream.

In the liver, beta oxidation breaks down fatty acids to acetyl coenzyme A. To this point everything appears to be proceeding smoothly, and we would expect that acetyl coenzyme A would be utilized via the Krebs cycle to produce ATP. Unfortunately, operation of the Krebs cycle is impaired in the diabetic. This impairment can be traced directly back to the lack of glucose metabolism.

Theoretically, each turn of the Krebs cycle regenerates oxaloacetic acid to begin the cycle over again. This process is not completely efficient, however, and some of the Krebs cycle's intermediates are lost to alternate metabolic pathways.

The loss of some of these intermediates means some oxaloacetic acid is not regenerated to initiate the next turn of the cycle. Thus, to keep the Krebs cycle active and continue to utilize acetyl coenzyme A, there must be a continuing supply of new oxaloacetic acid.

There are two major sources of oxaloacetic acid in animal tissue. One of these is the carboxylation (addition of CO_2) of pyruvic acid

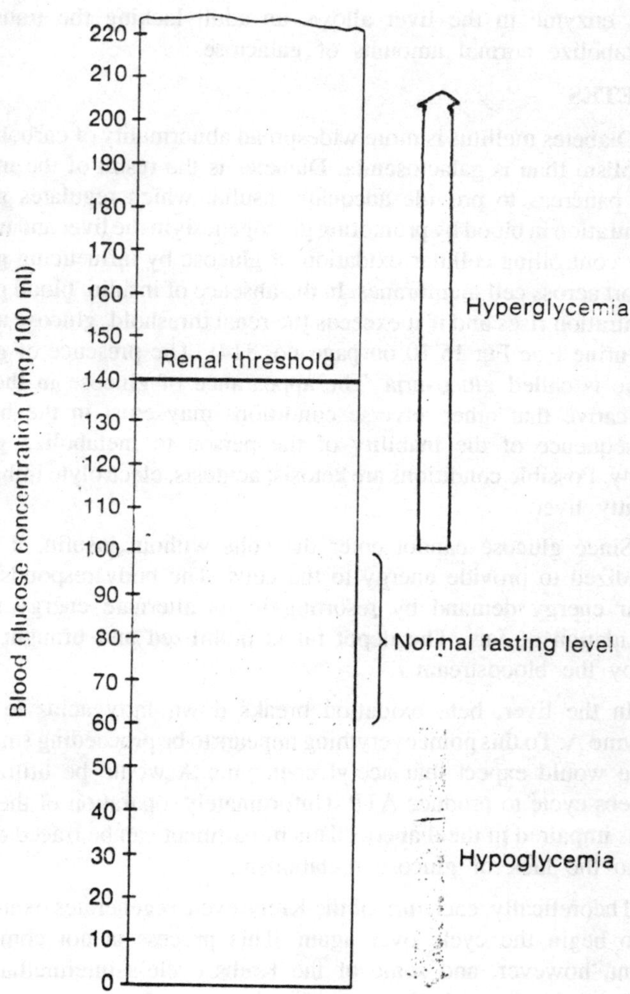

Fig. 15.10 : Summary of blood glucose levels and normal responses. During hypoglycemia, glycogenolysis and glycogenesis occur to increase glucose concentration. When glucose concentration is in the normal range, glycogenesis and glycogenolysis occur to maintain concentration. During hyperglycemia, glycogenesis and lipid synthesis occur to decrease glucose concentration. When glucose concentration exceeds the renal threshold, glucose is removed from the blood by the kidneys and appears in the urine.

to produce L-Malic acid:

$$\underset{\text{pyruvic acid}}{\begin{array}{c} CO_2H \\ | \\ C{=}O \\ | \\ CH_3 \end{array}} + CO_2 \xrightarrow{NADPH_2 \quad NADP} \underset{\text{L-malic acid}}{\begin{array}{c} CO_2H \\ | \\ HO{-}C{-}O \\ | \\ CH_2 \\ | \\ CO_2H \end{array}}$$

The L-malic acid is then oxidized to oxaloacetic acid by NAD.

The other major source of oxaloacetic acid is the carboxylation of phosphoenolpyruvic acid:

$$\underset{\text{phosphoenolpyruvic acid}}{\begin{array}{c} CO_2H \quad\quad\quad O \\ | \quad\quad\quad\quad \| \\ C{-}O{-}P{-}OH \\ \| \quad\quad\quad\quad | \\ CH_2 \quad\quad\quad OH \end{array}} + CO_2 \longrightarrow \underset{\text{oxaloacetic acid}}{\begin{array}{c} CO_2H \\ | \\ C{=}O \\ | \\ CH_2 \\ | \\ CO_2H \end{array}} + P_i$$

Notice that the source of both pyruvic acid and phosphoenolpyruvic acid is glycolysis (see Fig 15.9). In the diabetic, glycolysis is impaired, limiting the amount of pyruvic acid and phosphoenolpyruvic acid available to produce oxaloacetic acid. The resultant decreased amount of oxaloacetic acid leads to a slowdown in the rate of operation of the Krebs cycle. The combination of a decreased rate of acetyl coenzyme A utilization in the Krebs cycle and the increased rate of acetyl coenzyme A formation by beta oxidation of fatty acids in the liver leads to a buildup of acetyl coenzyme A concentration in the body.

In a normal individual, an increased acetyl coenzyme A concentration is a single for the liver to increase its rate of fatty acid synthesis and simultaneously to decrease the rate of fatty acid breakdown. In the diabetic, however, the energy demands of the body are increasing the rate of fatty acid breakdown. Consequently, there is a tendency in the diabetic for the buildup of acetyl coenzyme A to continue.

The Formation of Ketone Bodies in Diabetes

In an effort to cope with the increased amounts of acetyl coenzyme A, the liver converts acetyl coenzyme A to the so-called *ketone bodies,* which are released into the bloodstream and then eliminated via urine. The ketone bodies are acetoacetic acid, β-hydroxybutyric acid, an acetone.

$$\underset{\text{acetoacetic acid}}{CH_3\overset{O}{\overset{\|}{C}}CH_2CO_2H} \qquad \underset{\text{β-hydroxybutyric acid}}{CH_3\underset{OH}{\underset{|}{C}}HCH_2CO_2H} \qquad \underset{\text{acetone}}{CH_3\overset{O}{\overset{\|}{C}}CH_3}$$

Notice that one of these ketone bodies is not a ketone: β-hydroxybutyric acid is included among the ketone bodies since it arises from reduction of acetoacetic acid. In essence, acetoacetic acid arises from the coupling of two acetyl coenzyme A molecules:

$$\underset{\text{acetyl coenzyme A}}{2CH_3{-}C({=}C){-}S{-}CoA} \rightleftharpoons \underset{\text{acetoacetyl coenzyme A}}{CH_3{-}\overset{O}{\overset{\|}{C}}{-}CH_2{-}C({=}C){-}S{-}CoA} \quad +CoASH$$

The acetoacetyl coenzyme A undergoes hydrolysis to acetoacetic acid and coenzyme A:

$$\underset{\text{acetoacetyl coenzyme A}}{CH_3{-}\overset{O}{\overset{\|}{C}}{-}CH{-}C({=}O){-}SCoA} + H_2O \rightarrow \underset{\text{acetoacetic acid}}{CH_3{-}\overset{O}{\overset{\|}{C}}{-}CH_2{-}C({=}O){-}OH} \quad +CoASH$$

Acetoacetic acid readily undergoes two reactions in the body. One of these is reduction $NADH_2$:

The other reaction is decarboxylation to acetone:

$$CH_3\overset{O}{\overset{\|}{C}}CH_2C({=}O)SCoA \rightarrow CH_3\overset{O}{\overset{\|}{C}}CH_3 + CO_2$$

The formation and release of acetoacetic acid in small amounts is a normal continuing process. In a healthy individual the total amounts of ketone bodies is less than 1 mg/100mL of blood. This is because the ketone bodies can be degraded back to acetyl coenzyme A in muscle tissue and oxidize by the Krebs cycle. In the diabetic, because of the increased rate of acetoacetic acid production and the slowdown of the Krebs cycle, ketone body concentration rises (see Fig. 15.11 on page no. 318).

If the concentration of ketone bodies in the blood exceeds 1 mg/100mL, the condition of ketonemia is said to exist. The average daily excretion via urine of ketone bodies in the normal individual is 20 mg. Greater than normal amounts of ketone bodies in the urine is called *ketonuria.* The combination of ketonemia and ketonuria is called *ketosis.* A diabetic individual with severe ketosis may exhibit a blood concentration of ketone bodies as high as 90 mg/100mL and excrete as much as 5000 mg of ketone bodies per hour!

Diabetic Acidosis and Electrolyte Imbalance

Two of the ketone bodies are carboxylic acids. As the concentration of these acids in the blood increases, the pH of blood starts to decrease. A continued decrease of the blood pH could lead to *acidosis.* To prevent acidosis, the blood buffers react with the acids to neutralized them. Recall that one of the blood buffers is the H_2CO_3/HCO_3^-. buffer system. As acetoacetic acid and β-hydroxybutyric acid enter the bloodstream, they react with bicarbonate ions to produce H_2CO_3, which decomposes to water and carbon dioxide. The CO_2 is exhaled by the lungs.

Continued release of acetoacetic and β-hydroxybutyric acids into the blood begins to deplete the bicarbonate ion concentration. Depletion of HCO_3^- ion concentration reduces the ability of the blood to neutralize the acids and eventually, as more and more acid is released into the bloodstream, acidosis results. This in turn leads to still further complications. As the pH of the blood goes down, the ability of hemoglobin to transport oxygen decreases and an oxygen deficiency develop. Breathing may become rapid, or even difficult and painful.

As the plasma HCO_3^- concentration continues to fall, chloride ions from plasma enter the red blood cells in exchange for HCO_3^-. This makes more HCO_3^- available in plasma to neutralize the acids and to aid to the effort to return the plasma pH to normal levels. Unfortunately, this chloride/ bicarbonate shift is not sufficient to restore plasma pH and instead serves to initiate electrolyte imbalance. Electrolyte imbalance is further enhanced by kidney action. To counter act acidosis, the kidneys produce a more acidic urine. Even in the most acidic

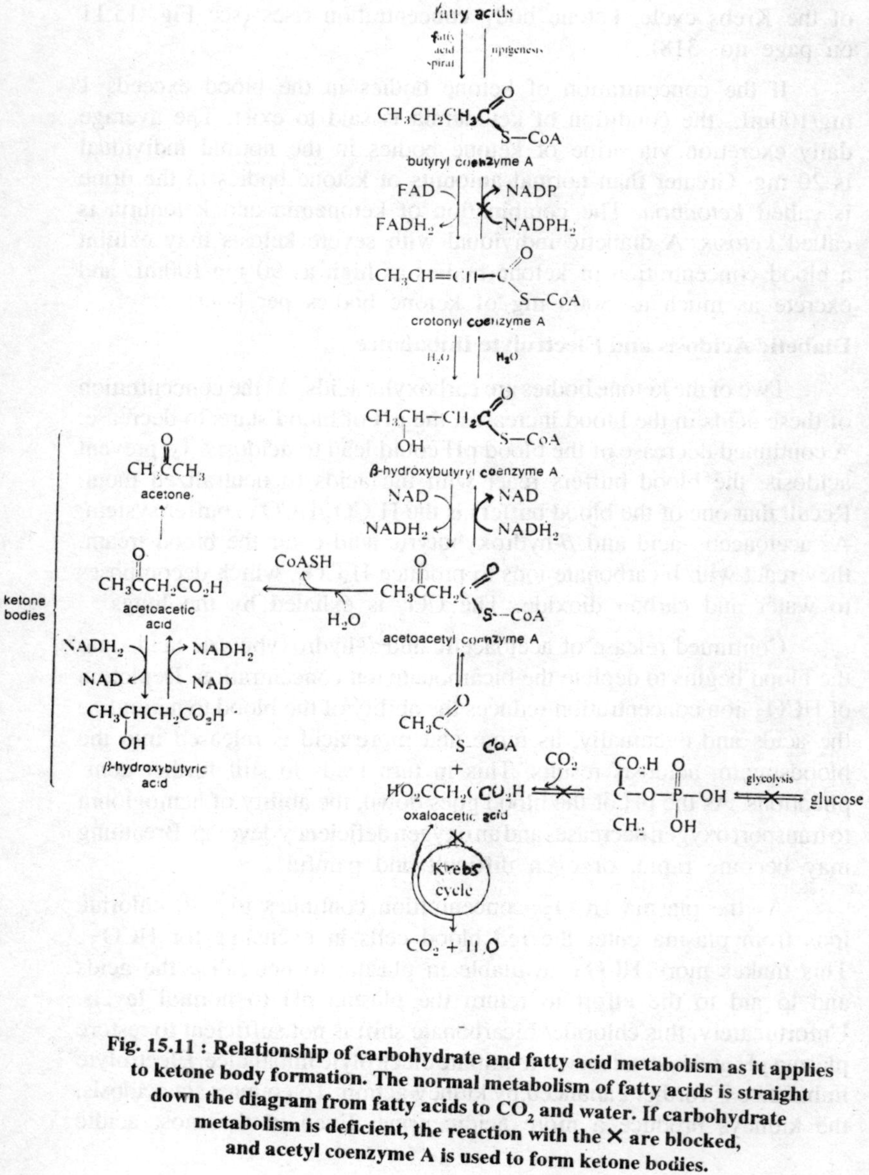

fatty acids

fatty acid spiral | lipigenesis

$CH_3CH_2CH_2C(=O)S-CoA$

butyryl coenzyme A

FAD → $FADH_2$; NADP ✕ $NADPH_2$

$CH_3CH=CH-C(=O)S-CoA$

crotonyl coenzyme A

H_2O | H_2O

$CH_3CH(OH)-CH_2C(=O)S-CoA$

β-hydroxybutyryl coenzyme A

NAD → $NADH_2$; NAD ← $NADH_2$

$CH_3CCH_2C(=O)S-CoA$ (with C=O)

acetoacetyl coenzyme A

CoASH ; H_2O

ketone bodies:

CH_3CCH_3 (with C=O)

acetone

$CH_3CCH_2CO_2H$ (with C=O)

acetoacetic acid

$NADH_2$ → NAD ; $NADH_2$ ← NAD

$CH_3CHCH_2CO_2H$ (with OH)

β-hydroxybutyric acid

$CH_3C(=O)S-CoA$

\+

$HO_2CCH_2CCO_2H$ ✕ CO_2 ; $CO_2H-C(CH_2)-O-P(=O)(OH)-OH$ ✕ glycolysis glucose

oxaloacetic acid

✕ Krebs cycle

$CO_2 + H_2O$

Fig. 15.11 : Relationship of carbohydrate and fatty acid metabolism as it applies to ketone body formation. The normal metabolism of fatty acids is straight down the diagram from fatty acids to CO_2 and water. If carbohydrate metabolism is deficient, the reaction with the ✕ are blocked, and acetyl coenzyme A is used to form ketone bodies.

urine the kidneys are capable of producing, however, acetoacetic acid and β-hydroxybutyric acid exist as anions. In order to maintain electrical neutrality of the urine, cations must also be excreted. These are primarily sodium ions and as they are excreted in urine, their concentration in plasma is diminished.

The situation in the plasma has now reached the point where HCO_3^-, Cl^- and Na^+ concentrations have decreased because of the release of ketone bodies into blood. The decreased concentration of ions in the plasma, if uncompensated, would lead to decreased osmotic pressure of the plasma. To prevent this and to dissolve the greater quantities of materials being excreted, the kidneys increase the urine volume. Thus, these factors further complicate acidosis by inducing a state of dehydration. Hence, metabolic ketosis eventually causes acidosis and dehydration that ultimately may lead to coma and finally death.

The aims of clinical treatment of diabetic acidosis become obvious from the discussion above. In the short-term treatment, acidosis, sodium loss, and water loss must be controlled. This can be accomplished by intravenous injection of an isotonic solution of sodium chloride to replace needed water, sodium ions, and chloride ions. In severe cases, an isotonic solution of sodium bicarbonate or a mixture of NaCl and $NaHCO_3$ maybe given intravenously. The addition of bicarbonate serves to give an immediate increase in the blood pH. In less severe cases, sodium lactate or the sodium salt of a Krebs cycle intermediate may be included in the intravenous solution. These materials allow the rate of the Krebs cycle operation to be increased. As the Krebs cycle operation increases, the normal concentration of HCO_3^- is restored by oxidation of acetyl coenzyme A and blood pH is controlled by natural means.

For the long term, treatment must establish normal carbohydrate metabolism to prevent a reoccurring accumulation of ketone bodies. For a diabetic individual, normal carbohydrate metabolism can usually be established by a combination of diet control and repeated insulin injections.

16

Amides, Amino Acids and Proteins

This chapter includes the reactions and properties of compounds such as amides, amino acids, and proteins.

AMIDES

Ammonia behaves as a base by accepting a hydrogen ion. It is able to do this because it has a pair of unshared electrons. Amides are related to the products of the reaction between ammonia and a carboxylic acid: Ammonia, acting as a base, can react with an acid to produce an ammonium salt.

$$H_3N: + HO{-}\overset{O}{\overset{\|}{C}}{-}R \longrightarrow \underset{\text{ammonium ion}}{[NH_4]^+} + \underset{\text{anion of the carboxylic acid}}{\left[HO{-}\overset{O}{\overset{\|}{C}}{-}R\right]^-}$$

At elevated temperatures, the salt decomposes by the elimination of water to form an amide:

$$(NH_4)^+ \left[O{-}\overset{O}{\overset{\|}{C}}{-}R\right]^- \xrightarrow{\Delta} \underset{\text{amide}}{R{-}\overset{O}{\overset{\|}{C}}{-}NH_2} + HOH$$

Thus, the *net effect* is that H from ammonia and —OH from the acid are split out as water and the remaining fragments combine to form an amide:

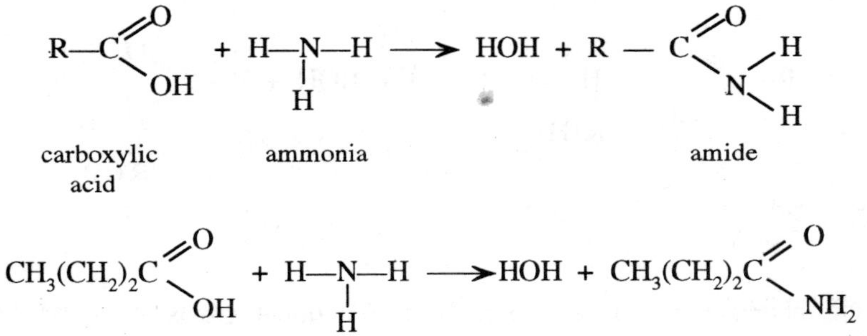

$$R{-}C(=O){-}OH + H{-}\underset{\underset{H}{|}}{N}{-}H \longrightarrow HOH + R{-}C(=O){-}N(H)(H)$$

carboxylic acid, ammonia, amide

$$CH_3(CH_2)_2C(=O){-}OH + H{-}\underset{\underset{H}{|}}{N}{-}H \longrightarrow HOH + CH_3(CH_2)_2C(=O){-}NH_2$$

AMINES

Amines are compounds related to ammonia by having one or more alkyl groups attached directly to the nitrogen atom.

Amines can be classified as primary, secondary, or tertiary. Primary and secondary amines have a hydrogen bonded directly to the nitrogen atom; tertiary amines do not (Table 16.1).

Table 16.1 : Classifications of amines

Primary Amines $\left(R-\underset{\underset{H}{\vert}}{N}-H\right)$	Secondary Amines $\left(R-\underset{\underset{R}{\vert}}{N}-H\right)$	Tertiary Amines $\left(R-\underset{\underset{R}{\vert}}{N}-R\right)$
$CH_3CH_2{-}\underset{\underset{H}{\vert}}{N}{-}H$	$CH_3{-}\underset{\underset{H}{\vert}}{N}{-}CH_3$	$CH_3{-}\underset{\underset{CH_3}{\vert}}{N}{-}CH_3$
$H{-}\underset{\underset{CH_2)_3CH_3}{\vert}}{N}{-}H$	$CH_3{-}\underset{\underset{H}{\vert}}{N}{-}$ —CH_3	$CH_3CH_2{-}\underset{\underset{CH_2}{\vert}}{N}{-}CH_3$
$CH_2{-}NH_2$	$CH_3{-}\underset{\underset{CH_2CH(CH_3)_2}{\vert}}{N}{-}H$	$-CH_3CH_2{-}\underset{\underset{CH_2CH(CH_3)_2}{\vert}}{N}{-}CH_2)_2-$

Amines chemically behave similarly to ammonia. To form an amide, an amine must have a hydrogen directly bonded to its nitrogen atom. Thus, primary amines and secondary amines as well as ammonia meet this requirement and form amides when reacted with carboxylic acids.

$$\underset{\text{carboxylic acid}}{R{-}C(=O){-}OH} + \underset{\text{amine}}{H{-}N(R'(H)){-}R'} \xrightarrow{\Delta} HOH + \underset{\text{amide}}{R{-}C(=O){-}N(R'(H)){-}R'}$$

The characteristic of an amide is its functional group:

$$-\overset{\displaystyle O}{\overset{\|}{C}}-\underset{\displaystyle R(H)}{\underset{|}{N}}-R(H)$$

Esters and amides are derivatives of carboxylic acids and a comparison of their formation points out this similarity:

$$\underset{\text{acid}}{R{-}C(=O){-}OH} + \underset{\text{alcohol}}{H{-}O{-}R'} \rightleftharpoons HOH + \underset{\text{ester}}{R{-}C(=O){-}O{-}R'}$$

$$\underset{\text{acid}}{R{-}C(=O){-}OH} + \underset{\text{primary or secondary amine}}{H{-}N(R'(H))(R'(H))} \rightleftharpoons HOH + \underset{\text{amide}}{R{-}C(=O){-}N(R'(H))(R'(H))}$$

Tertiary amines do not have a hydrogen directly bonded to the amine nitrogen and, consequently, do not form amides.

AMINO ACIDS AND PROTEINS

What do skin, fingernails, hair, milk, and egg whites have in common? They all contain protein, complex aggregates of amino acids.

Amino acids are important cellular components containing carboxylic acid and amine groups in the same molecule. The most important biochemical function amino acids perform is their reaction with each other to form proteins, materials present in all living cells. The key reaction in the buildup of a protein by the combination of amino acids is the formation of an amide linkage between amino acid units. This linkage is called a *peptide bond.* Peptide bond formation results from the reaction of an amine group on one amino acid molecule with the acid group on another amino acid molecule.

The combination of two amino acids joined by a peptide bond is called a *dipeptide.* Using glycine and alanine, let's write reactions for dipeptide formation between these two amino acids. Being amino acids, they both contain acid and amine groups. This means that two possibilities arise:

1. Glycine can act as the acid and alanine as the amine.

$$H_2N-CH_2-C(=O)OH + H-NH-CH(CH_3)-C(=O)OH \rightarrow HOH + H_2N-CH_2-C(=O)-NH-CH(CH_3)-C(=O)OH$$

peptide bond

(glycine) amino acid 1   (alanine) amino acid 2   glycylalanine, a dipeptide

2. Alanine can act as the acid and glycine as the amine.

$$H_2N-CH(CH_3)-C(=O)OH + H_2N-CH_2-C(=O)OH \rightarrow HOH + H_2N-CH(CH_3)-C(=O)-NH-CH_2-C(=O)OH$$

peptide bond

alanine   glycine   alanylglycine, a dipeptide

Apparently it appears that the two dipeptides are the same. Note, however, that in glycylalanine the peptide bond nitrogen is bonded directly to a carbon bearing a hydrogen and a methyl group; in alanylglycine this nitrogen is bonded directly to a carbon bearing two hydrogens. Therefore, these two compounds are structural isomers.

Since the dipeptide still contains uncombined amine and carboxyl groups, the process of peptide bond formation can continue. R represents the particular side chain of each alpha amino acid.

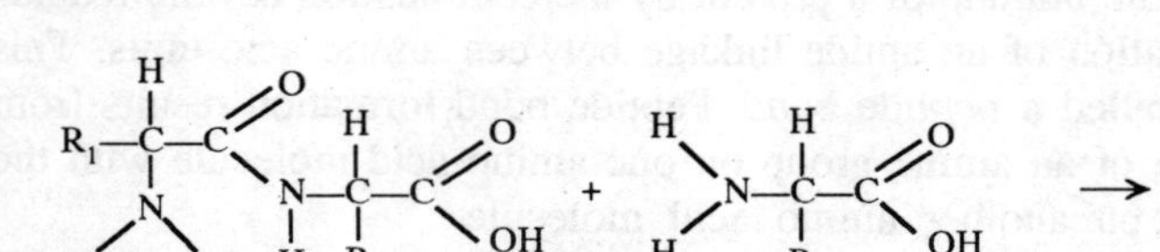

a dipeptide (amino acids 1 and 2) + amino acid 3 →

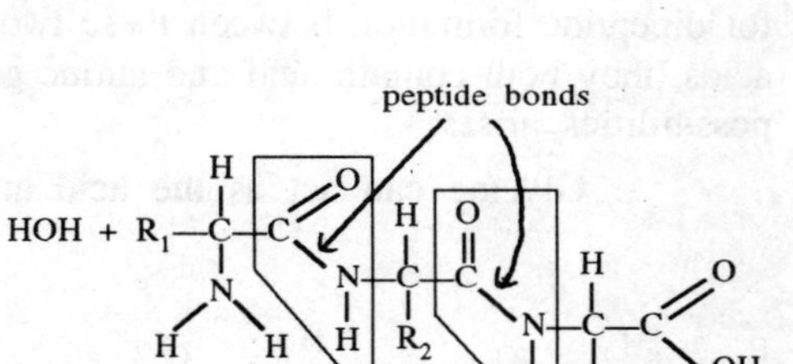

HOH + a tripeptide (peptide bonds)

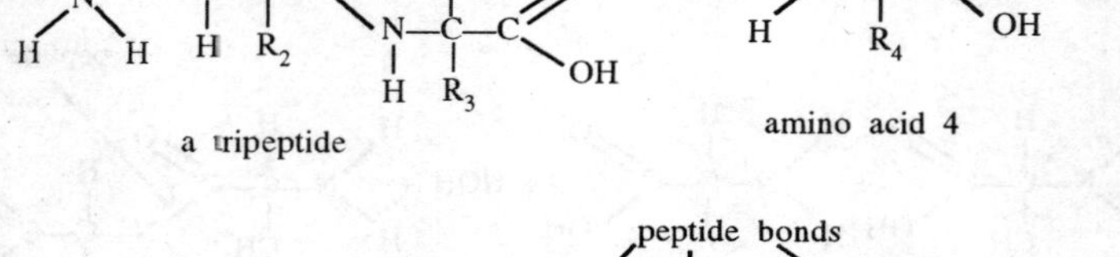

a tripeptide + amino acid 4 →

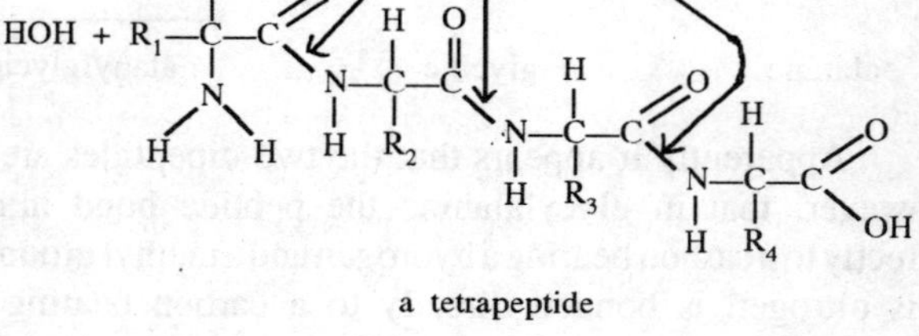

HOH + a tetrapeptide (peptide bonds)

Peptide bond formation can continue by the joining of additional amino acids to the lengthening peptide. Pentapeptides (five amino acids), hexapeptides (six amino acids), etc., can be formed. Combinations of more than 10 amino acids are called *polypeptides.* Large numbers of amino acid units can be incorporated in this way, eventually culminating in the formation of a protein:

N-terminal end

$$H_2N-\underset{R}{\overset{H}{C}}-\overset{O}{\overset{\|}{C}}-\left[\underset{H}{N}-\underset{R}{\overset{H}{C}}-\overset{O}{\overset{\|}{C}}\right]_x-\underset{H}{N}-\underset{R}{\overset{H}{C}}-\overset{O}{\overset{\|}{C}}-OH$$

C-terminal end

X is a large number, usually greater than 50. Thus, proteins have molecular weights ranging from approximately 6000 amu to over a million amu. It should also be noted that a protein or polypeptide contains an uncombined amine group at one end of the molecule, the so-called N-terminal end, and an uncombined carboxyl group at the other, the C-terminal end. By convention, peptides are named starting from the N-terminal amino acid unit and proceeding by naming all component amino acids, ending with the C-terminal amino acid, *e.g.*, glycylalanine and alanylglycine in the examples above.

Previously we saw that two different amino acids could combine to yield two different dipeptides. How many tripeptides could three different amino acids form? Let's play "molecular scrabble" and consider the possible combinations using letters to represent the amino acids glycine (G), alanine (A), and valine (V); six combinations are possible:

G—A—V A—V—G V—G—A

G—V—A A—G—V V—A—G

Each combination represents a different tripeptide.

As the number of different component amino acids increases, the number of possible combinations increases tremendously. For example, 20 different amino acids give 2.4×10^{18} theoretically possible combinations using each amino acid only once. Not all possible combinations are used or are present in a given species.

PRIMARY STRUCTURAL FEATURES OF PROTEINS

The peptide bond is the major structural factor in animal tissue. *The sequence of amino acids constitutes the primary structure of a protein or polypeptide.* Sickle cell anemia is an example of what may happen if improper sequencing of a protein's-amino acids occurs. The protein globin found in human hemoglobin consists of two pairs of protein chains. In normal individuals, one of the chains has the primary structure: valine—histidine—leucine—threonine—proline—glutamic acid—glutamic acid— for its first seven amino acids. In an individual suffering from sickle cell anemia, the seventh amino acid is valine instead of glutamic acid. When combined with oxygen, both normal and sickle cell hemoglobin are soluble in plasma. As the oxygenated blood passes form arteries into capillaries, oxygen in released to the surrounding tissues. When this occurs, the solubility of sickle cell hemoglobin is sharply reduced and, within the red blood sickle cell, hemoglobin crystallizes out of solution. This causes the red blood cells to become misshapen. The misshapen cells block capillaries and small blood vessels and reduce normal blood flow. The difference in solubility between normal and sickle cell hemoglobin has been traced to the difference in amino acid sequence. Glutamic acid's side chain is $—CH_2—CH_2—CO_2H$. While that of valine is $—CH—(CH_3)_2$. The presence of the carboxylic acid side-chain group in glutamic acid makes the group much more water (plasma) soluble than valine's hydrocarbon-like side chain. Carboxylic acid groups can hydrogen bond with water; hydrocarbon groups cannot. Thus, here on the molecular level is the basis for a disease brought on by changing only 1 out of 146 amino acid units. This is an outstanding example of how chemical structure molecularly determines biological function.

SECONDARY AND TERTIARY STRUCTURAL FEATURES

The primary structure of all proteins is dictated by number, kind, and order of their component amino acids. Many proteins can be isolated as crystalline solids. Observations like this led to the concept that additional factors are categorized as contributing to the secondary and tertiary features of a protein's structure. Hydrogen bonding between peptide bond groupings gives secondary structural features to protein molecules. Factors contributing to tertiary features include disulfide bonds, salt bridges and hydrogen bonds between amino acid side-chain groups, and hydrophobic interactions. *Hydrophobic* means water hating.

SECONDARY STRUCTURE AND HYDROGEN BONDING.

The protein or polypeptide chain is primarily linked together by peptide linkages. The

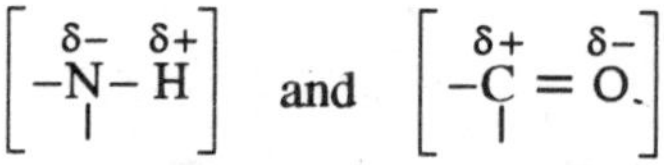

$$\left[\overset{\delta-}{-\underset{|}{N}}-\overset{\delta+}{H}\right] \quad \text{and} \quad \left[-\overset{\delta+}{\underset{|}{C}}=\overset{\delta-}{O}\right]$$

bonds of the peptide linkages are highly polar because of electronegativity differences between the bonded atoms. This creates a condition favorable for hydrogen bond formation. Linus Pauling an R.B. Corey proposed two possible ways by which hydrogen bonding could stabilize the folding of a protein in a particular way. One way is the pleated-sheet configuration, a network of adjacent polypeptide chains held together in a nearly planer arrangement by *intermolecular* hydrogen bonding. The hydrogen bonding occurs between atoms already involved in peptide bonds. Figure 16.1(a) and (b) (see page no. 328 and 329) shows this arrangement. Silk and muscles are natural proteins whose structural properties seem to be related to this type of arrangement.

The more commonly occurring configuration is the alpha helix arrangement in which hydrogen bonds exist intramolecularly between $-\underset{|}{N}-H$ groups and carbonyl-oxygen groups of the peptide linkages down the chain.

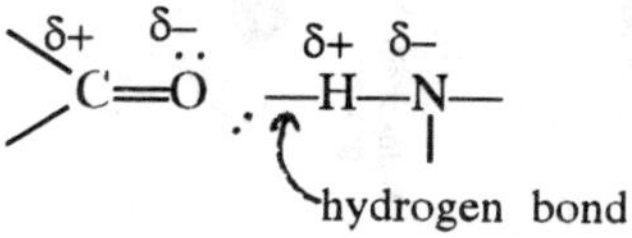

$$\rangle\overset{\delta+}{C}=\overset{\delta-}{\ddot{O}} \; \cdots \; -\overset{\delta+}{H}-\overset{\delta-}{\underset{|}{N}}-$$

hydrogen bond

Secondary structure of proteins results from hydrogen bonding between amide groups.

The curliness of hair and the flexibility of skin are attributed to the secondary structure of proteins.

This intramolecular hydrogen bonding causes the chain to coil into a helix (see Fig. 16.1(c) on page no. 330) because of the sequence of amino acids present in a protein, the primary structure dictates the coiling of the protein.

TERTIARY STRUCTURAL FEATURES

Tertiary structure refers to folding of a polypeptide chain into a unique three dimensional structure stabilized by covalent forces (disulphide

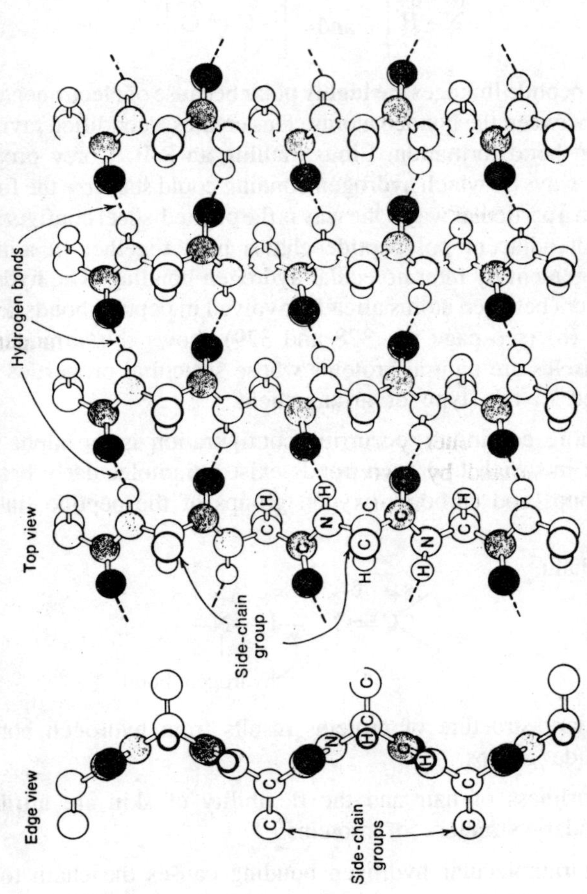

Figure 16.1(a) Pleated-sheet structure of extended polypeptide chains. Groups indicated by heavy lines are meant to project forward; those by light lines, backward. Dashed lines indicate hydrogen bonds that hold parallel polypeptide chains to each other and stabilize the configuration.

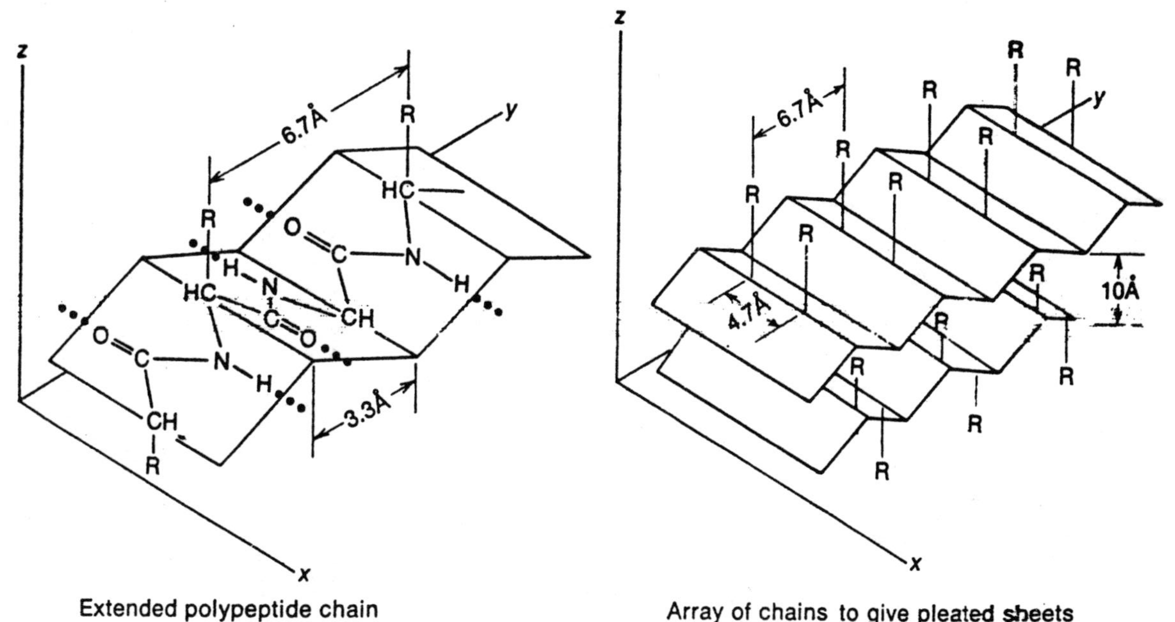

Figure 16.1(b) Schematic diagram of the pleated sheet model of β-keratin. The peptide bonds lie in the plane of the pleated sheet; the side chains lie above or below the sheet and alternate along the chain. The polypeptide chains are held together by hydrogen bonds.

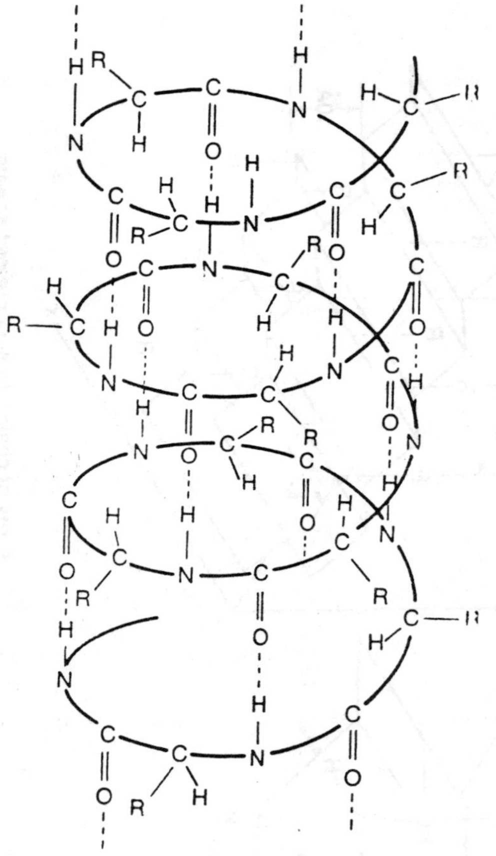

Fig 16.1(c). Secondary structure—the peptide chain twists into a helix, in which the C=O from one turn forms a hydrogen bond with the N-H of the turn below.

bonds), ionic forces (slat bridges), hydrogen bonds, and London interactions.

Cysteine is an amino acid containing a thiol group (—S—H) that is not involved in peptide bond formation. The presence of cysteine units in a protein provides the possibility for disulfide bond formation. Under mild oxidizing conditions, hydrogen is lost form the thiol and combination of sulfur atoms between cysteine units occurs. In this way, disulfide bond formation concrete cross-linking between protein chains or within a protein helix (see Fig. 16.2). Note from Fig 16.2 that disulphide bond formation is reversible.

Insulin is a protein containing several disulfide linkages between two polypeptide chains.

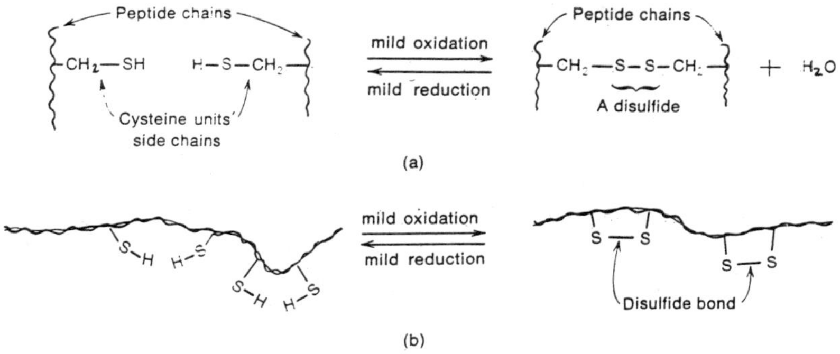

Peptide chains
$-CH_2-SH$ $H-S-CH_2-$
Cysteine units' side chains
mild oxidation
mild reduction
Peptide chains
$-CH_2-S-S-CH_2-$ + H_2O
A disulfide

(a)

mild oxidation
mild reduction
S—H H—S
S—H H—S
S — S
S — S
Disulfide bond

(b)

Fig. 16.2 : Disulfide bond formation. (a) Intermolecular, (b) Intramolecular.

The side chains of the acidic and basic amino acids in a protein contain free carboxyl and amine groups. These can undergo Bronsted—Lowry acid—base interactions that produce sites of opposite charges that attract each other. These attractions are termed salt bridges (Salt bridges result from the neutralization of the acid groups in the side chains of the acidic amino acids by the amine groups of basic amino acids making up the protein) and help further stabilize the molecule's tertiary structure (See Fig 16.3).

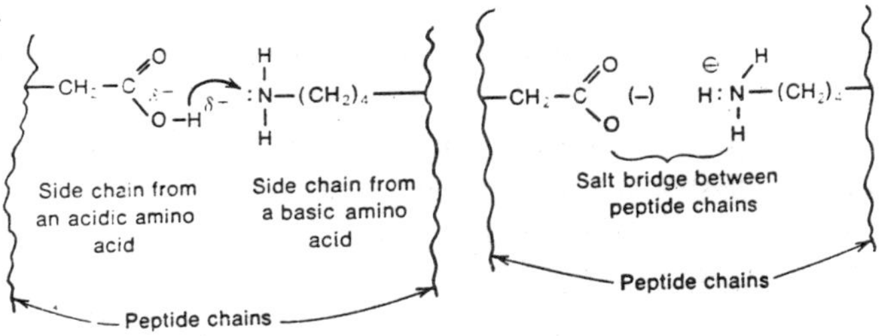

Side chain from an acidic amino acid
Side chain from a basic amino acid
Peptide chains
Salt bridge between peptide chains
Peptide chains

Fig. 16.3 : Salt bridge formation.

Several amino acids have side chains capable of hydrogen bonding with side chains of other amino acids in a polypeptide or protein.

Several examples are given in Fig 16.4. Note that this type of hydrogen bonding involves side chains, not peptide bond sites.

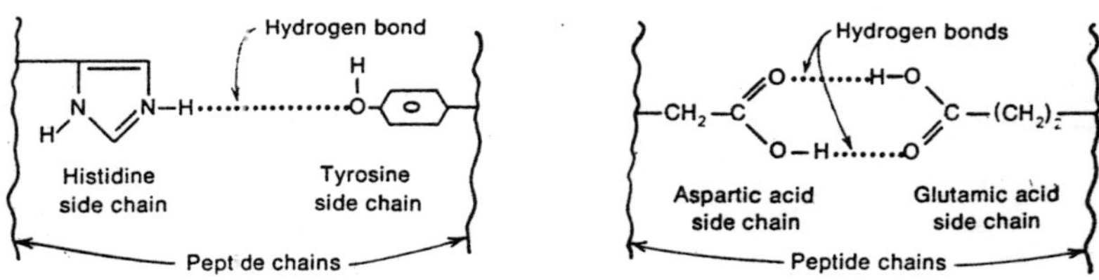

Fig. 16.4 : Hydrogen bonding between amino acid side chains.

Cellular activity occurs in an aqueous environment. Many proteins contain amino acids with hydrocarbon-like side chains. These side chains cannot hydrogen bond to water and thus do not promote the protein's water solubility. Instead, the protein molecule folds so that the alkane-like hydrophobic side chains are folded inward. This minimizes hydrophobic group contact with water while exposing more polar side chains for interaction with polar water molecules.

pH and Protein Solubility

Protein solubility is pH dependent. An example of this is the souring of milk. Bacteria in the milk produce lactic acid as a metabolic waste product. As lactic acid concentration increases, acidity increases and the pH of milk falls. When the pH reaches 4.6, casein (the major protein in milk) precipitates and forms a curd.

Proteins are least soluble at the *isoelectric point, the pH at which the protein molecule has no* **net** *charge.* Oppositely charged groups in a protein molecule cause the molecule to be attracted to neighboring molecules bearing opposite charges. This attraction is strong enough to cause the molecules to coagulate and precipitate from solution. At pH values below its isoelectric point, a protein molecule has a *net* positive charge; pH values greater than the isoelectric point cause protein molecules to have a *net* negative charge. In both cases, similarly charged molecules will repel each other, thus remaining in solution as a colloidal suspension.

Proteins and Buffer Action

The presence of acidic and basic amino acids in a protein provides side chains that can exhibit acid and base behavior toward added reagents. Depending on the pH, the carboxylic acid and amine groups in side chains of component amino acids may be in various forms. Generally, the amine group is in the form

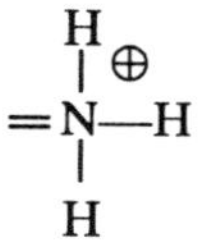

$$=\overset{\overset{\displaystyle H}{|}\;\oplus}{\underset{\underset{\displaystyle H}{|}}{N}}-H$$

The carboxylic acid is in the form

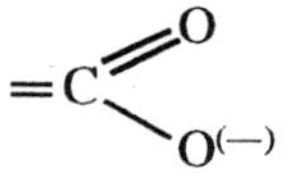

$$=C\begin{matrix}\nearrow O \\ \searrow O^{(-)}\end{matrix}$$

Thus, proteins can function as buffers (see Fig. 16.5). Blood proteins are the main buffers that help to maintain blood pH between 7.3 and 7.5.

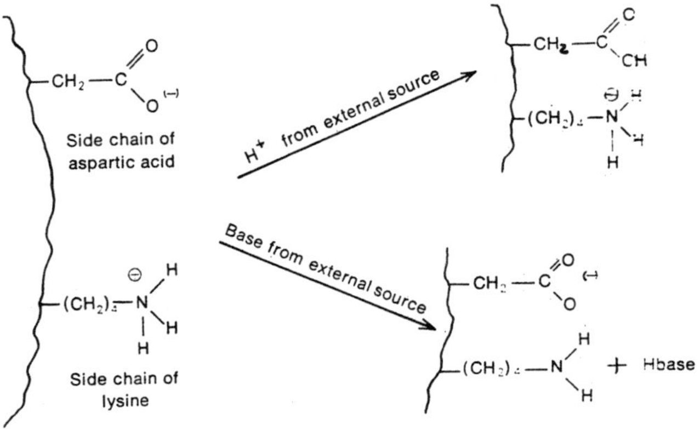

Fig. 16.5 : Protein buffering action.

DENATURATION

The structure of a protein, particularly its secondary and tertiary features, can be changed by a number of conditions. Changes in structure that cause a change in biological activity or function are called *denaturation*. This term applies to alterations of secondary and tertiary structural features. Denaturation can be reversible or irreversible depending on the extent and type of structural change. Denaturation is not usually used to describe changes in a protein's primary structure.

Common experiences give evidence that heat, sunlight, and / or chemical reagents can denature proteins. Egg whites contain the protein albumin. Frying an egg causes denaturation of the protein and converts the white form a liquid to a solid. Anyone who has suffered a painful sunburn can attest to the action of ultraviolet radiation on protein (skin). A 70 percent ethanol solution serves as a disinfectant because of its denaturing action of proteins in bacteria. Ethanol and various other polar organic solvents are capable of forming hydrogen bonds to proteins, thus interrupting the protein's normal hydrogen bonding pattern.

The (—S—H)- containing amino acid cysteine is a component of many proteins. Its ability to form disulfide bonds has been described. Unshared electron pairs on sulphur atoms in disulphide linkages make these atoms excellent electron pair donors for coordinate covalent bond formation. The carboxylate groups

$$—C\begin{matrix}\nearrow O \\ \searrow O^{(-)}\end{matrix}$$

of glutamic and aspartic acids also can form coordinate covalent bonds. Hg^{2+}, Ag^{+}, and Pb^{2+} ions (so -called heavy metal ions) form very strong, coordinate covalent bonds with these groups in a protein. This denatures the protein. Compounds containing these ions, if taken internally, act as poisons by combining with sulfur atom in disulfide linkages and carboxylate groups. If accidentally swallowed, these heavy metal ions can be kept from general circulation by coagulation in the stomach. Raw egg white or milk are given as antidotes for heavy metal poisoning because they contain proteins that readily combine with the heavy metal ions to form insoluble solids. An emetic is then given that removes the metal ion-containing precipitate before digestive juice action can destroy the protein and liberate the harmful ions.

Curiously, mercurials and other heavy metal compounds were among the earliest used antiseptics. In low concentration, their compounds are still used. Eyes of newborn babies are sometimes treated with a silver nitrate solution to prevent gonorrhea infection. Mercurochrome and the salicylic acid derivative merthiolate are topical, mercury-containing disinfectants.

Tannic acid and picric acid also denature proteins. They are common components of medication used in burn treatment. A physiological danger associated with a burn injury is loss of body fluids and attack by infectious bacteria at the burned region. Upon denaturation, proteins coagulate and form a protective, impervious scan over the burned area.

ENZYME ACTIVITY

Enzyme consist of a large protein part, the apoenzyme portion, and a smaller portion, the cofactor. The apoenzyme is nonactive catalytically unless combined with the cofactor. The cofactor usually is the active site in the enzyme. Other regions of the enzyme molecule act as binding sites to hold the substrate in the most favorable position for reaction to occur. This implies that a definite structural arrangement of the enzyme must be present for proper catalytic action.

In general, increasing the temperature causes an increase in reaction rate. Up to a point this is also true for enzymatic reactions.

As temperature increases, a point is reached at which the reaction proceeds at a maximum rate. This is called the optimum temperature. Most of our enzymes have an optimum temperature of 37°C.

Beyond the optimum temperature, an increase in temperature actually lowers the reaction rate. This might seem like strange behavior unless we remember that enzymes contain proteins that are denatured by heat. Heating alters the secondary and tertiary structural features of the enzyme. This causes a change in the orientation of the active site in relation to the position of the substrate. The change in orientation prevents the enzyme from catalyzing the reaction and thus reaction rate decreases.

Effect of pH of enzyme Activity

Experimentally it is found that an enzyme exhibits maximum activity over the narrow pH range in which its molecules possess the proper net charge for substrate binding.

Most body fluid enzymes have optimum pH values between 6 and 8. Gastric juice enzymes are notable exceptions. Pepsin is most active at pH 1.5—2.5 and rapidly loses its catalytic behavior above pH 6.9.

PROTEIN HYDROLYSIS

Amides will undergo hydrolysis. This means that peptide bonds in proteins can be hydrolyzed and the primary structure destroyed. The component amino acids are liberated if acidic or enzymatic hydrolysis is used. During digestion, both acidic and enzymatic hydrolysis occur and the water-soluble amino acids are absorbed into the general circulation for transport to metabolic sites. A molecule of water is required for the hydrolysis of each peptide bond in a polypeptide or protein.

17

Protein Metabolism

INTRODUCTION

Generally speaking, the use and treatment of proteins by the body is considerably different than that of carbohydrates and lipids. Proteins have very specific structures serving specific purposes. In addition to the insoluble proteins of epithelial, muscle, connective, glandular, and brain tissues, there are the soluble protein enzymes, plasma proteins, hormones, hemoglobin, antibodies, and nucleoproteins. In contrast to lipids and carbohydrates, the body possesses no storage depots for proteins nor are proteins used as a major energy source.

Excluding water, proteins are the main material of animal tissue and constitute about 75 percent of the dry weight of mammalian tissue. Table 17.1 gives the approximate percent composition of selected mammalian tissue.

Table 17.1 : Percent composition of selected mammalian tissues

Component	*Muscle*	*Whole Blood*	*Liver*	*Brain*	*Skin*
Water	72–78	79	60–80	78	66
Protein	18–20	19	15	8	25
Lipid	3	1	3–20	12–15	7
Carbohydrate	0.6	0.1	1–15	0.1	Trace

DIGESTION AND ABSORPTION

Hydrolysis of peptide bonds in proteins is called *digestion*. The proteins are hydrolyzed to their constituent amino acids in the gastrointestinal tract. Protein digestion is important for two reasons: (1)

High molecular weight compounds are converted to low molecular weight compounds that can be transported through cell membranes. (2) Biological specificity is destroyed, preventing antigenic reaction.

Digestion takes place in both the stomach and small intestine under the influence of proteolytic enzymes secreted in gastric and pancreatic juices. Those enzymes contained in gastric juice require a very acidic media for optimum activity. Pepsin, the primary enzyme in the stomach, has its maximum activity at a pH between 1.5 and 2.5. This low pH is attained by the secretion of 0.16 *M* HCl by the parietal cells of the stomach. Blood plasma is the source of the required hydrogen ions and chloride ions. The concentration of H^+ in the parietal cells is approximately 1 million times that of blood. Formation and maintenance of this large concentration difference requires energy. Calculations suggest the body expends a minimum of 1500 cal to produce 1L of this HCl solution. Secretion rates of gastric juices vary from 150 to 500 mL/hr, depending on both appetite and the presence of food in the stomach. Typically, secretion continues throughout and for 10 to 20 min. after eating.

Overproduction and secretion of gastric juice leads to the condition of hyperacidity, which causes heartburn and indigestion. Continued hyperacidity for prolonged lengths of time may lead to peptic or duodenal ulcers because of digestion of the stomach or duodenum mucosa.

Pernicious anemia is associated with a condition known as *anacidity* (without acidity) in which virtually no hydrochloric acid is produced. In this disease very little protein digestion can take place in the stomach since pepsin is inactive. The result is a nutritional deficiency of amino acids even though the individual is apparently on an adequate diet.

Following protein digestion in the stomach and small intestine, the resultant amino acids are absorbed directly by active transport into the bloodstream through the intestinal mucosa.

Amino Acid Pool and Nitrogen Balance

Body proteins are constantly undergoing degradation and resynthesis. Thus, a dynamic equilibrium exists for body proteins. The degradation and resynthesis rates vary widely for different proteins. These rates are referred to as *protein turnover rates.* Blood and liver proteins experience a rapid turnover rate with about one-half of their protein undergoing degradation and resynthesis every 6 days. Collagens (Collegens are the proteins of connective tissue) proteins apparently experience the slowest turnover rates, with a half-life of approximately 3 years.

The combination of amino acids from protein degradation and those from dietary sources are referred to as the body's *amino acid pool.* The amino acid pool may be considered as all the free amino acids in the body. They are *not* located in a specific depot. The amino acids in the pool are available to all tissues and circulate via the bloodstream. The normal amino acid concentration is 4 to 8 mg/100mL of plasma. Whenever amino acid concentration exceeds this range, the amino acids are rapidly removed from circulation by the liver. In the liver, they are degraded by removal of the *amine* group. The non-nitrogen-containing residue is usually converted to acetyl coenzyme A. The amine group may be used for synthesis of other nonprotein nitrogen-containing compounds or converted to urea and excreted via the urine.

A normal healthy adult on a correct diet will excrete as much nitrogen as taken in. Such an individual is said to be in *nitrogen balance.* A growing child or a person recovering from prolonged illness experiences a positive nitrogen balance in which less nitrogen is excreted than is included in the diet. Negative nitrogen balance may result from starvation, wasting diseases, or a lack of sufficient amounts of the essential amino acids in the diet.

OVERVIEW OF AMINO ACID METABOLISM

A wide variety of metabolic reactions exists for amino acids. Some of the reactions are specific for certain amino acids, while others are of a general type for all or most amino acids. The variety of reactions that a particular amino acid might undergo is independent of how it entered the amino acid pool. Amino acids from dietary sources are immediately mixed in the blood with those from tissue protein degradation with the result that no metabolic distinction can be made.

The general metabolic pathways for amino acid utilization can be summarized:

1. Synthesis of proteins.
2. Synthesis of nonprotein nitrogen-containing compounds such as heme, nucleic acids, and creatine.
3. Transamination to form required nonessential amino acids.
4. Oxidative deamination involving the removal of the amine group. The amine group is generally converted to urea via ammonia. The remainder of the amino acid molecule is frequently metabolized as an energy source.
5. Undergoing a specific reaction usually related to the presence of a particular element (sulphur) or a particular chemical group (aromatic or heterocyclic rings).

TRANSAMINATION

Transamination is the transfer of an amine group from an amino acid to an α-keto acid. This transfer results in the formation of a different amino acid and a different α-keto acid. The general equation for this reaction may be written

$$R{-}\underset{NH_2}{\underset{|}{C}}H{-}CO_2H + R'{-}\underset{O}{\underset{\|}{C}}{-}CO_2H \rightarrow R{-}\underset{O}{\underset{\|}{C}}{-}CO_2H + R'{-}\underset{NH_2}{\underset{|}{C}}H{-}CO_2H$$

This reaction gives the body the ability to synthesize the nonessential amino acids that it requires. With the exception of lysine and threonine, every natural amino acid has been found to undergo transamination reactions. The enzymes required for transamination are known as *transaminases* and required a derivative of the B vitamin pyridoxine.

A specific example of transamination is the reaction between L-glutamic acid and pyruvic acid to produce α-ketoglutaric acid and alanine.

$$\underset{\text{L-glutamic acid}}{HO_2CCH_2\underset{NH_2}{\underset{|}{C}}HCO_2H} + \underset{\text{pyruvic acid}}{CH_3\underset{O}{\underset{\|}{C}}CO_2H} \rightarrow \underset{\alpha\text{-ketoglutaric acid}}{HO_2CCH_2CH_2\underset{O}{\underset{\|}{C}}CO_2H} + \underset{\text{alanine}}{CH_3\underset{NH_2}{\underset{|}{C}}HCO_2H}$$

Most transamination reactions apparently involve L-glutamic acid as the source of amine groups. This is probably because the Krebs cycle may furnish abundant amounts of α-ketoglutaric acid and the liver can directly aminate α-ketoglutaric acid with ammonia to produce L-glutamic acid.

$$\underset{\alpha\text{-ketoglutaric acid}}{HO_2CCH_2CH_2\underset{O}{\underset{\|}{C}}CO_2H} + NH_3 \xrightarrow[\text{glutamic acid dehydrogenase}]{NADH_2 \;\; NAD} \underset{\text{L-glutamic acid}}{HO_2CCH_2CH_2\underset{NH_2}{\underset{|}{C}}HCO_2H} + H_2O$$

OXIDATIVE DEAMINATION

Oxidative deamination is the removal of the amine group from an amino acid to produce ammonia and an α-keto acid. The overall reaction can be written

The reaction takes place in two steps. In the first, either NAD or FMN is the oxidizing agent and requires the enzyme amino acid oxidase. The second step is a spontaneous hydrolysis to produce ammonia and the α-keto acid.

Step 1.

$$HO_2CCH_2CH_2\underset{\displaystyle NH_2}{\underset{|}{C}}H{-}CO_2H \xrightarrow[\text{amino acid oxidase}]{NAD \;\curvearrowright\; NADH_2} HO_2CCH_2CH_2\underset{\displaystyle NH}{\underset{\|}{C}}{-}CO_2H$$

Step 2.

$$HO_2CCH_2CH_2\underset{\displaystyle NH}{\underset{\|}{C}}{-}CO_2H + H_2O \rightarrow HO_2CCH_2CH_2\underset{\displaystyle O}{\underset{\|}{C}}{-}CO_2H + NH_3$$

The $NADH_2$ produced in step 1 is reoxidized to NAD via the cytochrome chain. The overall equation for this reoxidation is

$$NADH_2 + \tfrac{1}{2}O_2 \rightarrow NAD + H_2O$$

The reaction sequence involving NAD as shown above is specific for glutamic acid.

Oxidative deamination of other amino acids occurs primarily via transaminations with α-ketoglutaric acid to form glutamic acid. This is followed by oxidative deamination of the glutamic acid. Only glutamic acid undergoes oxidative deamination in large amounts.

Minor amounts of amino acids undergo direct oxidative deamination rather than via glutamic acid formation. In this case FMN is required as the oxidizing agent instead of NAD.

DISPOSAL OF AMMONIA

Ammonia is a toxic compounds and its concentration must not build up in cells. Besides being toxic, ammonia is a base: accumulation of large amounts of it could change the pH of body fluids. Therefore, if the ammonia produced via deamination of amino acids is not required for anabolic purposes, it must be excreted. The form in which the ammonia is excreted depends in part on where it is produced. Ammonia produced by kidney cells during metabolism may be released directly into the urine. Excess ammonia produced elsewhere in the body is converted to either urea or glutamine. Most deamination reactions occur in the liver. Any excess ammonia produced in the liver is transformed to urea and excreted via urine. The overall equation for urea formation may be written

$$2NH_3 + CO_2 \rightarrow H_2N\text{—}\underset{\underset{O}{\|}}{C}\text{—}NH_2 + H_2O$$

urea

Urea formation takes place in a series of reactions known as the ornithine cycle, shown in Fig. 17.1

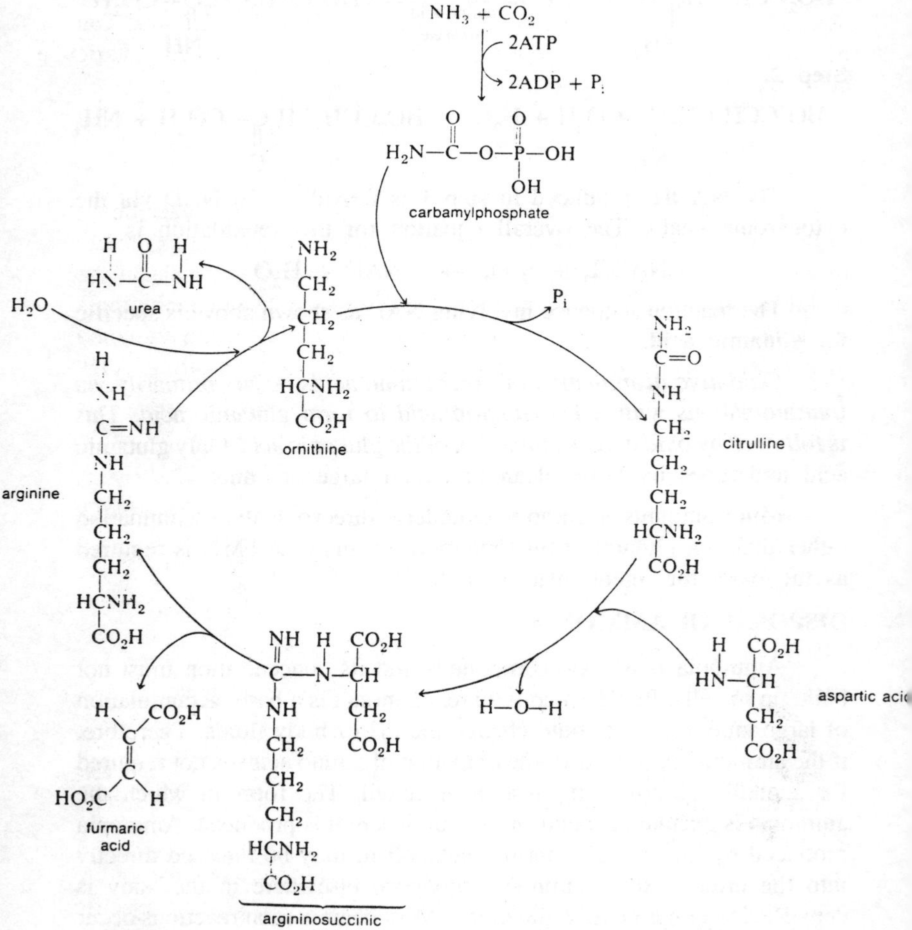

Fig. 17.1 : Urea formation via ornithine cycle.

Notice in Fig 17.1 that the second ammonia molecule required for urea formation comes from aspartic acid and not from free ammonia. Thus, the ornithine cycle uses ammonia, carbon dioxide, and aspartic acid to produce urea and fumaric acid.

The fumaric acid may be converted to oxaloacetic acid via a portion of the Krebs cycle. Figure 17.2 (see page no. 344) shows the use and regeneration of aspartic acid in relation to the ornithine cycle. Transamination between oxaloacetic acid and glutamic acid regenerates aspartic acid for use in the ornithine cycle. Since glutamic acid can be synthesized in the liver directly from ammonia and α-ketoglutaric acid, the overall equation for urea formation is

$$2NH_3 + CO_2 \rightarrow H_2N{-}\overset{\overset{\displaystyle O}{\|}}{C}{-}NH_2 + H_2O$$

Ammonia produced by the body elsewhere than in the kidneys or liver is principally used in glutamine synthesis. Glutamine apparently serves two functions: (1) as a means of transporting ammonia in the bloodstream in a nontoxic form and (2) as a temporary store of ammonia in the body. Glutamine is synthesized from glutamic acid and ammonia in a reaction catalyzed by glutamine synthetase and requiring ATP.

$$\underset{\text{glutamic}}{\begin{array}{c} CO_2H \\ | \\ H{-}C{-}NH_2 \\ | \\ CH_2 \\ | \\ CH_2 \\ | \\ CO_2H \end{array}} + NH_3 \xrightarrow[\text{ATP} \quad \text{ADP} + P_i]{\text{glutamine synthetase}} \underset{\text{glutamine}}{\begin{array}{c} CO_2H \\ | \\ H{-}C{-}NH_2 \\ | \\ CH_2 \\ | \\ CH_2 \\ | \\ C{-}NH_2 \\ \| \\ O \end{array}} + H_2O$$

For a healthy individual on a normal diet, most of the glutamine produced is transported to the liver via the bloodstream. In the liver, glutamine is hydrolyzed and the released ammonia is converted to urea and excreted.

We have examined the general degradation reactions for amino acids. In addition to these general reactions, there are many specific reactions for individual amino acids; however, the details of these are not of concern in this course. Some hereditary diseases result however

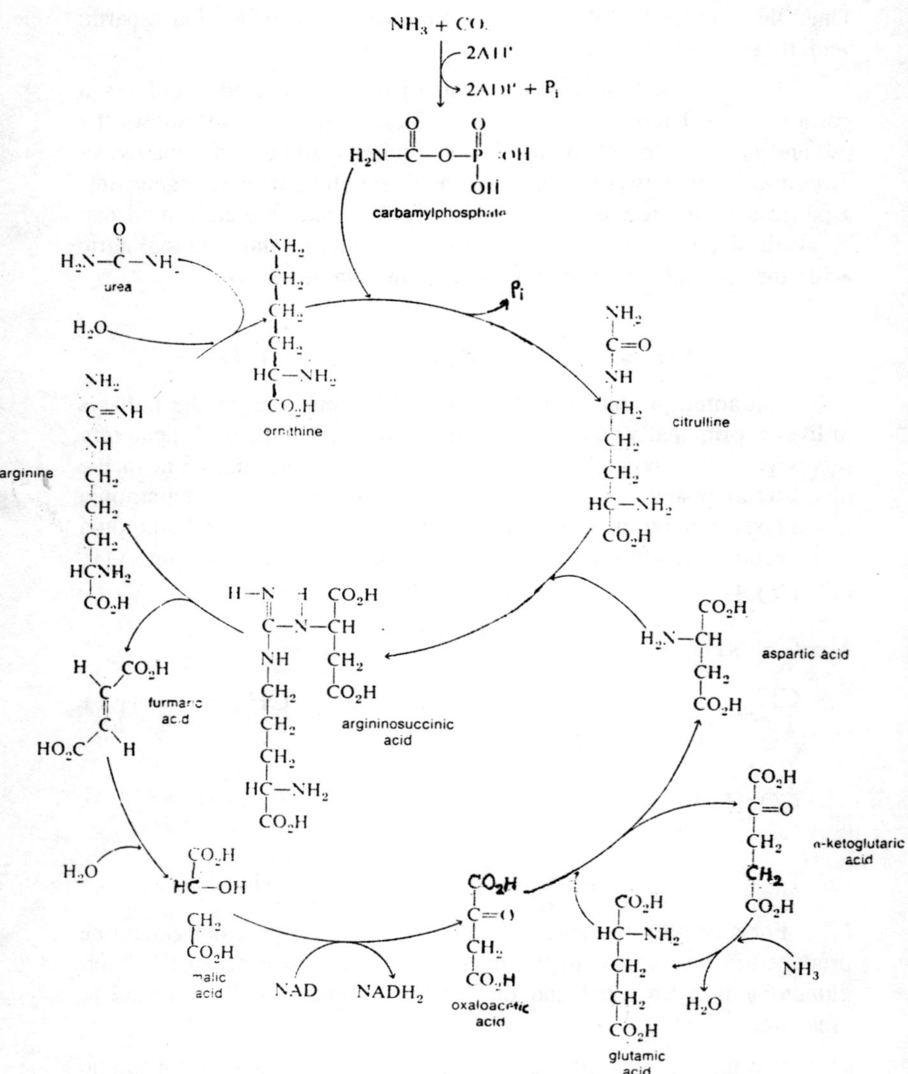

Fig. 17.2 : The use and regeneration of aspartic acid in relation to the ornithine cycle.

from the inability of the body to perform a specific reaction of a particular amino acid because of the lack of the required enzyme. Recall that enzymes are primarily proteins. Thus, the absence of an enzyme is in some way related to a malfunction in protein synthesis. We shall examine several hereditary diseases involving amino acid metabolism after our study of protein synthesis.

PROTEIN SYNTHESIS

For many years, the synthesis of proteins was thought to be the reversal of protein hydrolysis. However, this proposal could not explain how a cell was able to ensure the proper sequencing of the amino acid residues in the protein being synthesized.

As research on proteins continued, it was discovered that many hereditary diseases were associated with a malfunction of protein synthesis, most notably enzymes and proteinaceous hormones. Knowledge that hereditary diseases involve abnormalities in protein synthesis and that inheritance of traits is associated with genes carried on chromosomes in the cell nucleus led to the *one gene-one enzyme* theory. According to this theory, genes control metabolism by controlling synthesis of enzymes. A specific gene is present for each enzyme produced by a cell. Consequently, mutation of single gene results in an abnormality of a single specific enzyme.

Portions of DNA molecules are called genes. Thus DNA controls protein synthesis. The nuclei of cells have been found to contain *deoxyribonucleic acids* (DNA). The DNA is found only in the chromosomes and the amount of DNA present is proportional to the number of genes. This formation, coupled with additional unmentioned evidence, led scientists to conclude that *DNA is the genetic material of life* (expect for certain viruses) *and possesses the ability to control the formation of proteins, including enzymes.* DNA has the ability to replicate (form an exact replica of itself) and also is capable of undergoing mutation.

THE CHEMISTRY AND STRUCTURE OF DNA

DNA molecules are polymers with molecular weights exceeding 1×10^9. Thus, DNA is among the largest molecules known. Complete hydrolysis of DNA results in the isolation of six compounds: adenine, guanine, cytosine, thymine, β-2-deoxyribose, and phosphoric acid (see structure on page 346).

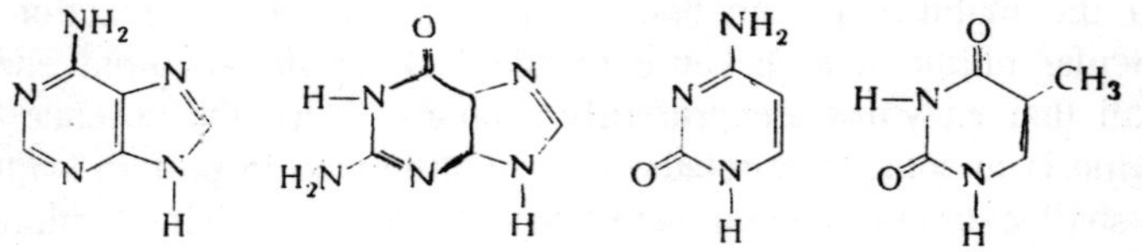

adenine guanine cytosine thymine

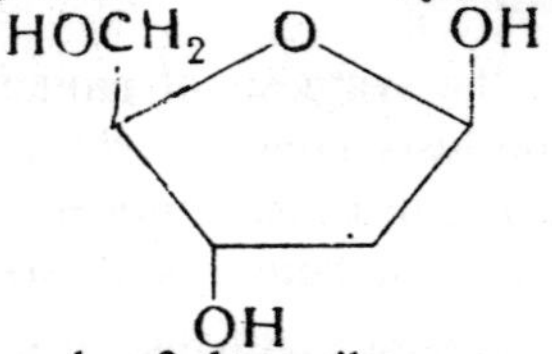

b-o-2-deoxyribose

Adenine, guanine, cytosine, and thymine are heterocyclic bases. Heterocycles are ring structures containing other atoms in addition to carbon in the ring. They are amines and consequently behave as bases because of the presence of unshared electron pairs on the nitrogen atoms. Deoxyribose is a carbohydrate. In DNA, deoxyribose is bonded to each of the heterocyclic bases via the No. 1 carbon of the sugar. The resultant molecules are known as *nucleosides*; their structural formulas are shown below.

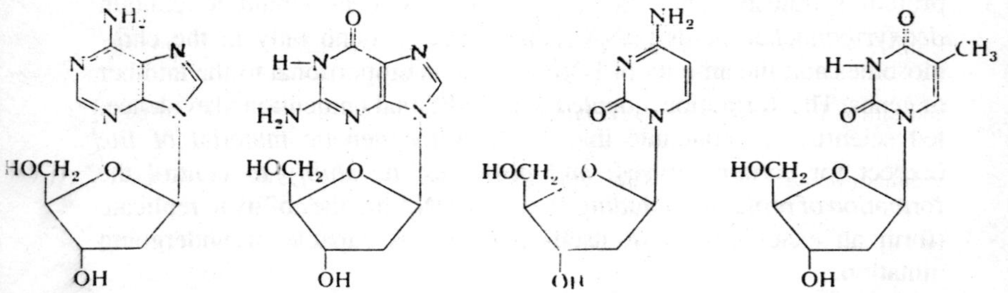

deoxyadenosine deoxyguanine deoxycytidine thymidine

In DNA, these nucleosides are joined together by phosphate ester linkages between the No. 5 carbon of the deoxyribose of one nucleoside and the No. 3 carbon of the deoxyribose of another nucleoside molecule. *The total number of nucleoside molecules and the order in which the four different nucleosides are bonded is what makes one gene different*

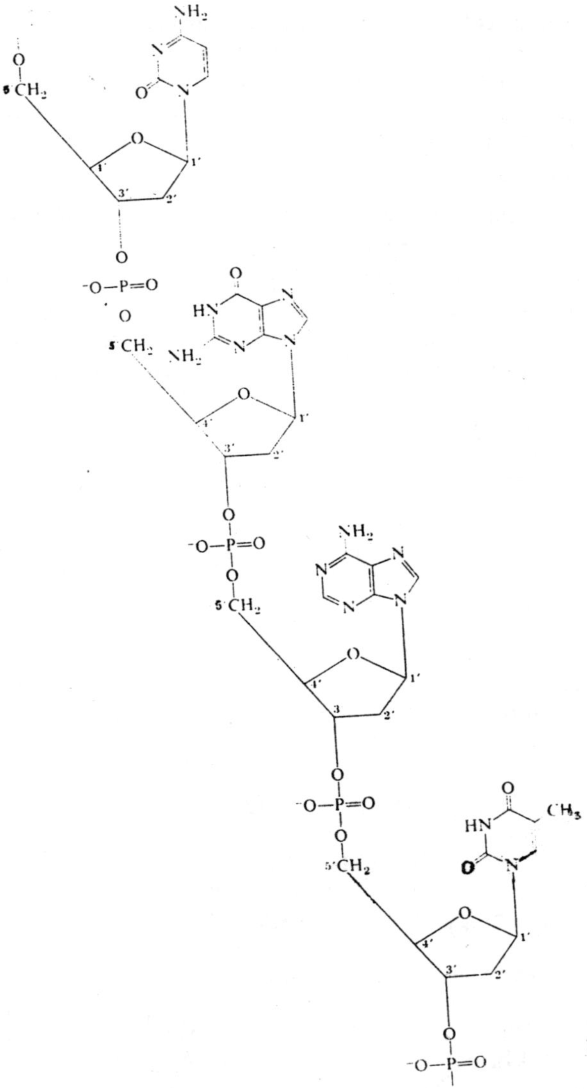

Fig. 17.3 : A portion of a strand of DNA showing the phosphate deoxyribose units forming the backbone. of the two strands in a DNA

from another gene. This is true whether we are comparing two different genes in a human cell or comparing a human gene to a plant gene. A portion of DNA molecule showing the phosphate ester bonds between nucleosides is depicted in Fig 17.3 (see on page 348). Notice in this figure that the alternating deoxyribose-phosphate sequence forms the backbone of the DNA molecule and the heterocyclic bases extend outward from the backbone.

The DNA molecules have a highly ordered structure that is maintained via hydrogen bonding between the heterocyclic bases. The hydrogen bonding is very specific: Adenine will hydrogen bond to thymine and cytosine will hydrogen bond to guanine. Other hydrogen bonding combinations do not normally occur in DNA. This specific hydrogen bonding between the heterocyclic bases is referred to as base pairing (Fig. 17.4).

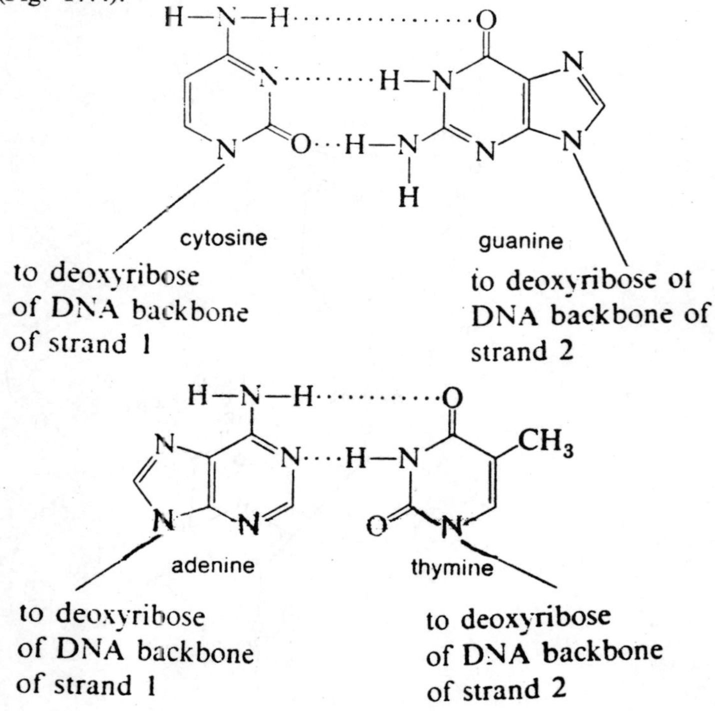

cytosine

guanine

to deoxyribose of DNA backbone of strand 1

to deoxyribose of DNA backbone of strand 2

adenine

thymine

to deoxyribose of DNA backbone of strand 1

to deoxyribose of DNA backbone of strand 2

Fig. 17.4 : Hydrogen bonding between heterocyclic bases

The hydrogen bonding between complementary base pairs does *not* occur between nucleosides within the same chain but occurs between nucleosides in two separate chains. Thus, the hydrogen bonding or base pairing serves to join two complementary chains.

The two long chains, joined by thousands of hydrogen bonds, are twisted or interwined to form a helical coil. This helical coiling of the paired chains produces the well-publicized *double helix* of DNA shown in Fig 17.5. This structural model for DNA was first proposed by M. Crick and J. Watson in 1954.

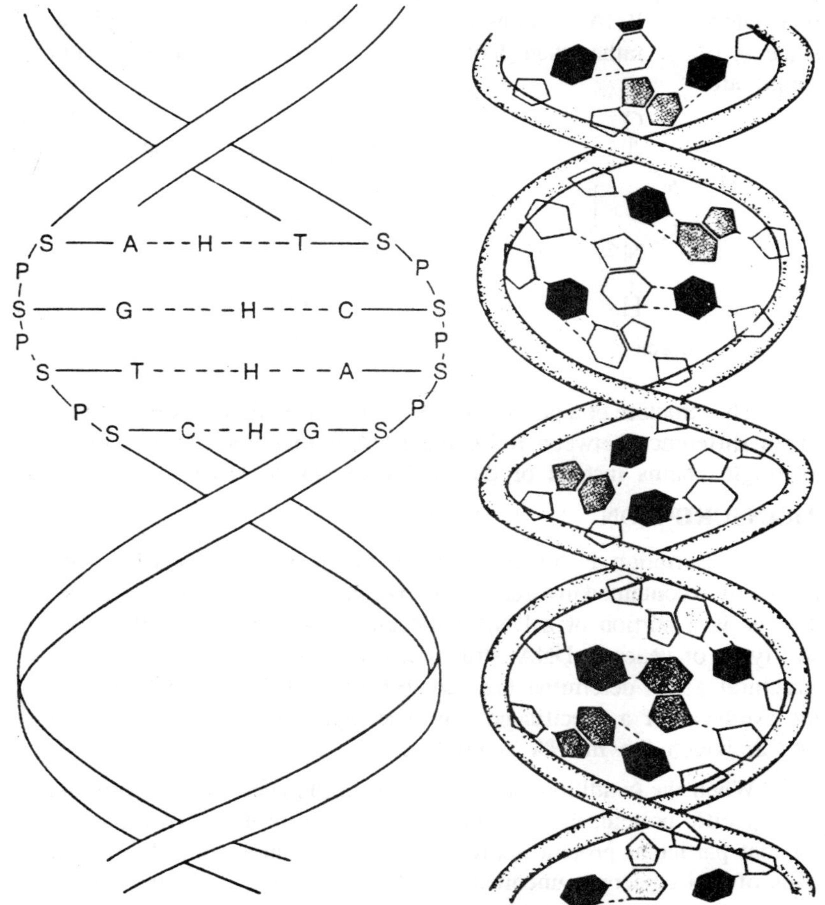

Fig. 17.5 : Double helix of DNA showing the hydrogen bonding between strands.

MESSENGER RNA

The DNA of the genes in the chromosomes is present in the nucleus of a cell. Protein synthesis occurs on cytoplasmic particles called *ribosomes* that are attached to the endoplasmic reticulum. A type of ribonucleic acid (RNA) known as messenger RNA or *m*RNA serves to carry the information contained in the DNA to the ribosomes.

Messenger RNA has many similarities to DNA and yet has its own characteristics. Like DNA, *m*RNA is a polymer made up of heterocyclic bases and a five-carbon sugar and is held together by phosphate ester bonds. The major differences are (1) *m*RNA is a much smaller polymer than DNA, (2) RNA contains uracil instead of thymine, and (3) the sugar is ribose rather than deoxyribose. The structures of uracil and ribose are

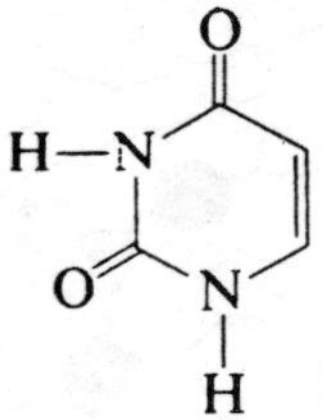

uracil

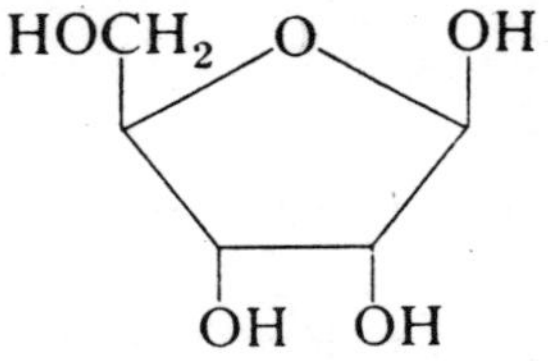

β-D-ribose

The structure of a portion of *m*RNA is shown in Fig. 17.6. Another major difference between RNA and DNA is that RNA exists largely as single chains instead of the double chains of DNA.

TRANSCRIPTION

Transcription is the name given to the process by which the information contained in a gene is incorporated into the *m*RNA molecule. A gene is a portion of a DNA molecule. It is thought that there are two types of genes in DNA: structural genes and regulator genes. The structural genes determine the structure of *m*RNA, thus determining the synthesis of a specific protein. The regulatory genes control the rate at which the *m*RNA is produced.

When the conditions in a cell indicate the need for the synthesis of a specific protein, the structural gene portion of the DNA molecule for that particular protein unravels. During this unraveling the helical coils of DNA become uncoiled and the two strands separate. One of the stands serves as a template (pattern) for the formation of *m*RNA.

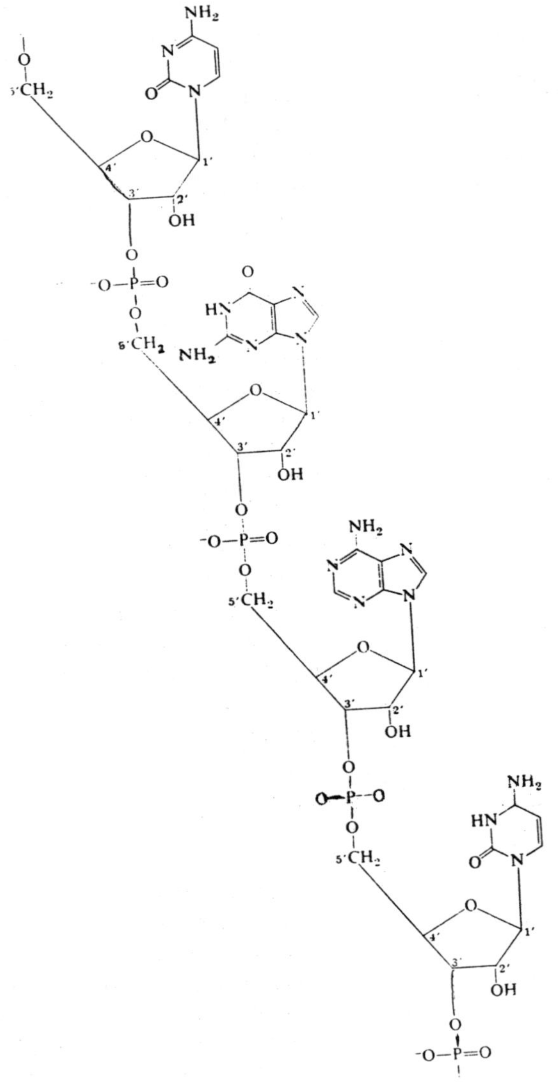

NH$_2$ N O N O 5'CH$_2$ O 4' 1' 3' 2' OH O $^-$O–P=O O 5'CH$_2$ O HN NH$_2$ N N N O 4' 1' 3' 2' OH O $^-$O–P=O O 5'CH$_2$ NH$_2$ N N N N O 4' 1' 3' 2' OH O O–P–O O 5'CH$_2$ NH$_2$ HN O N O 4' 1' 3' 2' OH O $^-$O–P=O

Fig. 17.6 : A portion of a strand of RNA showing the phosphate-ribose units forming the backbone.

This formation occurs in several steps. The nucleosides are phosphorylated and form nucleotides, also called nucleic acids. The nucleotides hydrogen bond to their complementary base in the DNA molecule. Under that action of an enzyme, the nucleotides are joined by phosphate ester bonds to form the *m*RNA molecule. When the *m*RNA molecule has been synthesized; it breaks away from the DNA strand and diffuses from the cell nucleus to the ribosomes. After the newly synthesized *m*RNA has diffused away, the two complementary strands of the DNA are rejoined into the normal DNA double helix. This process is depicted in Fig 17.7 (see p. 353).

TRANSLATION

The order of the nucleosides in the *m*RNA determines the sequence of the amino acids in the protein to be synthesized. The *m*RNA nucleoside order must be translated. The translation is accomplished by another type of RNA called transfer RNA (*t*RNA). Transfer RNA molecules serve both as amino acid carriers and interpreters of the genetic code. For each amino acid there is at least one specific *t*RNA. The *t*RNA molecules bond to the carboxyl group of their particular amino acid by the formation of an ester linkage.

Another portion of the *t*RNA molecule recognizes a code contained in the *m*RNA for the specific amino acid that the *t*RNA carries. The code consists of three adjacent heterocyclic bases in the *m*RNA. The three adjacent bases are known as a triplet or a codon. For example, it has been determined that the base sequence guanine, guanine, uracil in *m*RNA is specific for the *t*RNA molecule that carries glycine. Thus, when this sequence of bases occurs in a *m*RNA molecule, a glycine is incorporated into the protein molecule being synthesized. The recognition by a *t*RNA molecule for its proper triplet in the *m*RNA occurs by base pairing.

Table 17.2 contains the codons for the various amino acids. Notice that there are several codons for most of the amino acids.

Consider the beginning portion of a hypothetical *m*RNA molecule with nucleoside order UUUCCUACGGUAUUGUCG...... . Referring to Table 17.2, we see that the first triplet UUU is the code for phenylalanine; the second triplet CCU is for proline; the third, ACG, for threonine. Continuing in this manner we find the portion of the *m*RNA shown above would dictate the synthesis of the polypeptide phenylalanine-proline-threonine-valine-leucine-serine-.....

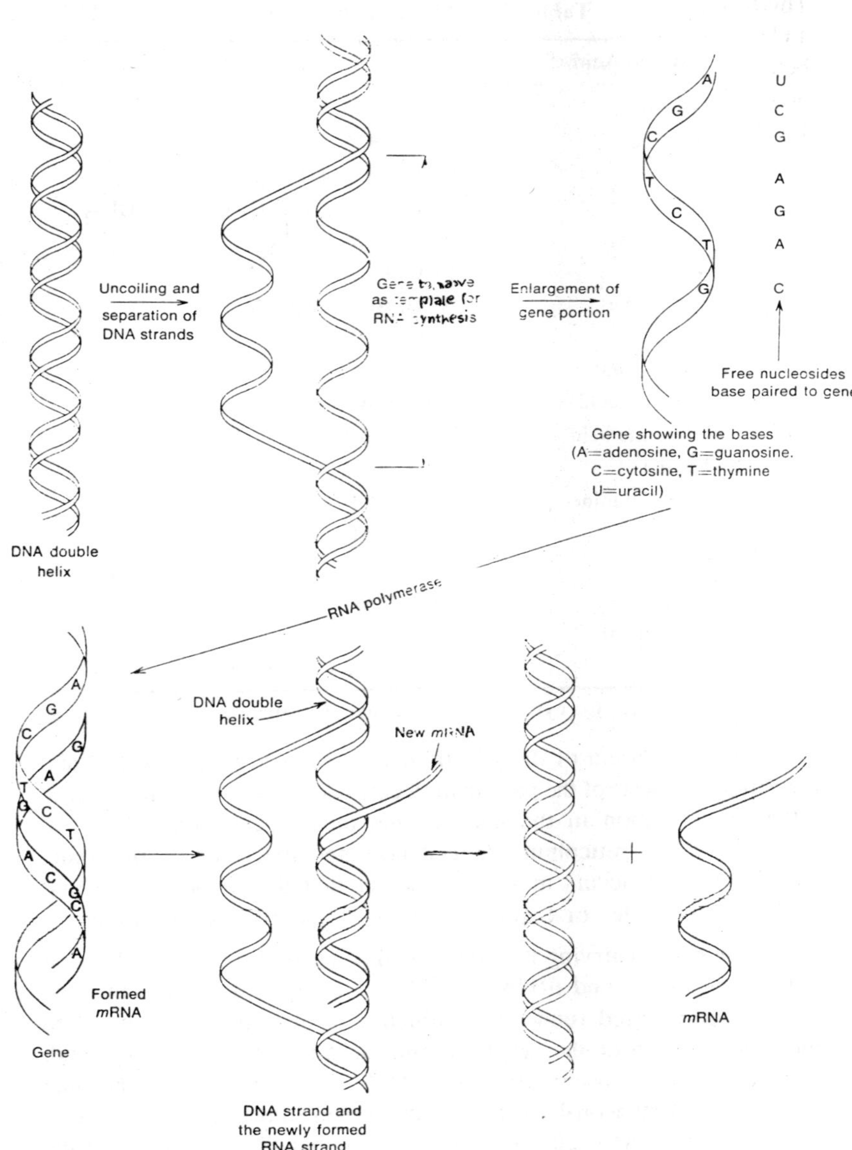

Fig. 17.7 : Schematic representation of *m*RNA formation.

Table 17.2 : Amino acid codons

Amino Acid	Codon*
Glycine	GGG, GGA, GGC, GGU
Alanine	GCG, GCA, GCC, GCU
Valine	GUG, GUA, GUC, GUU
Leucine	CUG, CUA, CUC, CUU, UUG, UUA
Isoleucine	AUA, AUC, AUU
Serine	UCG, UCA, UCC, UCU
Threonine	ACG, ACA, ACC, ACU
Cysteine	UGC, UGU
Methionine	AUG
Aspartic acid	GAC, GAU
Glutamic acid	GAG, GAA
Lysine	AAG, AAA
Phenylalanine	UUC, UUU
Tyrosine	UAC, UAU
Arginine	CGG, CGA, CGC, CGU, AGG, AGA
Histidine	CAC, CAU
Tryptophan	UGG
Proline	CCG, CCA, CCC, CCU

*U, uraci; C, cytosine; A, adenine; G, guanine.

The translation of the genetic code contained in the *m*RNA and the actual synthesis of the protein molecule takes place on the ribosomes. After its formation in the nucleus, the messenger RNA diffuses to the endoplasmic reticulum where it becomes attached to a ribosome. The attachment occurs in such a manner that the codon for the first *t*RNA carrying the first amino acid is at the ribosomal surface.

The *t*RNA carrying the amino acid bonds to the codon. Following bonding of the second *t*RNA to *m*RNA, the amino acid from the first tRNA is transferred (under the influence of appropriate enzymes) to the amino group of the second amino acid to form a peptide bond. After this transfer occurs, the first *t*RNA leaves the *m*RNA-ribosome complex. Simultaneously, the ribosome moves to the third codon. Now the process repeats itself: attachment of another *t*RNA, transfer of the growing protein chain, loss a *t*RNA molecule, and movement of the ribosome (see Fig 17.8)

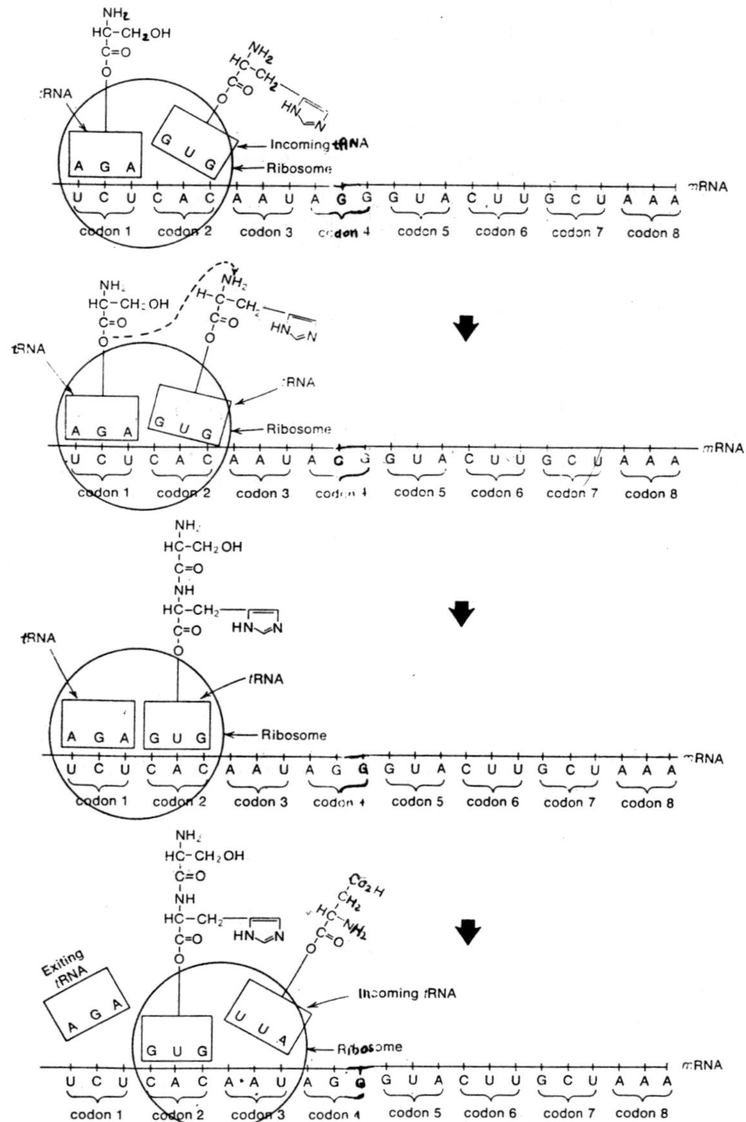

Fig. 17.8 : Protein synthesis by translation of the genetic code.

MUTATION

Mutations in cells can cause a change in the structure of proteins. This change can occur in several ways, such as the substitution of one amino acid for another. An example of this type of change is the substitution of a valine for a glutamic acid in hemoglobin. This particular substitution leads to sickling of red blood cells. Another possible change is the omission of an amino acid in a protein; yet another is the substitution of several amino acids in a sequence. This substitution occurs because the *m*RNA contains a GUG codon instead of the correct GAG codon. This error arises because the DNA has an adenine in place of a thymine.

The opportunities for a change in a protein structure are numerous. Changes can result as a consequence of an error in translation or transcription or because of a change in DNA. Errors in translation or transcription result in a change of an individual or several protein molecules and do not result in inheritable traits; hence, these two errors do not lead to mutations.

A change in DNA structure will cause an abnormality in all the molecules of at least one particular protein synthesized by the cell as long as the cell lives. This lasting variation in protein structure arises because a change in a DNA molecule results in the continuous production of altered *m*RNA molecules. Note only will the protein in this cell be abnormal, but this characteristic of producing abnormal proteins will be genetically passed on to new generations of cells during abnormal proteins will be genetically passed on to new generations of cells during reproduction. Thus, we observe an inheritable alteration; *mutation* has occurred.

To understand this phenomenon we must investigate the synthesis of DNA during the processes leading to mitosis. Before doing so, it must be emphasized that mutations are not necessarily bad. Mutation is a part of the revolutionary process and may aid the organism in surviving in a slowly changing or new environment. In the overall scheme of life we must view mutations as nature's experiments. If the mutation has positive survival value, it will be retained and the mutants will flourish, eventually replacing the *normal* individuals. If the mutation offers negative survival value, the mutants will die out. Some mutations apparently have neither a significant negative nor positive survival value and are simply carried along through generations providing differences between individuals of the same species.

The process by which DNA is synthesized is called *replication*. In many respects replication of DNA is very similar to the synthesis of *m*RNA. During the process, the two strands of the double-stranded DNA molecules become unraveled and separated. *Each* strand serves as a template for the synthesis of a new DNA strand. The new strand being synthesized is not identical to the template; rather, it is complementary to the parent strata. The parent strand and the newly synthesized strand join in a double helix coil to form a new DNA molecule. Since each strand of the original DNA molecule serves as a template, two new double-stranded DNA molecules result from the original molecule. This model of DNA replication is depicted in Fig 17.9.

Mutations occur whenever the nucleotide sequence in DNA undergoes a change. One way the nucleotide sequence might be changed occurs during base pairing, resulting in the substitution of one nucleotide by another during replication. High-energy radiation, such as X-rays, ultraviolet light, and radioactivity, also cause mutations. Certain chemicals are capable of causing mutations. Particularly effective chemicals are those that react with amino groups. Further, it appears that at least in certain species the cell itself has a built-in mechanism for causing mutations. For example, it has been found that the bacteria *Escherichia coli* contains mutator genes. One of these mutator genes causes the adenine-thymine base pair to be replaced by the cytidine-guanine base pair and vice versa.

HEREDITARY DISEASES

In previous chapters a number of hereditary disorders have been discussed including diabetes, galactosemia, and sickle cell anemia. Other hereditary diseases of sufficient incidence to warrant discussion are phenylketonuria, albinism., and hemophilia.

Phenylketonuria results from a lack of the enzyme phenylalanine hydroxylase, which catalyzes the conversion of phenylalanine to tyrosine. Although phenylalanine is an essential amino acid, the requirement for its is primarily to satisfy the body's need for tyrosine.

$$C_6H_5{-}CH_2{-}\underset{\displaystyle NH_2}{\underset{|}{CH}}{-}CO_2H \xrightarrow[\text{hydroxylase}]{\text{phenylalanine}} HO{-}C_6H_4{-}CH_2{-}\underset{\displaystyle NH_2}{\underset{|}{CH}}{-}CO_2H$$

phenylalanine　　　　　　　　　　tyrosine

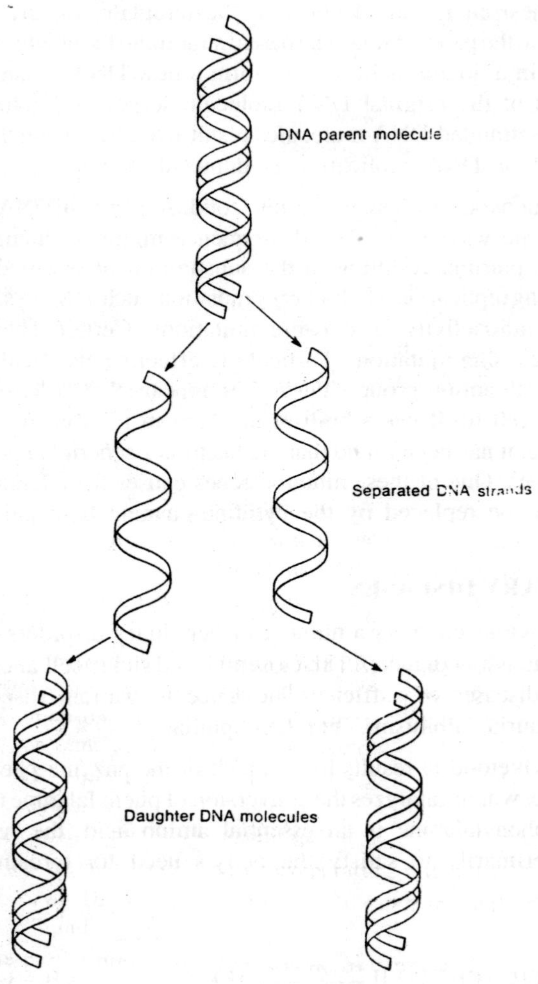

Fig. 17.9 : DNA replication. (The original strands are shown in color).

In the absence of this enzyme the normal metabolic pathway for phenylalanine is blocked. As the concentration of phenylalanine increases, the body resorts to utilizing it for transamination reactions. Transamination involving phenylalanine produces phenylpyruvic acid.

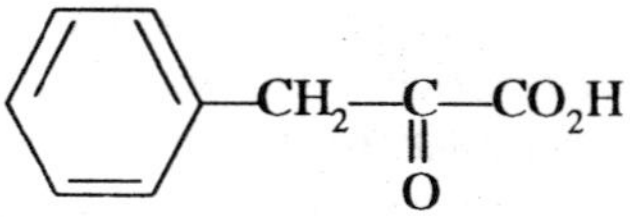

$$C_6H_5{-}CH_2{-}\underset{\underset{O}{\|}}{C}{-}CO_2H$$

As much as 1 to 2 g/day of phenylpyruvic acid may be excreted via the urine. Phenylpyruvic acid is a keto acid, hence the name phenylketonuria (PKU). Accumulation of phenylpyruvic acid rapidly leads to severe mental retardation. Approximately 1 in every 80 people of European descent carry this trait as a recessive gene. Fortunately, a simple blood test for phenylketonuria has been developed and is routinely done in most hospital 4 or 5 days after birth. Infants displaying this disease are placed on a diet containing very little phenylalanine. This diet must be maintained throughout periods of rapid brain growth and development if these individuals are to be given a chance for a normal adult life.

Albinism is another hereditary disorder associated with abnormal amino acid metabolism. Tyrosine is a precursor of melanin, a dark, polymeric pigment, which is synthesized in melanoblasts located in the basal layer of the skin. Skin color depends on the number and distribution of melanoblasts and melanin concentration. Melanin is also found in the retina, adrenal medulla, and portions of the brain. The absence of this pigment result albinism characterized by an extremely pale skin color and pink irises in the eyes.

The initial step in the conversion of tyrosine to melanin is the oxidation of tyrosine to dopa, catalyzed by the enzyme *tyrosinase*.

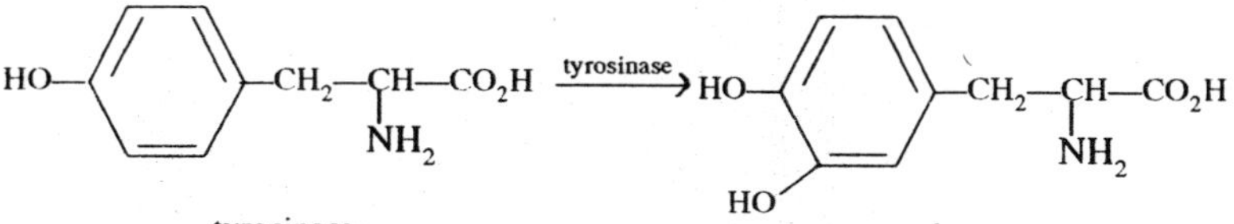

$$HO{-}C_6H_4{-}CH_2{-}\underset{\underset{NH_2}{|}}{CH}{-}CO_2H \xrightarrow{\text{tyrosinase}} (HO)_2C_6H_3{-}CH_2{-}\underset{\underset{NH_2}{|}}{CH}{-}CO_2H$$

tyrosinase   dopa

In albinos the enzyme is missing.

Hemophilia is a sex-linked hereditary disease, occurring primarily in males but transmitted solely through females. The disorder is characterized by a marked increase in the time required for blood clotting.

The essential feature of the extremely complex process of clotting is the conversion of fibrinogen to soluble fibrin that reacts to form insoluble fibrin in the presence of Ca^{2+} ions. The conversion of fibrinogen to fibrin is catalyzed by the enzyme thrombin. In order to prevent coagulation in the bloodstream, thrombin itself is not present in blood but is present as inactive prothrombin. Individuals with hemophilia lack the activators necessary for the conversion of prothrombin to thrombin.

18

Fatty Acid Metabolism and the Krebs Cycle

INTRODUCTION

In this chapter, the chemical changes and the parallel energy changes a metabolite undergoes will be described. The production of acetyl coenzyme A by fatty acid catabolism and its subsequent chemical processing in the Krebs cycle will be discussed. General type reactions studied previously will be applied in understanding specific reactions in the catabolic sequences.

TRIGLYCERIDE CATABOLISM

Triglycerides are hydrolyzed to fatty acids and glycerol. The glycerol is used in carbohydrate metabolism. The fatty acids are *exoergically* oxidized to carbon dioxide and water. Although a small amount of fatty acids is oxidized in skeletal muscle cells and heart muscle cells, the liver is the principal site of fatty acid oxidation. Mitochondria in liver cells contain high concentrations of the enzymes necessary for fatty acid oxidation.

The oxidation of a fatty acid proceeds through a series of steps called the *fatty acid spiral* in which two-carbon fragments are split off and combine with coenzyme A to produce acetyl coenzyme A. This compound is incorporated into a sequence called the Krebs cycle in which carbon dioxide is given off. Hydrogens from metabolite fragments in the Krebs cycle are passed along the respiratory chain enzymes producing water, energy, and ATP (Fig. 18.1).

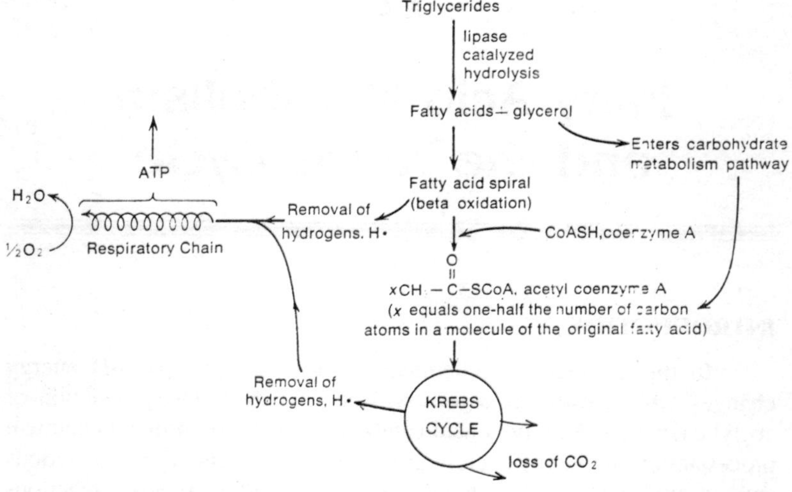

Fig. 18.1 : Metabolic sequence for triglycerides.

FATTY ACID SPIRAL—BETA OXIDATION

A summary of the steps in the fatty acid spiral (also called *beta oxidation*) is given in Fig. 18.2. The beta oxidation of a particular fatty acid, palmitic acid, is given in Fig. 18.3. Our discussion will focus on these two figures and they should be studied together. Numbered steps refer to the figures.

Palmitic acid is a 16-carbon acid and the goal of its beta oxidation is to convert the acid to an "active" derivative, which can subsequently lose the 2-carbon fragmens, $H\!-\!\overset{\displaystyle H}{\underset{\displaystyle H}{\overset{|}{\underset{|}{C}}}}\!-\!C\begin{smallmatrix}\nearrow O \\ \searrow SCoA\end{smallmatrix}$

Because the long hydrocarbon chain of the fatty acid is chemically inactive, it must be activated prior to oxidation. This activation involves

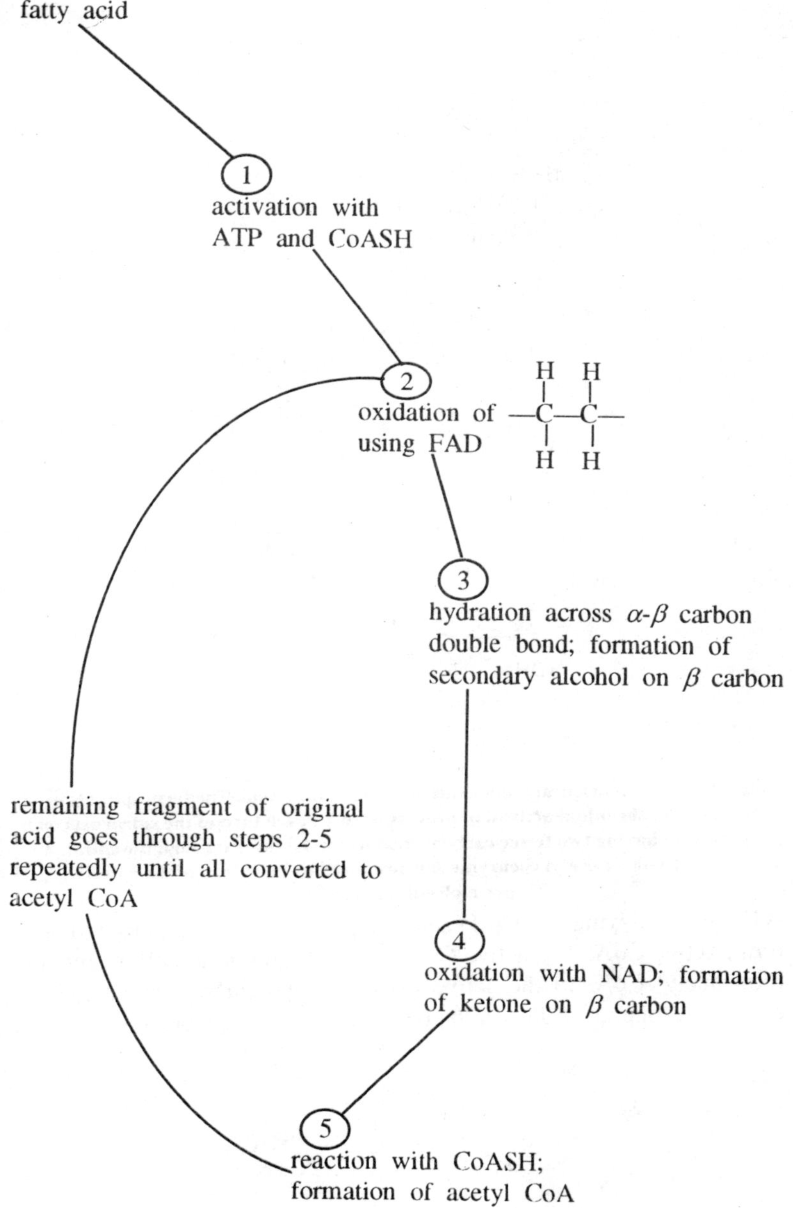

Fig. 18.2 : Summary of steps in fatty acid spiral (beta oxidation).

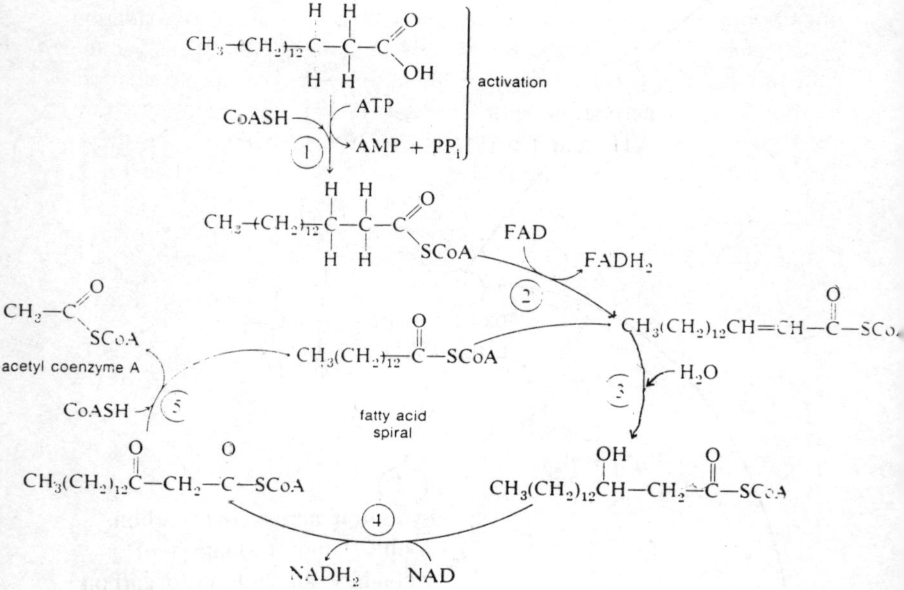

$$CH_3-(CH_2)_{12}-\overset{H}{\underset{H}{C}}-\overset{H}{\underset{H}{C}}-C(=O)OH$$

activation: CoASH, ATP → AMP + PP_i (1)

$$CH_3-(CH_2)_{12}-\overset{H}{\underset{H}{C}}-\overset{H}{\underset{H}{C}}-C(=O)SCoA$$

FAD → $FADH_2$ (2)

$$CH_3(CH_2)_{12}CH=CH-\overset{O}{\overset{\|}{C}}-SCoA$$

H_2O (3)

$$CH_3(CH_2)_{12}\overset{OH}{\overset{|}{C}}H-CH_2-\overset{O}{\overset{\|}{C}}-SCoA$$

NAD → $NADH_2$ (4)

$$CH_3(CH_2)_{12}\overset{O}{\overset{\|}{C}}-CH_2-\overset{O}{\overset{\|}{C}}-SCoA$$

CoASH (5)

$$CH_3-C(=O)SCoA$$ acetyl coenzyme A

$$CH_3(CH_2)_{12}-\overset{O}{\overset{\|}{C}}-SCoA$$

fatty acid spiral

Fig. 18.3 Fatty acid spiral (beta oxidation) of palmitic acid. The term spiral is used because after the initial activation process (step 1), each turn of the spiral acts on a substance having two fewer carbons than its predecessor through the spiral. A molecule of acetyl coenzyme A is produced by each turn of the spiral per molecule of fatty acid.

ATP and coenzyme A (step 1) and changes the fatty acid to its activated form, active CoA. The active CoA ends in a group that will eventually form acetyl CoA. In the active CoA, the two carbons preceding the carbonyl group are called *alpha* (α) and *beta* (β) carbons, respectively.

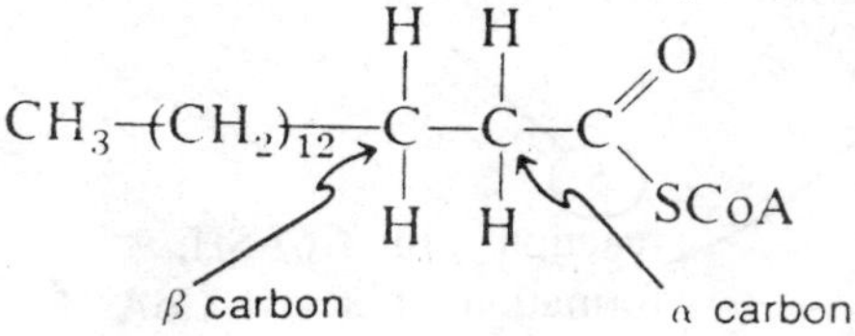

$$CH_3-(CH_2)_{12}-\overset{H}{\underset{H}{C}}-\overset{H}{\underset{H}{C}}-C(=O)SCoA$$

β carbon $\qquad$ α carbon

Steps 2 to 4 change the beta carbon from a $—CH_2—$ group to a ketone. This is accomplished by the following sequence: oxidation of the alpha-beta carbons using FAD (step 2), hydration of the resultant carbon-to-carbon double bond to form a secondary alcohol (step 3), and oxidation of the alcohol to a ketone (step 4). Because oxidation occurs at the beta carbon, the fatty acid spiral is alternatively called *beta oxidation.* In step 5, the alpha-to-beta carbon bond breaks and hydrogen from coenzyme A is transferred to the alpha carbon-containing fragment. The remainder of the coenzyme a bonds to the beta carbon.

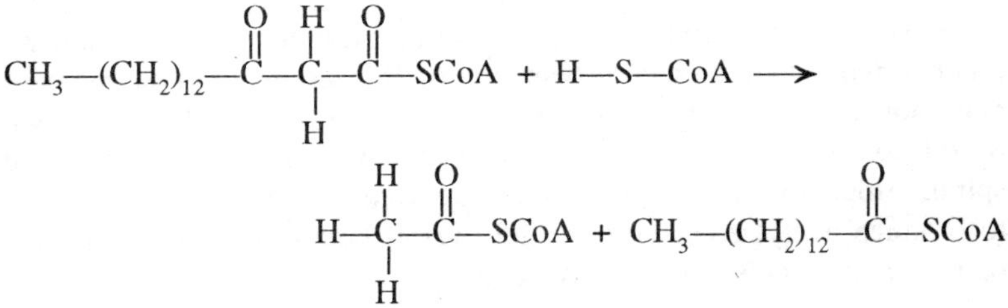

$$CH_3—(CH_2)_{12}—\overset{\overset{\large O}{\|}}{C}—\underset{\underset{\large H}{|}}{\overset{\overset{\large H}{|}}{C}}—\overset{\overset{\large O}{\|}}{C}—SCoA + H—S—CoA \longrightarrow$$

$$H—\underset{\underset{\large H}{|}}{\overset{\overset{\large H}{|}}{C}}—\overset{\overset{\large O}{\|}}{C}—SCoA + CH_3—(CH_2)_{12}—\overset{\overset{\large O}{\|}}{C}—SCoA$$

As a result of step 5, three important changes take place:

1. Acetyl coenzyme A is produced.
2. Two carbons (shaded portion above) from the original fatty acid chain are removed from the chain and produce acetyl coenzyme A.
3. The fatty acid, now minus two carbons, is in the form for step 2 to occur. No further activation or consumption of ATP is required. The compound then undergoes step 2 to 5.

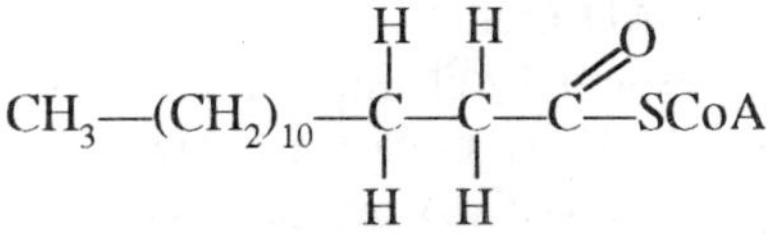

$$CH_3—(CH_2)_{10}—\underset{\underset{\large H}{|}}{\overset{\overset{\large H}{|}}{C}}—\underset{\underset{\large H}{|}}{\overset{\overset{\large H}{|}}{C}}—\overset{\nearrow O}{C}—SCoA$$

After going through six turns of the spiral, palmitic acid is shortened to

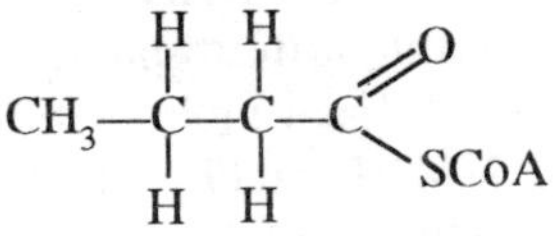

$$CH_3—\underset{\underset{\large H}{|}}{\overset{\overset{\large H}{|}}{C}}—\underset{\underset{\large H}{|}}{\overset{\overset{\large H}{|}}{C}}—C\begin{matrix}\nearrow O \\ \searrow SCoA\end{matrix}$$

This compound goes through step 2 to 5.

$$CH_3-CH_2-CH_2-C(=O)SCoA \xrightarrow[\text{(2)}]{FAD \rightarrow FADH_2} CH_3-CH-CH-C(=O)SCoA \xrightarrow[\text{(3)}]{HOH} CH_3-CH(OH)-CH_2-C(=O)SCoA$$

$$\xrightarrow[\text{(4)}]{NAD \rightarrow NADH_2} CH_3-C(=O)-CH_2-C(=O)SCoA \xrightarrow[\text{(5)}]{CoASH} CH_3-C(=O)-SCoA + H-CH_2-C(=O)SCoA$$

Note that in step 5 the last two-carbon fragment of palmitic acid is converted directly to acetyl coenzyme A. This last two-carbon fragment does not have to go through steps 2 to 5. The net result is that two acetyl coenzyme A's are produced in the final step of the fatty acid spiral. Since two-carbon fragments are removed from palmitic acid to produce acetyl coenzyme A, the overall beta oxidation of this 16-carbon acid yields eight acetyl coenzyme A's:

$$\text{16-carbon acid} \times \frac{\text{1 acetyl CoA}}{\text{two-carbon fragment from aci}} = \text{8 acetyl coenzyme A's}$$

FATTY ACID SPIRAL ENERGETICS

The activation step in the oxidation of palmitic acid converts ATP to AMP. Obviously, cells can't keep converting ATP to AMP or ADP without needing to generate new ATP. The initial investment of an ATP to activate the fatty acid spiral is paid back many times through oxidative phosphorylation.

The FAD and NAD used during the beta oxidation of palmitic acid are part of the respiratory chain; 1 mole of palmitic acid reduces 7 moles of FAD to $FADH_2$; 7 moles of NAD are reduced to $NADH_2$. Although 8 moles of acetyl CoA are produced per mole of palmitic acid, the last step of beta oxidation does not require FAD or NAD. Thus, only 7 moles of $FADH_2$ and 7 moles of $NADH_2$ are produced per mole of palmitic acid. In the respiratory chain, these reduced coenzymes are oxidized back to FAD and NAD. The hydrogens acquired during beta oxidation are passed along the respiratory chain eventually producing water, energy, and ATP. Each mole of $FADH_2$ generates 2 moles of ATP; each mole of $NADH_2$ yields 3 moles of ATP. The production of ATP by beta oxidation of a mole of palmitic acid can be summarized

Credit	ATP production
7 moles $FADH_2 \times \frac{2 \text{ moles ATP}}{1 \text{ mole } FADH_2}$	= 14 moles
7 moles NADH2 $\times \frac{3 \text{ moles ATP}}{1 \text{ mole } NADH_2}$	= 21 moles
	35 moles ATP
	—1 mole ATP used to activate the process
Net production :	34 moles ATP/mole palmitic acid.

ACETYL COENZYME A CONVERSION—THE KREBS CYCLE

We have just seen that ATP production accompanies acetyl coenzyme A formation by beta oxidation of fatty acids. Additional energy can be derived form acetyl coenzyme A by processing it in the mitochondria through a sequence of reactions called the *Krebs cycle. Acetyl coenzyme A from any source can enter the cycle.* Lipids, carbohydrates, and proteins produce acetyl coenzyme A during their metabolism and the acetyl CoA can be further catabolized in the Krebs cycle.

A metabolic cycle is a series of reactions having a substance common to its initial and final steps. The formation of citric acid from oxaloacetic acid and acetyl CoA begins the cycle. Citric acid undergoes a series of reactions in the cycle, the last of which produces oxaloacetic acid so that the cycle may continue (Fig 18.4).

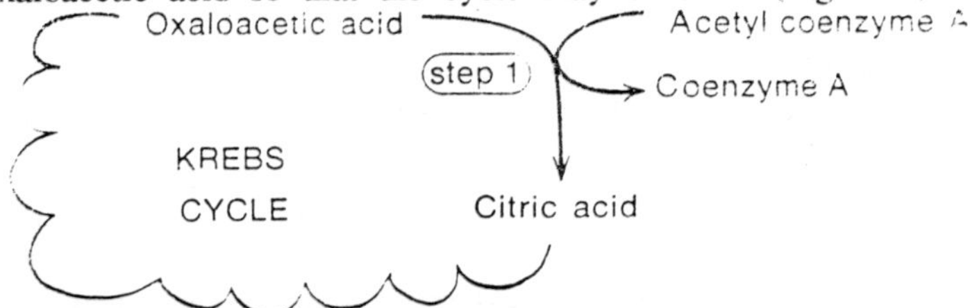

Fig. 18.4 : Relationship of oxaloacetic acid and acetyl coenzyme A to the Krebs cycle.

Because the subsequent reactions in the Krebs cycle involve citric acid and its derivatives, the name *citric acid cycle* is also used for the Krebs cycle. It is also called the *tricarboxylic acid cycle.* All three names are commonly used and the choice of name is largely one of personal preference. We will use the term *Krebs cycle.*

Figure 18.5 and 18.6 should be studied together. References made in the text to certain steps refer to the enumerated steps in the figures. Fig 18.5 is a diagram that presents a narrative overview of the cycle's reactions. Figure 18.6 uses structural formulas to parallel the descriptions given in Fig. 18.5.

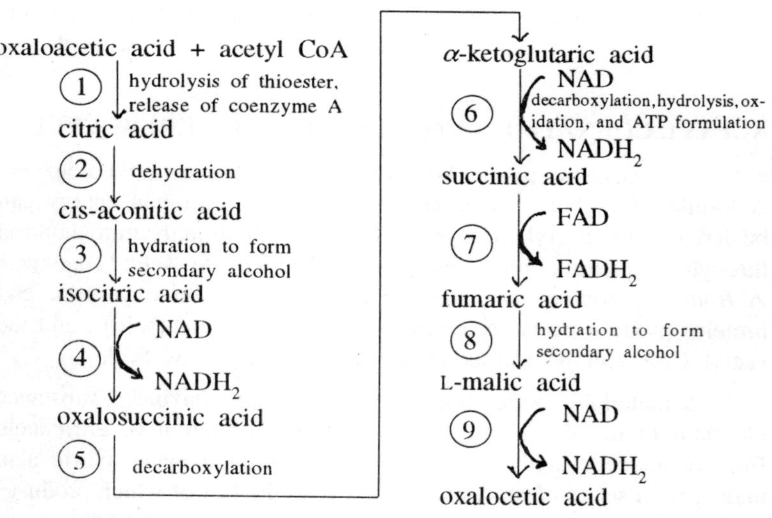

Fig. 18.5 : Overview of steps in the Krebs cycle. Steps 7–9 also occur in the fatty acid spiral.

The Krebs cycle also works cooperatively with the respiratory chain. NAD and FAD in the respiratory chain remove hydrogens from citric acid derivatives formed in the Krebs cycle (step 4, 6, 7 and 9). The hydrogens are passed along the respiratory chain where they eventually combine with oxygen to form water. During the respiratory chain processes, oxidative phosphorylation occurs producing ATP (Fig. 18.1). Step 6 *directly* produces ATP from ADP and P_i without the respiratory chain.

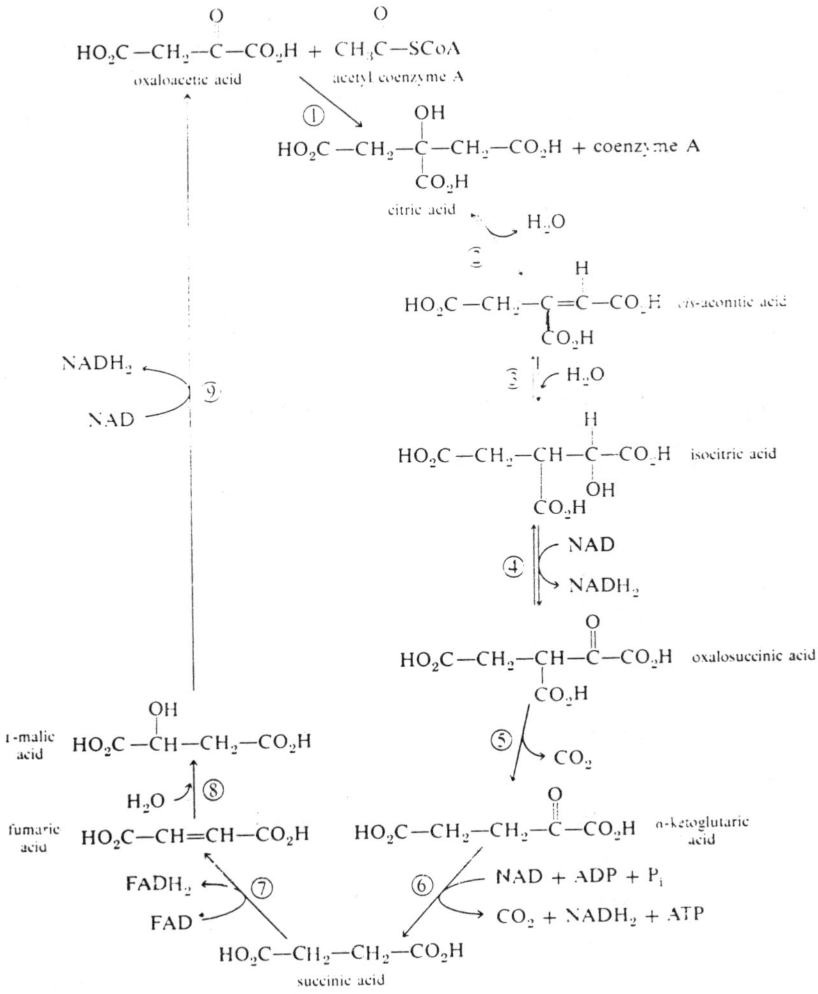

$$HO_2C-CH_2-\overset{O}{\overset{\|}{C}}-CO_2H + CH_3\overset{O}{\overset{\|}{C}}-SCoA$$

oxaloacetic acid, acetyl coenzyme A

① →

$$HO_2C-CH_2-\underset{CO_2H}{\overset{OH}{C}}-CH_2-CO_2H + \text{coenzyme A}$$

citric acid

② H_2O

$$HO_2C-CH_2-\underset{CO_2H}{C}=\overset{H}{C}-CO_2H$$ *cis*-aconitic acid

③ H_2O

$$HO_2C-CH_2-\underset{CO_2H}{CH}-\underset{OH}{\overset{H}{C}}-CO_2H$$ isocitric acid

④ NAD → $NADH_2$

$$HO_2C-CH_2-\underset{CO_2H}{CH}-\overset{O}{\overset{\|}{C}}-CO_2H$$ oxalosuccinic acid

⑤ → CO_2

$$HO_2C-CH_2-CH_2-\overset{O}{\overset{\|}{C}}-CO_2H$$ α-ketoglutaric acid

⑥ NAD + ADP + P_i → CO_2 + $NADH_2$ + ATP

$$HO_2C-CH_2-CH_2-CO_2H$$ succinic acid

⑦ FAD → $FADH_2$

$$HO_2C-CH=CH-CO_2H$$ fumaric acid

⑧ H_2O

$$HO_2C-\overset{OH}{CH}-CH_2-CO_2H$$ L-malic acid

⑨ NAD → $NADH_2$ (regenerates oxaloacetic acid)

Fig. 18.6 : The Krebs cycle. Note the conversion of acetyl CoA to citric acid and the regeneration of oxaloacetic acid in step 9.

Using Fig. 18.5 and 18.6 several other features should be noted;

1. The production of isocitric acid from citric acid (step 2 and 3) converts the nonoxidizable tertiary alcohol group in citric acid to an oxidizable secondary alcohol group in isocitric acid.
2. Step 4 is the oxidation of a secondary alcohol group to a ketone.
3. Step 6 is complex and only the net reaction is shown. The step involves decarboxylation. Hydrolysis, and oxidation. The overall changes produce enough energy to form ATP *directly* from ADP and P_i.
4. Steps 7 to 9 are the same three-step sequence used in steps 2 to 4 of beta oxidation: oxidation with FAD to form a double bond (step 7), hydration of the double bond to form a secondary alcohol (step 8), and oxidation of the alcohol to a ketone using NAD (step 9).

ATP PRODUCTION AND THE KREBS CYCLE

A mole of acetyl CoA going through the Krebs cycle produces 3 moles of $NADH_2$, 1 mole of $FADH_2$, and *directly* 1 mole of ATP. $NADH_2$ and $FADH_2$ are part of the respiratory chain and are oxidized back to NAD and FAD. The hydrogens are passed by the respiratory chain to form water eventually by combination with oxygen. ATP's are produced from ADP and P_i with energy made available form the respiratory chain reactions 3 ATP/$NADH_2$ and 2 ATP/$FADH_2$. We can summarize the amount of ATP produced from 1 mole of acetyl CoA being processed through the Krebs cycle.

Step 4.	1 $NADH_2$ × 3ATP/$NADH_2$	= 3ATP
Step 6.	1 $NADH_2$ × 3 ATP/ $NADH_2$	= 3ATP
Step 6.	1 mole ATP *directly* from ADP + Pi	= 1ATP
Step 7.	1 $FADH_2$ × 2 ATP/ $FADH_2$	= 2 ATP
Step 9.	1 $NADH_2$ × 3 ATP/ $NADH_2$	= 3 ATP
	Net result:	12 ATP/acetyl CoA

In the section on the fatty acid spiral, we saw that a mole of palmitic acid yields 34 moles of ATP and 8 moles of acetyl CoA upon beta oxidation. After combining with oxaloacetic acid to form citric acid, the acetyl CoA proceeds through the Krebs cycle with a production of 12 ATP/acetyl CoA. Therefore, the overall amount of

ATP produced from a mole of palmitic acid catabolized to carbon dioxide and water is

1 mole palmitic acid through fatty acid spiral → 34 moles ATP

In Krebs cycle:

$$8 \text{ moles acetyl CoA from palmitic acid} \times \frac{12 \text{ moles ATP}}{1 \text{ mole acetyl CoA}} \rightarrow 96 \text{ moles ATP}$$

Net production : 130 moles ATP/mole palmitic acid

Note that most of the ATP production occurs via energy released during the Krebs cycle and respiratory chain processes.

19

Hormones and Vitamins

Introduction

With aid of chemical messengers, higher organisms integrate many chemical processes within and among a vast number of cells. Messages are transmitted in two ways:

1) Organ-to-organ transfer of information through nerve action or
2) humoral communication, information via body fluids, mainly blood plasma. Endocrine glands are specialized organs that synthesize and secrete hormones *directly* into the bloodstream. These substances function as chemical messengers. In this chapter, the roles hormones, vitamins, and body fluids play in the maintenance of good health will be examined.

HORMONES

Vertebrate hormones, their classes, target cells, and effects are listed in Table 19.1 (see p. 374 and 375).

Hormones may be chemically classified as proteinoid or lipoid (Fig 19.1). Many hormones control metabolism by either stimulating or reducing enzyme activity in primary metabolic pathways.

The hypophysis (pituitary gland) is a small two-lobed structure. Each lobe functions essentially as a separate endocrine gland secreting specific hormones. Fig 19.2 indicates the secretion of each lobe in mammals.

Humoral, from Latin humor, fluid. Endocrine, from Greek endon, within; krino, to separate. Hormone, from Greek hormon, arouse, excite.

Growth Hormone (Somatotropin)

Growth hormone (GH), also called somatotropin, stimulates skeletal growth and an increase in body weight. Dwarfism results from undersecretion of GH during periods when skeletal growth should normally occur. Oversecretion of growth hormone prior to adulthood results in giantism; oversecretion after full skeletal growth is complete causes acromegaly, a condition in which enlargement of the hands, feet, jaws, and face occurs.

Human growth hormone functions by altering the permeability of cell membranes toward amino acids, thereby affecting the rate of protein-synthesis. Somatotropin also effects carbohydrate and lipid metabolism. It decreases the rate at which carbohydrates are utilized

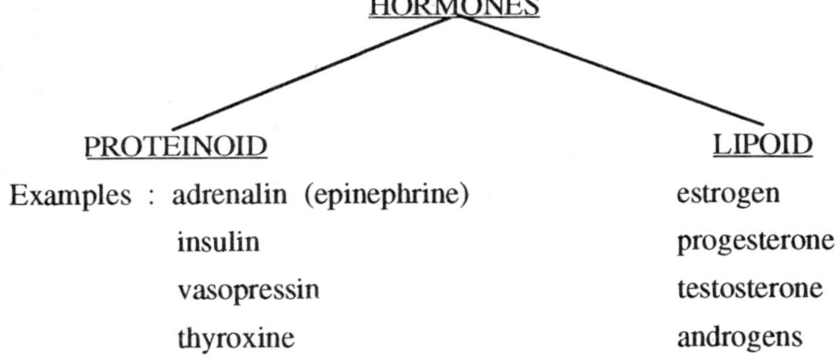

Fig. 19.1. : Hormone classification

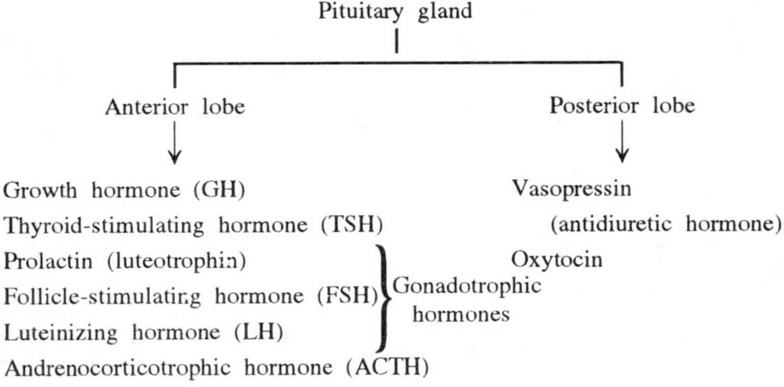

Fig. 19.2 : Hormones secreted by the anterior and posterior lobes of the pituitary gland.

Table 19.1 : Vertebrate Hormones

Hormone	*Target*	*Effects*
I. Peptides, protein hormones	All tissues	Growth of tissues (easily seen in long bones, metabolism of protein, mobilization of fat)
A. Hypophysis—anterior lobe		
Growth hormone (GH, somatotropin)		Lipolysis
Adrenocorticotrophin (ACTH, corticotrophin)	Adrenal cortex	Synthesis and release of glucocorticoids
Thyroid-stimulating hormone (TSH, thyrotrophin)	Adipose tissue	Lipolysis
	Thyroid gland	Synthesis and secretion of thyroxine and triiodothyronine
Male	Seminiferous tubules	Production of sperm
Follicle-stimulating hormone (FSH)		
Luteinizing hormone, LH	Testes	Synthesis and secretion of androgens
Female	Ovary (follicles)	Follicle maturation
FSH		
LH	Ovary	Final maturation of follicle, estrogen secretion, ovulation, corpus luteum formation, progesterone secretion
Prolactin	Mammary glands (alveolar cell)	Milk production in prepared gland
Melanocyte-stimulating hormones	Melanophores	Pigment dispersal in melanophores (darkening of skin)
B. Hypophysis—posterior lobe		
Oxytocin (letdown factor, milk-ejection factor)	Uterus	Contraction of smooth muscle, milk ejection
	Mammary glands	
Vasopressin (antidiuretic hormone, ADH)	Kidney	Reabsorption of water
	Arteries	Contraction of smooth muscle

C. Pancreas		
Insulin	All cellls	Carbohydrate, fat, and protein metabolism, hypoglycemia
Glucagon	Liver	Hyperglycemia
D. Ovary		
Relaxin	Pelvic ligaments	Separation of pelvic bones
E. Thyroid		
Thyrocalcitonin (calcitonin)	Bones, kidney	Excretion of calcium and phosphorus, inhibited calcium released from bones, decreased blood calcium levels
II. Amino acid derivatives		
A. Thyroid		
Thyroxin Triiodothyronine	Most cells	Increased metabolic rate, growth, and development
B. Adrenal medulla		
Norepinephrine Epinephrine	Most cells	Increased cardiac activity, elevated blood pressure, glycolysis, hyperglycemia
III. Steroids and lipids		
A. Testes		
Androgen (testosterone)	Most cells	Development and maintenance of masculine characteristics and behaviour
B. Ovary estrogen	Most cells	Development and maintenance of feminine characteristics and behaviour
C. Corpus luteum Progesterone	Uterus, mammary glands	Maintenance of uterine endometrium and stimulation of mammary duct formation
D. Adrenal cortex		
Hydrocortisone Cortisone	Most cells	Balanced carbohydrate, protein, and fat metabolism; antiinflammatory action
Aldosterone	Kidney	Reabsorption of Na^+ from urine
E. Prostate, seminal vesicles, brain, nerves		
Prostaglandins	Uterus	Contraction of smooth muscle

by the body, thus causing hyperglycemia and glucosuria (diabetogenic effect). Somatotropin increases the release of lipids from fat depots. This increases fat metabolism; excess acetyl CoA is produced; and ketosis results causing ketonemia and, in severe cases, ketonuria. This is referred to as the ketogene effect of growth hormone.

THYROID STIMULATING HORMONE

Thyroid-stimulating hormone (TSH) acts on two sites- fat depots and the thyroid gland. The hormone causes release of fatty acids from fat depots and thus acts as a regulator for fatty acid metabolism. TSH also controls thyroid function by stimulating iodine uptake, formation of iodine-containing hormones such as thyroxine, and release of these hormones into circulation. Using injected radioactive I-131 as a tracer, an animal's rate of iodine uptake can be followed.

VASOPRESSIN AND OXYTOCIN

Vasopressin and oxytocin are hormones having similar chemical structures: They are octapeptides, each containing a disulfide bond. They differ by two amino acid residues (Fig 19.3). Vasopressin's main function is to regulate water loss from the body via urine.

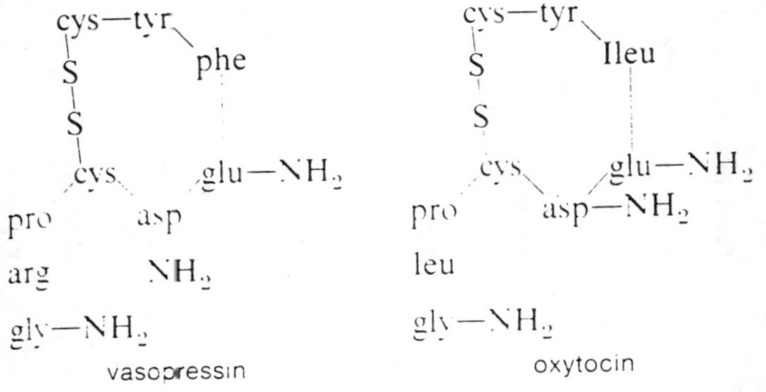

```
   cys—tyr                      cys—tyr
   |      \                     |      \
   S       phe                  S       Ileu
   |                            |
   S                            S
   |                            |
   cys      glu—NH2             cys      glu—NH2
pro      asp                 pro      asp—NH2
arg      NH2                 leu
gly—NH2                      gly—NH2
     vasopressin                  oxytocin
```

Fig. 19.3 : Symbolized structural formulas of vasopressin and oxytocin. Glu-NH$_2$ and asp-NH$_2$ represent glutamic acid and aspartic acid residues whose side chain carboxyl groups have been converted to the amide form. In gly-NH$_2$, the terminal carboxylic acid group has been converted to the amide.

Oxytocin causes powerful contractions of the uterus during childbirth. This hormone is frequently administered to stimulate uterine contractions during childbirth. Oxytocin also causes milk in the mother's breasts to be released into ducts where it can be obtained by the sucking infant.

TYROSINE-DERIVED HORMONES

The amino acid tyrosine is the substance from which two important groups of hormones are synthesized — catecholamine-type hormones and thyroid hormones (Fig. 19.4).

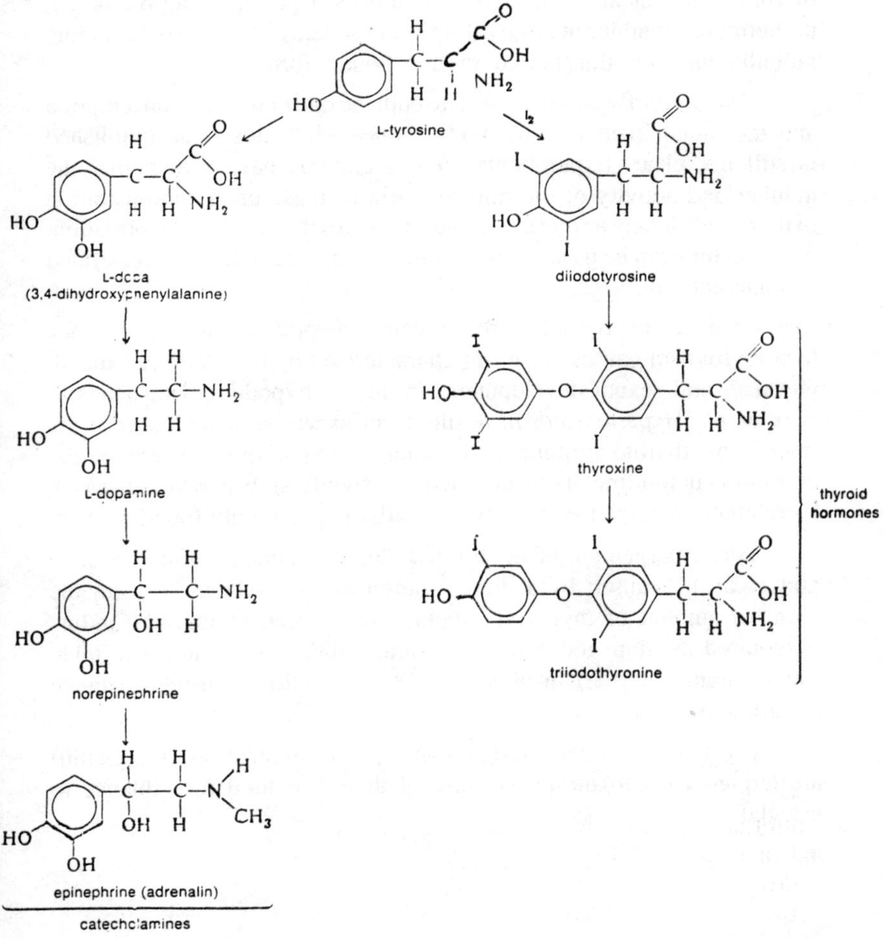

Fig. 19.4 : Hormone derivatives of tyrosine

It is known that at least four physiologically active, structurally related, iodine-containing compounds act as thyroid hormones. Under normal conditions, the thyroid gland rapidly takes up iodine for the synthesis of these hormones. The two most important ones are thyroxine and triiodothyronine. These hormones are stored in the thyroid gland by combination with a protein to form thyroglobulin. Hydrolysis of thyroglobulin releases the thyroid hormones. Upon entering the blood, the hormones combine mainly with a plasma protein called thyroxinebinding globulin and are transported in this bound form.

The thyroid hormones regulate both the rate of oxygen consumption and the metabolism in most body tissues. How this is accomplished is still not clear. Experiments on test animals have shown that the number and activity of the mitochondria increase upon administration of thyroxine. It may be that thyroxine stimulates the rate of ATP formation, which in turn can be used to provide the energy necessary for accelerated cellular activity.

Insufficient thyroid secretion causes *hypothyroidism*. In infants, hypothyroidism causes cretinism characterized by retardation of metal, physical, and sexual development. In adults, hypothyroidism leads to myxedema. *Hyperthyroidism* results from excessive stimulation of the thyroid by thyroid-stimulating hormones (TSH). Enlargement of the thyroid occurs and metabolic rate increases significantly (Graves' disease). Exephthalmos (protrusion of the eyeballs) is commonly found as well.

Any enlargement of the thyroid gland is termed a goiter. Simple goiter can be caused by a dietary deficiency of iodine. To form the required amount of thyroxine, about 1 mg./ week of ingested iodine is required. Sodium iodide or potassium iodide can be added to table slat (sodium chloride) in about a 1: 10.000 ratio to ensure adequate dietary iodine.

L-Dopa, dopamine, norepinephrine, and epinephrine (adrenalin) are termed *catecholamines* because of their structural resemblance to catechol

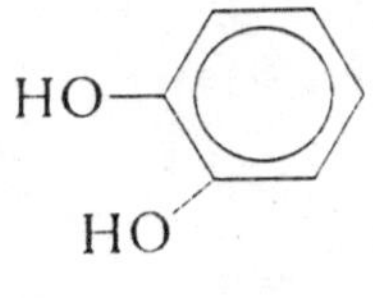

HO
HO

They are produced by the medulla (inner portion) of the adrenal glands. Many individuals suffering form Parkinson's disease are aided by treatment with L-dopa. This material reduces Parkinson symptoms such as tremor and lack of muscle control, although exactly how the drug does this is still not known.

Norepinephrine and epinephrine are sometimes called *emergency hormones* because of their effectiveness in rallying the body during times of physical or emotional stress. Epinephrine is in higher concentration and is more effective than norepinephrine. During stress, epinephrine stimulates a rapid increase in blood sugar through liver glycogenolysis. Release of fatty acids from fat depots also occurs. Accompanying these changes are increases in heartbeat, blood flow in limb muscles, and oxygen consumption. Phosphorylase is the enzyme that catalyzes the release of glucose from glycogen. When the body is resting, most of the phosphorylase is inactive so that glucose is not released from glycogen. Epinephrine indirectly activates phosphorylase and glycogenolysis rapidly occurs.

INSULIN AND GLUCAGON

Insulin is secreted by the beta cells of the islets of Langerhans in the pancreas. Insulin increases the rate of glucose transport across cell membranes especially in muscle and fatty tissues. A person suffering from untreated diabetes mellitus experiences abnormally high glucose concentration in the blood because of failure to transport glucose into cells.

Glucagon is a protein hormone produced by the alpha cells of the pancreas. Recent evidence indicates it can be produced by newly identified gastrointestinal tract A cells as well. New experimental data suggest that the development of severe diabetic hyperglycemia requires the presence of glucagon as well as a lack of insulin. The recently discovered hormone somatostatin controls the release of glucagon and insulin. It has been suggested that glucagon and insulin counterbalance each other with insulin primarily concerned with the storage of sugar and amino acids ingested at mealtime and with glucagon, by stimulating glycogenolysis, maintaining glucose level between meals. If too much glucagon is released, glucose concentration rapidly increases.

Diagnosis of diabetes mellitus is facilitated by administration of a glucose tolerance test. After fasting, a patient ingests 50g of glucose in solution. Blood samples are taken periodically and the blood glucose levels are determined. These concentrations are compared to those of a normal individual and a diabetic. Fig 19.5 illustrates typical glucose tolerance test results for a normal individual and a diabetic.

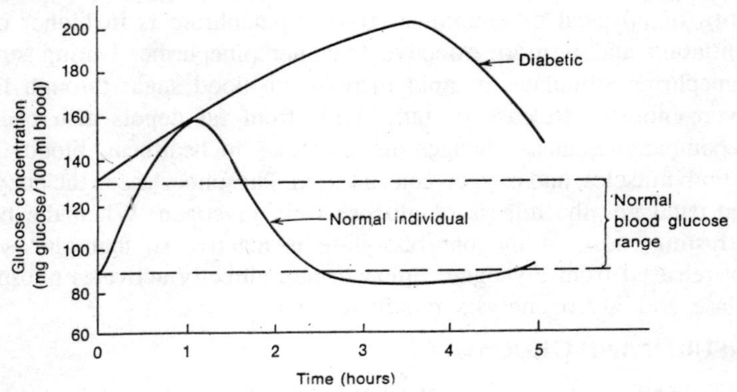

Fig. 19.5 : Glucose tolerance test curves for a normal person and a diabetic.

Following ingestion of the glucose solution, a normal individual's blood glucose concentration quickly rises from its normal fasting value of less than 100 mg. of glucose per 100 mL of blood (100 mg%) to about 160 mg%. It then drops to below normal level within 3 hr. In this individual, an increase in insulin secretion follows glucose intake and facilitates rapid transport of glucose into cells for metabolism.

A diabetic usually has a fasting blood glucose level of at least 120 mg%. Upon ingestion of 50 g of glucose, a gradual rise in blood glucose level occurs in the diabetic during the next 2 to 3 hr. The blood sugar level slowly decreases during the third through fifth hours. The slowness of the decline of glucose level in the diabetic following glucose ingestion indicates that an increase in insulin secretion does not occur in the diabetic. Note that the diabetic's glucose level does not fall below the diabetic's normal fasting level during the third to fifth hour following ingestion. This is in contrast to the normal individual's behavior during this period in which glucose level may actually fall below the normal fasting level. In normal individuals, glucose level returns to normal fasting level values after the fifth hour due to glucose release by glycogenolysis.

STEROIDS

Cholesterol is the most commonly occurring steroid in animal and is the parent compound for the synthesis of the major classes of steroid hormones: estrogens, androgens, progestogens, and the steroid hormones of the adrenal gland. The structural formula of cholesterol is shown below.

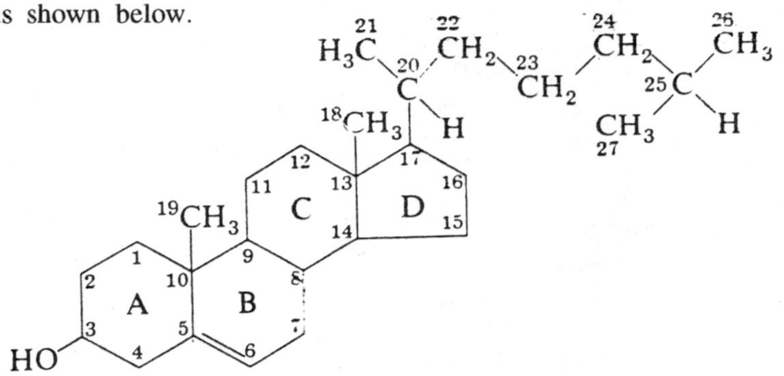

cholesterol

The rings of steroids are generally identified as A,B,C and D.

Follicle-stimulating hormone (FSH) and luteinizing hormone (LH) regulate production and activity of sex hormones, e.g., testosterone and progesterone. Their roles in females and males are noted below.

Males		Females	
FSH	LH	FSH	LH
Regulation of sperm production	Sperm production	Estrogen production	Ovarian follicle rupture
Regulation of testosterone production		Ovarian follicle development	Corpus luteum development

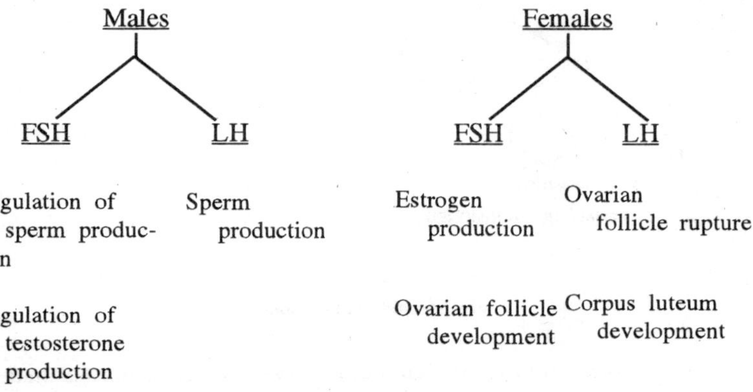

The term *estrogen* refers to a group of compounds having similar structures. Estradiol and estrone are the major estrogens. In males, androgens and testosterone are produced by specialized cells in the tests (Fig 19.6).

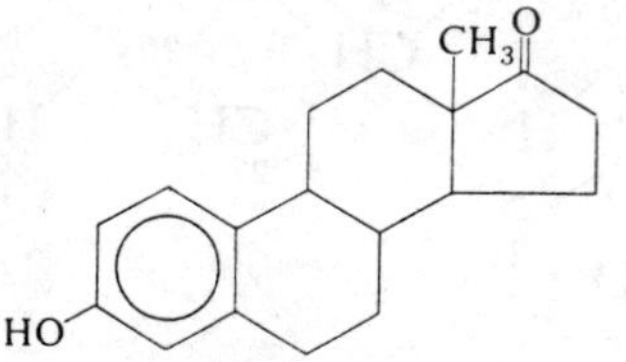

estrone

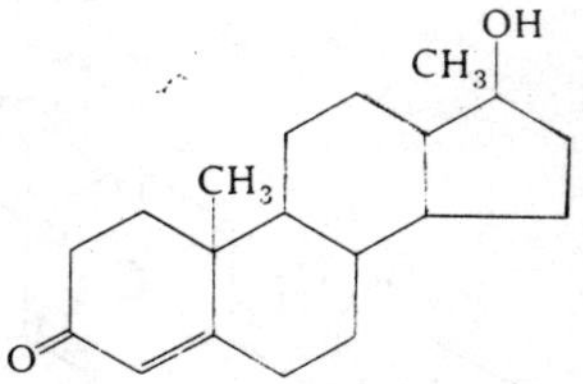

testosterone

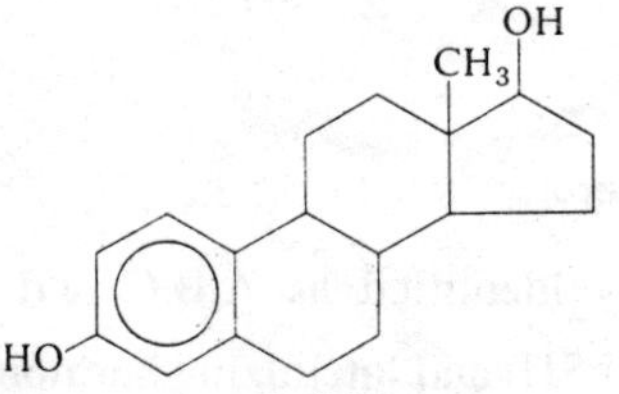

estradiol

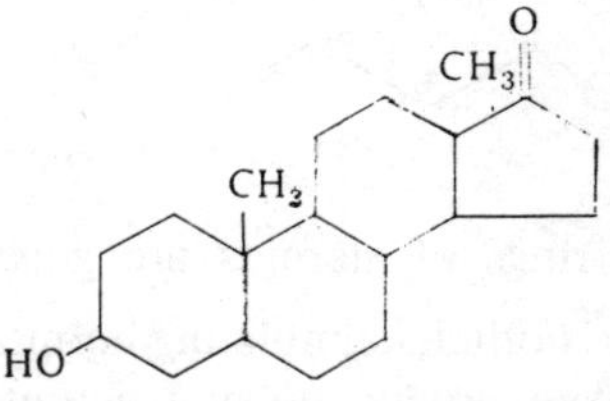

androsterone

MALE SEX HORMONES

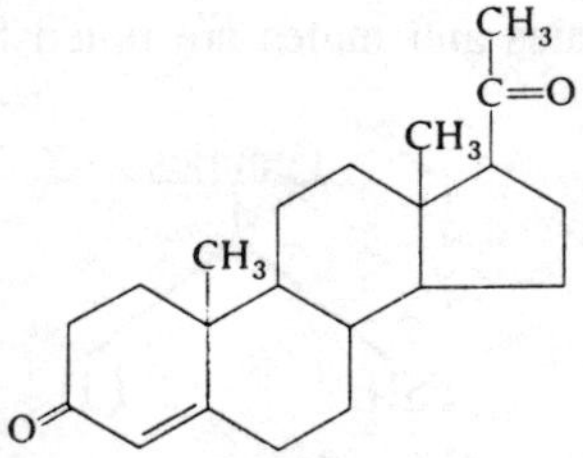

progesterone

FEMALE SEX HORMONES

Fig. Steroid sex hormones.

Low concentrations of female hormones occur in normal males: low concentrations of male hormones also are present in normal females. It is the higher concentrations of the testosterone in males and estrogens and progesterone in females that causes development and maintenance of appropriate secondary sexual characteristics.

FEMALE SEX HORMONES

In human females, changing rates of sex hormone secretion cause a periodic change in the ovaries, uterus, and associated sex organs. This periodic pattern of change is called the *menstrual cycle*. Beginning at puberty and continuing through menopause, the cycle normally is repeated approximately every 28 days except during pregnancy. Hormonal control of the cycle regulates the release of a mature egg cell and preparation of the uterine lining for the implantation of a fertilized egg cell. This is illustrated in Fig 19.7.

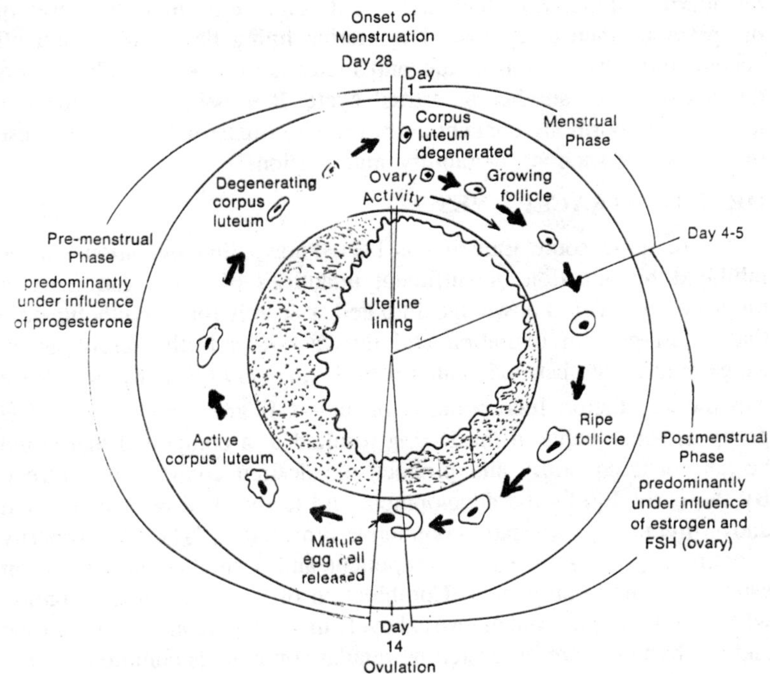

Fig. 19.7 : The human female menstrual cycle.

Following menstrual flow, an ovarian follicle, under the influence of FSH, progressively develops and matures in one of the ovaries. As it matures, the developing follicle secretes increasing quantities of estrogen that aid in the preparation of the uterine lining for arrival of a fertilized egg. Approximately 14 days after the onset of menstrual flow, the influence of luteinizing hormone (LH) causes rapture of the mature follicle, releasing a mature egg cell (ovulation). LH and prolactin promote formation of the corpus luteum within the ruptured follicle. The corpus luteum grows and secretes progesterone, which further promotes uterine lining growth. Progesterone also inhibits development of other ovarian follicles, inhibits uterine contraction, and promotes changes within the breasts.

If the egg is not fertilized, the corpus luteum degenerates causing progesterone secretion to stop and estrogen production to decline. The decrease in concentrations of these hormones stops the thickening of the uterine lining. By about the twenty-eight day after the onset of the previous menstrual flow. The uterine lining begins to slough off accompanied by hemorrhaging and menstruation begins. This marks the beginning of another menstrual cycle. It is believed that the low level of sex hormone concentration prior to menstruation is the cause of symptoms such as irritability and tension.

ORAL CONTRACEPTIVES

For quite some time it has been known that ovulation can be inhibited by injection of sufficient quantities of either estrogens or progesterone. This knowledge initiated a search for a synthetic drug that could prevent ovulation and thus act as a birth control agent. Research indicated that modification of the D-ring substituents in progesterone was the key factor. Incorporation of an —OH group and a —C≡CH group as substituents in the D ring resulted in a compound that could be administered orally and produce the desired contraceptive effect. By the early 1960's the development and testing had been completed and birth control pills became commercially available. "The Pill" primarily contains a progesterone-like component and a smaller amount of an estradiol-related component. Combination of these prevents ovulation while allowing a normal menstrual cycle in most woman. Norethindrone and northynodrel are progesterone-related compounds commonly used; mestranol is the common estradiol-like component.

Adrenal Steroids

The cortex (outer layers) of the adrenal glands produce two major types of hormones—mineralocorticoids and glucocorticoids. The mineralocorticoids regulate concentrations of mineral ions (mainly sodium and potassium ions). Glucocorticoids principally influence carbohydrate metabolism and are controlled by ACTH secretion. Emotional stress, temperature extremes, and drugs are among stimuli and influence ACTH secretion.

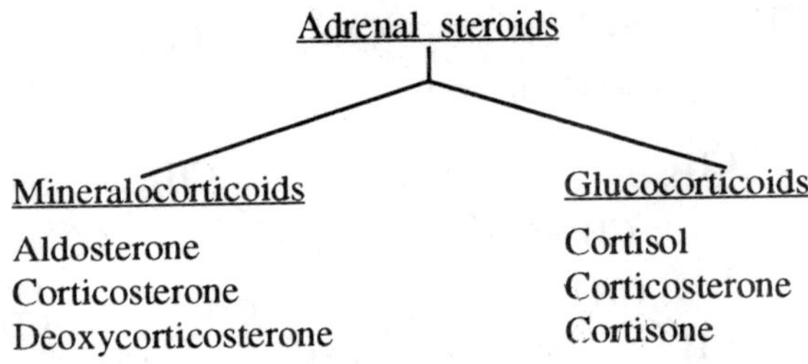

Feedback and Control of Hormone Concentration

ACTH, produced by the anterior pituitary, controls the level of glucocorticoid secretion. This control exemplifies what is called a *feedback relationship* in which as change in concentration of one substance causes an alternation of the concentration of a second substance. For example, a decrease in glucocorticoid level in the blood stimulates an increase in ACTH secretion. In turn, the ACTH secretion causes the adrenal cortex to secrete glucocorticoids. As glucocorticoid levels rise, this inhibits further ACTH secretion by the pituitary. Because an increase in secretion of one substance causes a decrease in secretion of another substance, the term *negative feedback* is used to describe this action. Thus, some hormones can regulate secretion of other hormones.

PROSTAGLANDINS

Prostaglandins are 20-carbon fatty acids derived from essential fatty acids such as arachidonic acid. These essential fatty acids are converted to prostanoic acid whose basic structure is common to all prostaglandins (see structure on page no. 386). The known prostaglandins in humans are classified into A,B,E, and F families.

Prostaglandins, although present in minute amounts throughout body tissues, are hormone-like compounds that act on localized target sites. Prostaglandins are probably formed in the same cells in which they influence activity. This is in contrast to the action of hormones, which effect target cells a long distance from the point of hormone secretion. Prostaglandins are rapidly inactivated in the bloodstream and the lungs. They have been found to participate in a broad spectrum

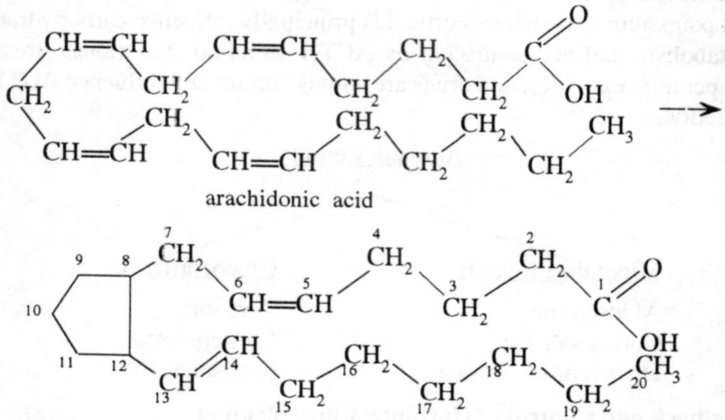

arachidonic acid

prostanoic acid

of activities and are useful in (1) treatment of high blood pressure, peptic ulcers, and asthma; (2) induction of labor; (3) therapeutic abortion; (4) prevention of blood clotting; and (5) controlling inflammation. Their physiological actions seem to depend on the type of organism and target tissue. Prostaglandins apparently function by regulating hormone action and recent research indicates that they influence the activity of the secondary messenger *c*-AMP. Prostaglandins interact with adenyl cyclase in a variety of tissues. This may be the reason for their diversity of effects. They may act as messengers between classical hormones and *c*-AMP.

VITAMINS

Since ancient times, a relationship between proper diet and good health has been recognized. Correct nutrition requires eating more than just carbohydrates, proteins, and lipids. In 1912 the vitamin theory was proposed. According to this theory, the lack of particular nutritional factors in the diet cause certain specific diseases such as scurvy, rickets, and pellagra, Shortly thereafter, specific substances were isolated from foods. When given to test animals suffering from dietary deficiency diseases, these substance secured the condition. These vital food factors were called vitamines, now spelled *vitamins*. Currently, a vitamin is defined as a naturally occurring organic compound found in foods that is necessary for normal body maintenance and growth. The vitamin may be present in foods either as the vitamin itself or as a compound that can be readily converted into the vitamin.

Vitamins are classified according to both their functional groups and solubility. Some are soluble in water, while others are water-insoluble but soluble in fats (the lipid-soluble vitamins). Table 19.2 classifies vitamins according to their solubility.

Table 19.2 : Classification of vitamins according to solubility

Vitamins

Water soluble	Lipid soluble
Thiamine (vitamin B_1)	Vitamin A
Riboflavin (vitamin B_2)	Vitamin D
Niacin (nicotinic acid)	Vitamin E
Vitamin B_6 (pyroxidal)	Vitamin K
Pantothenic acid	
Biotin	
Folic acid	
Vitamin B_{12}	
Ascorbic acid (vitamin C)	

Because they are fat-soluble, the lipid-soluble vitamins accumulate in body fat. Their availability from this type of storage may prevent deficiencies even when an individual is on a vitamin-deficient diet of a fat-soluble vitamin for some time. In contrast, water-soluble vitamins are not stored and therefore must be ingested daily.

Table 19.3 and 19.4 list their recommended daily allowance, and conditions caused by their deficiency.

Water-Soluble Vitamins

Thiamine was the first vitamin was the first vitamin isolated. In the body, it is found mainly as thiamine pyrophosphate, which functions as a coenzyme in decarboxylation reactions of α-keto acids. Because of its involvement in metabolism, thiamine is required by all body cells. During thiamine deficiency, glucose utilization by the central nervous system is significantly reduced: this can lead to muscle weakness, cardiac difficulty, and intestinal tract disorders.

Riboflavin is a component of the coenzymes FAD and FMN, both vital to respiratory chain function. Thus, riboflavin is required by all body cells and a deficiency reduces the rate of energy production during cellular metabolism. The general symptoms are similar to those noted below for niacin deficiency.

Table 19.3 : Vitamins Important for Human well-being

Water-soluble Vitamins

Vitamin	Recommended (Adult) Daily Allowance	Dietary Sources	Deficiency Effect(s)
Thiamine (B_1)	1-2 mg	Whole grams, liver, peas, beans	Beriberi
Riboflavin (B_2)	1-2 mg	Wheat germ, liver green leafy vegetables, milk, eggs	Reduction of rate of energy production from metabolism
Niacin	15-20 mg (estimated)	Wheat germ, meats	Pellagra
Pyridoxal (B_6)	Approximately 700 μg (estimated)	Egg yolks, liver, wheat germ, yeast	Convulsions in infants, dermatitis in adults
Pantothenic acid	10 mg	Yeast, liver, eggs	Not known for humans
Biotin	Approximately 200 μg (estimated)	Yeast, liver peanuts, chocolate	
Folic acid		Wheat germ, yeast, spinach	Anemia, growth failure, intestinal tract disorders
Cobalamin (B_{12})	Approximately 1 μg	Liver, meats, eggs	Pernicious anemia
Ascorbic acid (C)	75 mg	Citrus fruits, green vegetables	Scurvy

Table 19.4 : Vitamins Important for Human well-being

Lipid-soluble Vitamins

Vitamin	Recommended (Adult) Daily Allowance	Dietary Sources	Deficiency Effect(s)
Vitamin A	5000 I.U. (excess can cause hypervitaminism) 1 I.U.=0.3 μg retinol	Cod liver oil, yellow and green vegetables	Eye disorder (xerophthalmia)
Vitamin D_2	400 I.U. (avoid hypervitaminism) 1 I.U. = 0.025 μg cholecalciferol		Poor bone formation (hypervitaminism can cause brittle bones)
Vitamin D_3		Irradiation of compounds in the skin produces the vitamins fish liver oils	

(Contd....)

Table 19.4 : (Contd..)

Lipid-soluble Vitamins

Vitamin	Recommended (Adult) Daily Allowance	Dietary Sources	Deficiency Effect(s)
Vitamin E	10-30 μg (estimated)	Wheat germ oil, cottonseed oil, rice, green leafy vegetables	Required by many animals, human requirement and deficiency effe-cts uncertain
Vitamin K	Not known	Manufactured by intestinal bacterial in humans	Impairs blood clotting

Niacin (nicotinic acid) is present in the body as the nicotinamide portion of NAD and NADP. These coenzymes are essential for metabolism and therefore all body cells require niacin. All body functions are reduced during nacin deficiency; dermatitis, diarrhea, and mental disorders can results. These symptoms are characteristic of individuals suffering from pellagra.

Pyridoxine (vitamin B_6) exists in cells mainly as pyridoxal phosphate and pyridoxamine. These three substances as a group are termed vitamin B_6. The vitamin serves as a coenzyme during decarboxylation and transamination reactions. Thus, it is of importance to amino acid and protein synthesis.

Cobalamin (vitamin B_{12}) is a complex molecule containing molecule

containing pyrrole rings bonded to a central cobalt ion. Pernicious anemia results from inadequate absorption of vitamin B_{12}. Gastric juices contain proteins that bind that vitamin and facilitate its absorption. If these proteins (gastric intrinsic factor) are lacking due to defective gastric secretion, pernicious anemia may result even when a sufficient dietary supply of the vitamin is ingested.

Biotin is another member of the vitamin B complex of vitamins. Its main role is in fatty acid synthesis where it acts as a carrier of carbon dioxide. By accepting CO_2 from biotin, acetyl CoA is converted to malonyl CoA, the first step in fatty acid synthesis. It also is involved in the carboxylation of pyruvic acid to malic acid, which is then oxidized to oxaloacetic acid for use in the Krebs cycle.

Lipid-Soluble Vitamins

β-Carotene is the parent substance for the production of vitamin A. An enzymatically catalyzed cleavage of a β-carotene molecule yields two molecules of vitamin A.

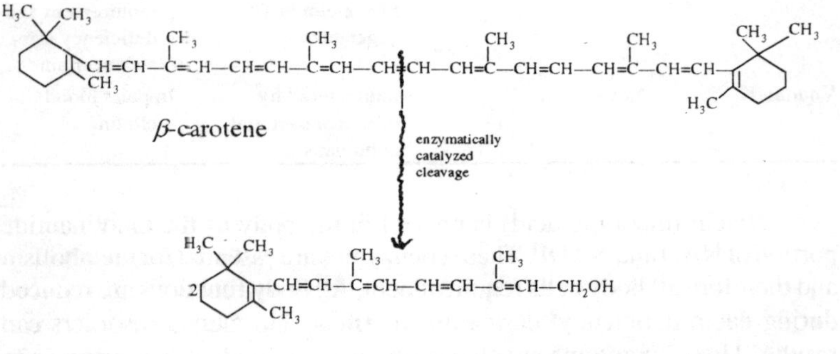

H_3C CH_3 ... $CH=CH-C(CH_3)=CH-CH=CH-C(CH_3)=CH-CH=CH-CH=C(CH_3)-CH=CH-CH=C(CH_3)-CH=CH-$ ... CH_3 CH_3 H_3C

β-carotene

enzymatically catalyzed cleavage

2 H_3C CH_3 ... $-CH=CH-C(CH_3)=CH-CH=CH-C(CH_3)=CH-CH_2OH$ ... CH_3

vitamin A (alcohol form, retinol)

Because of the system of alternating single and double bonds, both β-carotene and vitamin A absorb light in the visible region of the spectrum and; consequently, are colored. β-Carotene is purple and vitamin A is yellow. The alcohol form of vitamin A can be oxidized to retinene, the aldehyde form. A major function of vitamin A is in vision. Retinene reversibly combines with the protein opsin to form rhodopsin (visual purple), one of the pigments of vision.

Because of its involvement in vision, a dietary deficiency of vitamin A creates a lack of rhodopsin, which in turn causes a condition known as night blindness. Vitamin A is also necessary for normal growth and proper skin condition.

Vitamin D is sometimes called *the sunshine vitamin.* This is because 7-dehydrocholesterol, a steroid normally found in human skin, is converted to vitamin D_3 when irradiated by ultraviolet light. Ergosterol, found in molds and yeasts, is converted to vitamin D_2 (calciferol) when irradiated by sunlight. The natural vitamin D content of milk and other foods is supplemented by addition of irradiated ergosterol (see structure on page no. 391).

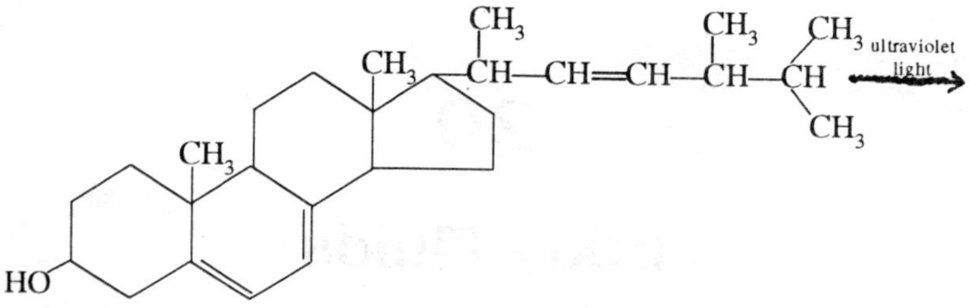

ergosterol

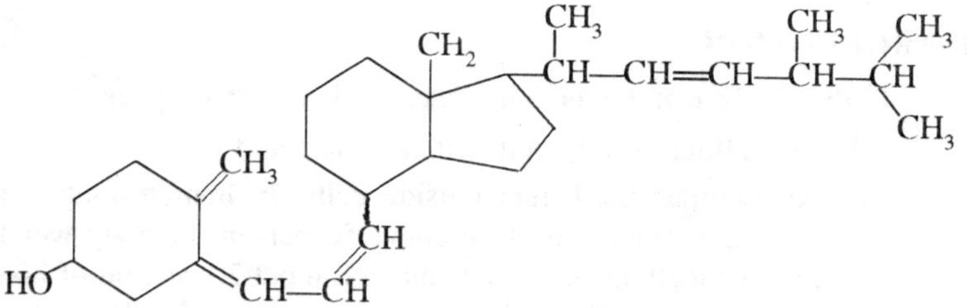

vitamin D_2 (calciferol)

Occurring in several forms, vitamin K is required for adequate prothrombin and factor VII formation by the liver. Both of these are necessary for proper clotting of the blood. Lack of vitamin K causes clotting time to increase and hemorrhaging occurs. Normally the widespread occurrence of vitamin K in common foods and its production by intestinal bacteria result in an adequate vitamin level.

20

Body Fluids

INTRODUCTION

Body fluids can be broadly classified into two types:

1) intracellular fluid, that *within* cells; and
2) extracellular fluid, that outside cells. In human adults, intracellular fluid makes up about 50 percent of body weight and extracellular fluid accounts for about 20 percent of body weight. Extracellular fluid is composed mainly (75 percent) of *interstitial* fluid consisting of water and soluble materials. Interstitial fluid bathes the cells and makes and exchanges of solutes between the blood and cells occur more readily. The bloodstream accounts for most of the remaining 25 percent of extracellular fluid. Digestive juice, synovial fluid, cerebrospinal fluid, semen, milk, and lymph are other extracellular fluids.

BLOOD

A human adult's body contains approximately 6 L of whole blood, which is a mixture consisting of *plasma* (55-60 percent by volume) and formed elements—red cells (erythrocytes), white cells (leukocytes), and platelets. Plasma can be separated form the formed elements by centrifuging whole blood.

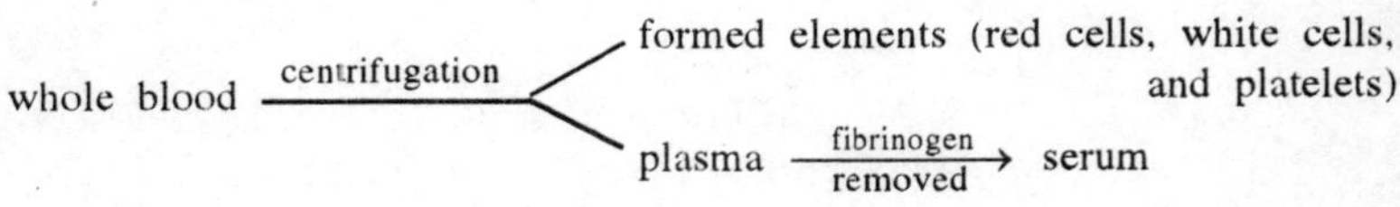

Plasma, the liquid portion of circulating or uncoagulated blood, is an aqueous solution containing plasma proteins and various cations and anions. Other solutes such as glucose, uric acid, urea, and polysaccharides are also present. Trace amounts of metal ions (Mn^{2+}, Cu^{2+}, Fe^{2+}, Co^{2+}, etc.) are also present. Table 20.1 lists the major organic solutes in plasma and their normal concentration ranges. The main plasma proteins are albumin, other globular proteins, and fibrinogen. The main ions present are sodium, potassium, calcium, magnesium, chloride, bicarbonate, monohydrogen and dihydrogen phosphates, and sulphate. If fibrinogen is removed from plasma, the resulting liquid is called *serum*.

Table 20.1 : Some organic solutes in blood plasma

Carbohydrates and Carbohydrate Metabolites (fasting levels)		Lipids, Ketone Bodies, and bile acids (fasting levels)		Selected Non-protein Nitrogen-containing Compounds (NPN) (fasting levels in venous blood)	
Substance	Concentration (mg/100 mL blood)	Substance	Concentration (mg/100 mL blood)	Substance	Concentration (mg/100 mL blood)
Glucose	60–100	Total lipid	385–675	Total NPN	16–40
Fructose	5–9	Neutral fat	0–260	Urea	10–23
Lactose (during lactation)	0–2	Unesterified		Creatinine	0.8–1.0
Polysaccharides	90–140	Fatty acids	8–31	Creatine	0.2–0.7
Glucosamine	50–90	Phospholipids	110–250	Uric acid	2.5–6
Lactic acid (resting)	5–20	Cholesterol	140–260		
Pyruvic acid	0.4–2	Ketone bodies	0.2–0.9		
Citric acid	1.5–3.2	Cholic acid	0.2–3.0		

Albumin is the most abundant plasma protein. Its main functions are to aid in maintaining osmotic pressure balance between plasma and interstitial fluid and to transport substances that are normally water insoluble such as fatty acids.

BLOOD CLOTTING

When blood vessels are cut or broken, bleeding occurs. Obviously bleeding cannot continue very long without severe physiological consequences. The body's major defence against loss of blood is blood clot formation. Clotting requires the participation of many substances and only a brief overview is presented here. Fibrinogen is a soluble plasma protein responsible for clotting. It converts into fibrin, an insoluble, fibrous protein whose strands form a network that traps the formed elements during clotting.

Many plasma proteins that participate in the clotting process have been discovered in the last 20 years. In individuals with clotting diseases, these proteins are either absent or inactive. Coagulation proteins participate in clotting through two related mechanisms called *intrinsic* and *extrinsic* pathways. The former refers to reactions that lead to thrombin formation using only factors present in plasma. The extrinsic pathway involves plasma factors as well as those extracted from tissue components. The two mechanisms seem to function simultaneously.

The conversion of fibrinogen to fibrin is catalyzed by the enzyme thrombin. Why, then, doesn't the blood continuously clot? Thrombin exists in the blood in an inactive form, prothrombin, and conversion of inactive prothrombin to active thrombin requires many factors such as calcium ions and proteins (collectively called *thromboplastins*). At least 13 factors are required to convert prothrombin to fibrin. Some of these factors are always present in the blood but others are only released when blood vessels are cut. Hemophilia is a hereditary disease in which an individual (usually male) lacks one or more of the enzymes or enzyme activators necessary for the conversion of prothrombin to thrombin.

$$\underset{\text{(inactive form)}}{\text{prothrombin}} \xrightarrow{\text{Vitamin K, } Ca^{2+}\text{, thromboplastins, other factors}} \underset{\text{(active form)}}{\text{thrombin}}$$

$$\underset{\text{(soluble)}}{\text{fibrogen}} \xrightarrow[Ca^{2+}]{\text{thrombin}} \underset{\text{(insoluble)}}{\text{fibrin}}$$

Normal prothrombin production in the liver depends on adequate dietary intake of vitamin K.

Factor 13, which is activated by thrombin, stabilizes fibrin. Recent experimental evidence indicates that fibrin formation results from the step-wise interaction of a large number of plasma proteins. For example, a protein called *kallikrein* has been shown to activate factor 12 and has been proposed to be part of the factor 12 molecule. Factor 12, in contact with kallikrein and collagen, forms the enzyme, activated factor 12. This activates factor 11 to activated factor 11 (an enzyme), which, in the presence of Ca^{2+} ions, activates factor 10, etc., and the sequence continues until fibrin formation is completed.

The requirement of calcium ions for clotting is easily demonstrated. The addition of a solution of either sodium oxalate or sodium fluoride to whole blood precipitates calcium oxalate or calcium fluoride. Because calcium ions are removed by this treatment, the blood will not clot.

$$\underset{\text{in blood}}{Ca^{2+}\ (aq)} + \underset{\text{from } Na_2C_2O_4}{C_2O_4^{2-}(aq)} \longrightarrow CaC_2O_4\ (s)$$

Sodium citrate solutions are also used to prevent clotting by removing calcium ions. Citrate ions form strong coordinate covalent bonds to calcium ions and prevent calcium for participating in the clotting process.

URINE

The kidneys filter a prodigious amount of blood daily. The kidney tubules dialyze waste products, electrolytes, metabolites, and water from the blood. These substances are either reabsorbed by the blood or released to form urine (diuresis), depending on physiological needs. In this way, the kidneys regulate the composition of body fluids.

The broad fluctuation of urine composition can be seen from specific gravity data. The normal range of urine's specific gravity is 1.008 to 1.030. Variation is due to factors such as fluid intake, muscular activity, digestion, and emotional condition. The normal range of urine production by adults in a 24-hr period is 600 to 2500 mL, averaging about 1500 mL. Clinically, 24- to 48-hr samples are collected for a compete urinalysis so as to minimize variation in composition caused by the factors mentioned above.

Urine is normally a pale amber color due to the presence of pigments that are normal by-products of hemoglobin decomposition. The pigments are collectively called *urbilinogen.* When freshly voided, urine is clear, containing no appreciable sediment. Upon standing, cloudiness may develop due to alkaline conditions caused by bacterial action that precipitates phosphates.

Urine pH can range between the extremes of 4.5 to 8.4 but ordinarily is 5.0 to 7.0. These ranges mainly reflect the ratio of $H_2PO_4^-$ to HPO_4^{2-} ions present, hydrogen ions being exchanged during conversion of these phosphate forms.

$$H_2PO_4^- \rightleftharpoons H^+ + HPO_4^{2-}$$

Diet is normally the chief influence on urine acidity. Meats and other high protein foods yield acids as waste products of their metabolism, causing urine to become more acidic. Most fruits and vegetables contain organic acids such as oxalic and citric acids that are metabolized via the Krebs cycle to bicarbonate ions, causing urine to become less acidic.

Blood Plasma and Urine Relationships

Plasma and urine are both aqueous solutions. The solutes that are neither retained by plasma nor released into cells are passed into urine by kidney action. Thus, the kidneys

1) eliminate metabolic wastes;
2) control water balance in intracellular and extracellular fluids;
3) control pH by either retaining or releasing hydrogen ions;
4) maintain osmotic pressure of cells by adjusting water content;
5) maintain electrolyte balance by reabsorbing or releasing inorganic ions according to cellular demands; and
6) retain substances such as glucose, proteins, and hormones that are necessary for normal cellular activities. Table 20.2 compares concentrations of solutes common to both urine and plasma.

Table 20.2 : Comparison of main components of urine and serum

Component	Concentration		Urine/Serum Ratio
	Urine (mg/100 mL)	Serum (mg/100 mL)	
Sodium	355	335	1.06
Potassium	160	20	8
Calcium	15	10	1.5
Magnesium	20	2.4	8.3
Chloride	500	350	1.4
Inorganic sulfate	8	2.5	4.0
Inorganic phosphorus	160	4	40
Urea	1800	30	60
Creatinine	75	1.5	50
Uric acid	60	3	20

When water retention by the kidneys occurs and normal amounts of water and salts are ingested, the total amount of interstitial fluid increases. As fluid retention continues, abnormally large amounts of fluid accumulate in interstitial regions and swelling occurs. The condition is called *edema.*

Obviously, the body can't allow fluid buildup to continue unchecked. For cells to receive their require amounts of water and solutes, the flow of these materials between intracellular fluid and extracellular fluid must be regulated. The concentration of electrolytes in body fluids is regulated by hormone-controlled kidney action. Hormone action controls and volume of water reabsorbed by the kidneys and the movement of ions across renal membranes.

Figure 20.1 compares the concentrations of body fluid components in blood plasma. Interstitial fluid, and cellular fluid. Except for protein concentrations, notice the similarity between blood plasma and interstitial fluid and their marked dissimilarity with cellular fluid. Extracellular fluid has relatively high concentrations of Na^+, HCO_3^-, and Cl^- ions; compared to intracellular fluid, K^-, SO_4^{2-}, HPO_4^{2-}, and Mg^{2+} are found in relatively low concentrations.

Since sodium ions make up about 90 percent of all positive ions in extracellular fluid, regulation of sodium ion concentration is mandatory and is controlled mainly by aldosterone, which increases the rate of sodium ion reabsorption in the kidney tubules. In the presence of excess aldosterone, few sodium ions are excreted; in its absence, almost no reabsorption of sodium ions occur. Thus, this system determines the amount of sodium ions lost via urine.

Equally important to the regulation of sodium ion level is the regulation of water. Most of the reabsorption of water occurs in the proximal tubules. Final adjustment of the release or reabsorption of water occurs in the distal tubules and is regulated by the action of minute amounts of the antidiuretic hormone, vasopressin. The rate at which vasopressin is released is governed by changes in the blood's osmotic pressure. *Vasopressin promotes increased reabsorption of water* from both the distal tubules and collecting ducts of the kidneys. Let's consider some examples of its action.

When blood contains a higher than normal concentration of solutes. It is said to be *hypertonic.* This stimulates release of vasopressin and also stimulates thirst. The increased secretion of vasopressin causes

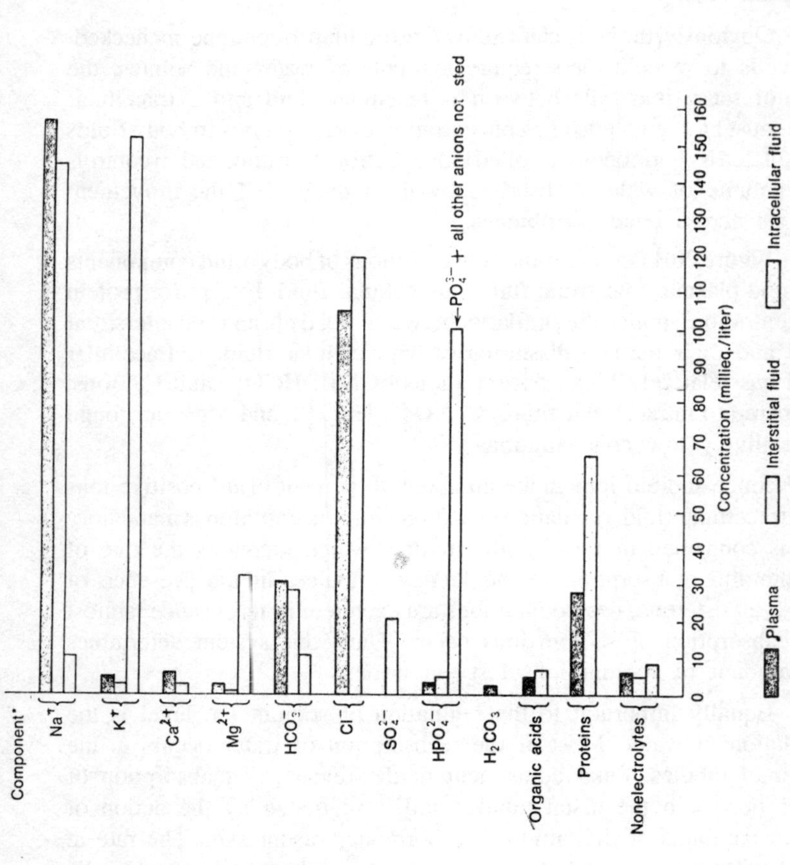

Fig. 20.1 : Comparison of concentrations of components of plasma, interstitial fluid and intracellular fluid.

increased reabsorption of water from the distal tubules into the bloodstream. Water intake also aids in returning the osmotic pressure of the blood to normal levels.

In a hypotonic condition, the blood has been diluted by the presence of excess water. To correct this condition, vasopressin secretion is reduced. Reduction of vasopressin levels decreases water reabsorption from the distal tubules and collecting ducts and causes an increase in the volume of urine excreted.

Electrolyte Balance

Kidney function regulates electrolyte levels by either reabsorbing electrolytes into the bloodstream or releasing them into urine. This regulation is termed *electrolyte balance.*

Solute concentration of plasma is in equilibrium with interstitial fluid that, in turn, is in equilibrium with intracellular fluid solutes. Therefore, interstitial fluid acts as an intermediary between the other two fluids. The principal ions in extracellular fluid are sodium and chloride. Potassium, phosphates, and ionized proteins are the chief intracellular ions. Sodium ions are actively transported from cells into interstitial fluid and plasma; potassium ions are actively transported into cells. Cellular and capillary membranes are impermeable to large molecules like protein sand such molecules are not exchanged directly between fluids.

In the cells lining the kidney tubules and collecting ducts, carbonic anhydrase catalyzes the reaction of carbon dioxide with water to produce carbonic acid. Dissociation of carbonic acid yields bicarbonate ions and hydrogen ions. The hydrogen ions are released into the urine. Normally, urine pH is 5.0 to 7.0; however, release of hydrogen ions can continue until the tubular fluid pH falls to about 4.5, the lower limit of the tubule lining's ability to release hydrogen ions. A pH of 4.5 represents an increase in hydrogen ion concentration of about 800 times that present at a pH of 7.4 the normal pH of blood.

As acidosis develops, the blood's hydrogen ion concentration increases. Additional bicarbonate ions are made available to buffer the additional hydrogen ions.

During alkalosis, a greater than normal concentration of bicarbonate ions are present in blood. The excess bicarbonate ions are then dialyzed into the urine where they are excreted along with sodium ions and other positive ions. Loss of bicarbonate ions via the urine decreases blood bicarbonate concentration to normal level and normal blood pH results.

The four major disturbances of plasma and urine acid-base balance are metabolic acidosis, respiratory acidosis, metabolic alkalosis, and respiratory alkalosis. The following equilibria are pertinent:

$$CO_2 + H_2O \underset{\text{anhydrase}}{\overset{\text{carbonic}}{\rightleftharpoons}} H_2CO_3 \rightleftharpoons H^+(aq) + HCO_3^-(aq)$$

Table 20.3 summarizes the net effects of these conditions on urine pH, plasma bicarbonate ion concentration, and plasma carbonic acid concentration.

Respiratory acidosis results from hypoventilation. Decreased breathing causes CO_2 concentration to increase. This causes the equilibria above to shift to the right and additional hydrogen ions are excreted via the urine. Hyperventilation causes the opposite effect. The overproduction of acidic metabolic waste products causes metabolic acidosis. These substances release hydrogen ions that combine with plasma bicarbonate to produce carbonic acid. The carbonic acid is then decomposed to carbon dioxide and water. Hydrogen ions move from distal tubule cells into urine and bicarbonate ions move from the cells into plasma to compensate for reduced bicarbonate levels in plasma. Metabolic alkalosis causes opposite effects.

Table 20.3 : Effects of pH change of plasma and urine

Disturbance	Urine pH	Plasma $[HCO_3^-]$ (meq/L)	Plasma $[H_2CO_3]$ (meq/L)
Normal level	5.5–7.0	25	1.25
1. Respiratory acidosis	↓	↑	↑
2. Respiratory alkalosis	↑	↓	↓
3. Metabolic acidosis	↓	↓	↓
4. Metabolic alkalosis	↑	↑	↑

↑ indicates increase ; ↓ indicates decrease

MILK

Milk is produced by the mammary glands of mammals for the nourishment of their young. During pregnancy, the high level of estrogens causes growth and branching of the ducts within the mammary glands.

By the end or pregnancy, the mother's breasts are fully developed for nursing. Prolactin initiates milk production and oxytocin ("letdown factor") stimulates release of milk from the breasts to the sucking infant.

Milk contains triglycerides, lactose, calcium, vitamin A, phosphorus, and the proteins casein and lactoglobulin. Because of its composition, milk is often called *nature's perfect food*, however, it is deficient in iron, copper, most of the B vitamins (except riboflavin and pantothenic acid), vitamin C, and vitamin D. Pasteurization further reduces the vitamin C content. Human milk contains more lactose and less protein than cow's milk.

Lactose is found exclusively in mammalian milk. Because galactose is not present in significant amounts in other body tissues or fluids, it must be formed from blood glucose in the mammary glands. The use of carbon- 14 radioisotope tracer experiments on test animals has demonstrated this to be true. Glucose containing radioactive carbon-14 was injected into test mammals. Lactose samples withdrawn later had equal amounts of C-14 carbon in their glucose and galactose components. The net overall production of lactose by mammary glands can be summarized:

$$\text{glucose in mammary blood vessels} \xrightarrow{\text{isomerizes}} \text{galactose}$$

$$\begin{array}{c} \text{H}\quad\text{O} \\ \diagdown\;\diagup\!\!\diagup \\ \text{C} \\ | \\ \text{H—C—OH} \\ | \\ \text{HO—C—H} \\ | \\ \text{H—C—OH} \\ | \\ \text{H—C—OH} \\ | \\ \text{CH}_2\text{OH} \end{array} \quad \xrightarrow{\text{isomerizes}} \quad \begin{array}{c} \text{H}\quad\text{O} \\ \diagdown\;\diagup\!\!\diagup \\ \text{C} \\ | \\ \text{H—C—OH} \\ | \\ \text{HO—C—H} \\ | \\ \text{HO—C—H} \\ | \\ \text{H—C—OH} \\ | \\ \text{CH}_2\text{OH} \end{array}$$

BILE AND BILE SALTS

Bile is secreted by the liver as a yellow fluid, pH 7.0-8.5, at a rate of between 500 and 700 mL daily. Its major constituents are cholesterol; bile acids, such as cholic and deoxycholic acids; bile salts; and bilirubin, a yellow bile pigment produced from the breakdown of heme. As bile accumulates, it is stored in the gallbladder where it is concentrated by water reabsorption and also acidified. During storage, bilirubin is oxidized to biliverdin, a green bile pigment.

Glycocholic and taurocholic acids are cholesterol derivatives. The combination of glycine and cholic acid produces glycocholic acid; taurine plus cholic acid yield taurocholic acid.

At the normal pH of bile, glycocholic and taurocholic acids occur as anions. Cations such as sodium and calcium are present to maintain electrical neutrality. The combination of these anions and cations produce bile salts. The bile salts facilitate fat digestion by emulsifying fats into micelles. The small droplets have a very large surface area exposed to the surrounding water. This enhances the action of pancreatic lipase in catalyzing the hydrolysis of fats. Free (unesterified) fatty acids and monoglycerides are also emulsified by bile salt action.

A certain amount of bile acids must be present in relation to the amount of cholesterol to keep cholesterol in solution. Failure of the gallbladder to acidify alkaline bile from the liver, excessive absorption of bile salts and fatty acids from the liver, or excessive secretion of cholesterol in the bile can cause the bile acid/ cholesterol ratio to fall below a critical level. When this occurs, precipitation of cholesterolic mixtures occurs and forms gallstones. Most commonly, gallstones consist of calcium-bilirubin and cholesterol.

Glossary of Terms

Absolute entropy—(of a substance) The increase in the entropy of a substance as it goes from a perfectly ordered crystalline form at 0 K (where its entropy is zero) to the temperature in question.

Absolute zero—The zero point on the absolute temperature scale; –273.15°C or 0 K; theoretically, the temperature at which molecular motion ceases.

Absorption spectrum—Spectrum associated with absorption of electromagnetic radiation by atoms (or other species) resulting from transitions from lower to higher energy states.

Acetal—A compound containing the $\mathrm{H{-}C(H)(O{-}R){-}O{-}R}$ group.

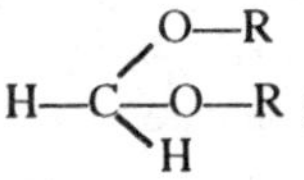

Acetyl coenzyme A (acetyl CoA)—An important intermediate in the Krebs cycle; combines with oxaloacetic to form citric acid.

Acid—A species that dissociates in water to liberate hydrogen ions; a hydrogen ion donor.

—A substance that produces H^+ (aq) ions in aqueous solution. Strong acids ionize completely or almost completely in dilute aqueous solution. Weak acids ionize only slightly.

Acid anhydrides—A group of compounds that forms acids when reacted with water.

—Compound produced by dehydration of a carboxylic acid; general formula is $\mathrm{R{-}\overset{O}{\overset{\|}{C}}{-}O{-}\overset{O}{\overset{\|}{C}}{-}R}$.

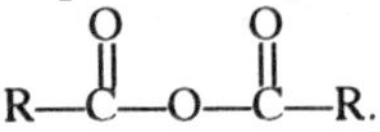

Acid (carboxylic acid)—A weak acid containing the $\mathrm{{-}C({=}O){-}O{-}H}$ group.

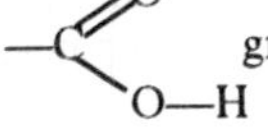

Acid halide—See Acyl halide.

Acidic salt—A salt containing an ionizable hydrogen atom; does not necessarily produce acidic solutions.

Acidosis—A condition in which blood pH drops from 7.4 toward 7.0.

Actinides—Elements 90 to 103 (after actinium).

Activated complex—The species produced when two molecules collide with enough energy to react.

Activation energy—The minimum amount of energy molecules must possess in order for a reaction to occur.

Activator—A metal ion that combines with the apoenzyme to form the active enzyme.

Active metal—Metal with low ionization energy that loses electrons readily to form cations.

Active series—A listing of metals (and hydrogen) in order of decreasing activity.

Active site—The portion of an enzyme molecule where the reaction with the substrate takes place.

Activity (of a component of an ideal mixture)—A dimensionless quantity whose magnitude is: equal to molar concentration in an ideal solution; equal to partial pressure in atm. in an ideal gas mixture; and defined as 1 for pure solids or liquids.

Activity series—See *Electromotive series.*

Actual yield—Amount of a specified pure product actually obtained from a given reaction. Compare with *Theoretical yield.*

Acyl group—Group of atoms remaining after removal of an —OH group of a carboxylic acid.

Acyl halide—Compound derived from a carboxylic acid by replacing the —OH group with a halogen (X), usually —Cl; general formula is $R{-}\overset{\overset{\large O}{\|}}{C}{-}X$.

Addition reaction—A reaction in which all the atoms of one molecule are added to another molecule.

Adhesive forces—Forces of attraction between a liquid and another surface.

Adsorption—Adhesion of species onto surfaces of particles.

Aerobic glycolysis—The metabolism of glucose to produce pyruvic acid which is converted to CO_2 and H_2O by the series of reactions known as the Krebs cycle.

Aerobic process—A process requiring oxygen.

Alcohol—A molecule with an —OH group bonded to a carbon atom.

Aldehyde—A compound containing the $-C\begin{matrix}{\nearrow}^{\!O}\\ \searrow_{H}\end{matrix}$ group.

—Compound in which an alkyl or aryl group and a hydrogen atom are attached to a carbonyl group; general formula, $R{-}\overset{\overset{\large O}{\|}}{C}{-}H$.

Aliphatic compound—A collective term of alkanes, alkenes, and alkynes.

Alkali metals—Metals of Group IA.

Alkaline earths—Elements of Group IIA.

Alkaline (basic) solution—A solution with a pH greater than 7.0.

Alkaline battery—A dry cell in which the electrolyte contains KOH.

Alkalosis—A condition when blood pH rises from 7.4 toward 7.8.

Alkanes—Hydrocarbons containing only single bonds.

Alkenes (olefins)—Hydrocarbons containing carbon-carbon double bonds.

Alkylbenzene—A compound containing an alkyl group bonded to a benzene ring.

Alkyl groups—An unstable hydrocarbon with one vacant bonding site on a carbon atom.

Alkynes—Hydrocarbons containing carbon-carbon triple bonds.

Allotropes—Different forms of the same element in the same physical state.

Allotropic modifications (allotropes)—Different forms of the same element in the same physical state.

Alpha particles—A helium atom's nucleus; two protons and two neutrons; symbolized $^{4}_{2}He$.

Amide—The product of a reaction between a carboxylic acid and a primary or secondary amine; functional group—$\overset{\overset{\displaystyle O}{\|}}{C}$—NR(H), with R(H) on the carbon.

Amine—Compound that can be considered a derivative of ammonia in which one or more hydrogens are replaced by alkyl or aryl groups.

Amines—A group of compounds having one to three alkyl groups and two to no hydrogen atoms directly bonded to a nitrogen atom.

—Derivatives of ammonia in which one or more hydrogen atoms have been replaced by organic groups.

Amino acid—A group of compounds in which a carboxylic acid group and an amine group are present in the molecule; functional group —CH(—NH_2)—CO_2H.

Amino group—The —NH_2 group.

Amorphous solid—A noncrystalline solid with no well-defined ordered structure.

Ampere—Unit of electrical current; one ampere equals one coulomb per second.

Amphiprotism—Ability of a substance to exhibit amphoterism by accepting or donating protons.

Amphoterism—The ability to react with both acids and bases.

Anabolism—The synthesis of cell components by the organism.

Anaerobic glycolysis—The metabolism of glucose by microorganisms to produce lactic acid.

Anaerobic process—A process not requiring oxygen.

Angstrom (Å)—10^{-10} meter.

Anhydrous—Without water.

Anion—a negatively charged atom or group of atoms.

—A negative ion; an atom or group of atoms that has gained one or more electrons.

Anode—In a cathode ray tube, the positive electrode.

Antibonding orbital—A molecular orbital higher in energy than any of the atomic orbitals from which it is derived; lends instability to a molecule or ion when populated with electrons; denoted with a star (*) superscript on symbol.

Apoenzyme—The protein portion of an enzyme.

Aromatic compound—Any compound containing a six-membered ring with the equivalent of three double bonds.

Aromatic hydrocarbons—Benzene and its derivatives.

Aryl group—Group of atoms remaining after a hydrogen atom is removed from an aromatic system.

Associated ions—Short-lived species formed by the collision of dissolved ions of opposite charge.

Atmosphere—A unit of pressure; the pressure that will support a column of mercury 760 mm high at 0°C.

Atom—The smallest particle of an element.

Atomic mass unit (amu)—One twelfth of the mass of an atom of the carbon-12 isotope; a unit used for stating atomic and formula weights also called dalton.

Atomic number—The number of protons in an atom's nucleus.

Atomic orbital—Region or volume in space in which the probability of finding electrons is highest.

Atomic radius—Radius of an atom.

Atomic weight—An atom's weight relative to the weight of a carbon-12 atom.

—Weighted average of the masses of the constituent isotopes of an element; the relative masses of atoms of different elements.

Aufbau ("building up") Principle—Describes the order in which electrons fill orbitals in atoms.

Autoionization—An ionization reaction between identical molecules.

Avogadro's Law—At the same temperature and pressure, equal volumes of all gases contain the same number of molecules.

Avogadro's number—See *Mole.*

Band—A series of very closely spaced, nearly continuous molecular orbitals that belong to the crystal as a whole.

Band theory of metals—Theory that accounts for the bonding and properties of metallic solids.

Barometer—A device for measuring pressure.

Base—Any species that decreases the hydrogen ion concentration when added to an aqueous solution; a hydrogen ion acceptor.

—A substance that produces $OH^-(aq)$ ions in aqueous solution. Strong soluble bases are soluble in water and are completely dissociated. Weak bases ionize only slightly.

Base pairing—The hydrogen bonding between adenine and thymine (or uracil) and between cytosine and guanine.

Basic anhydride—The oxide of a metal that reacts with water to form a base.

Basic salt—A salt containing an ionizable OH group.

Beta radiation—Negatively charged, resembling high-speed electrons; symbolized ${}^{0}_{-1}e$.

Binary acid—A binary compound in which H is bonded to one of the more electronegative nonmetals.

Binary compound—A compound consisting of two elements; may be ionic or covalent.

Biodegradability—The ability of a substance to be broken down into simpler substances by bacteria.

Boiling point—The temperature at which the vapor pressure of a liquid is equal to the applied pressure; also the condensation point.

Boiling point elevation—The increase in the boiling point of a solvent caused by dissolution of a nonvolatile solute.

Bomb calorimeter—A device used to measure the heat transfer between system and surroundings at constant volume.

Bond energy—The amount of energy necessary to break one mole of bonds of a given kind (in gas phase).

Bond order—Half the number of electrons in bonding orbitals minus half the number of electrons in antibonding orbitals.

Bonding orbital—A molecular orbital lower in energy than any of the atomic orbitals from which it is derived; lends stability to a molecule or ion when populated with electrons.

Bonding pair—Pair of electrons involved in a covalent bond.

Born-Haber cycle—A series of reactions (and accompanying enthalpy changes) which, when summed, represents the hypothetical one-step reaction by which elements in their standard states are converted into crystals of ionic compounds (and the accompanying enthalpy changes).

Boyle's Law—At constant temperature the volume occupied by a definite mass of a gas is inversely proportional to the applied pressure.

Bronsted-Lowry acid—A proton donor.

Bronsted-Lowry base—A proton acceptator.

Buffer solution—Solution that resists change in pH; contains either a weak acid and a soluble ionic salt of the acid or a weak base and a soluble ionic salt of the base.

Bugger—A solution containing a weak acid/conjugate base pair that can offset addition of limited amounts of base or acid to it without a significant pH change.

Buret—A piece of volumetric glassware, usually graduated in 0.1-mL intervals, that is used to deliver solutions to be used in titrations in a quantitative (dropwise) manner.

Calorie—The amount of heat required to raise the temperature of one gram of water from 14.5°C to 15.5°C, 1 calorie=4.184 joules.

Calorimeter—A device used to measure the heat transfer between system and surroundings.

Canal ray—Stream of positively charged particles (cations) that moves toward the negative electrode in cathode ray tubes; observed to pass through canals in the negative electrode.

Capillary action—The drawing of a liquid up the inside of a small-bore tube when adhesive forces exceed cohesive forces, or the depression of the surface of the liquid when cohesive forces exceed adhesive forces.

Carbohydrate—A polyhydroxyaldehyde or ketone, most of which have the general formula $C_x(H2O)_y$.

Carbonyl group—The $-\overset{\overset{\displaystyle O}{\|}}{C}-$ group.

Carboxylation—The addition of carbon dioxide to an organic molecule to produce an acid.

Carboxylic acid—Acids containing the $-C(=O)-O-H$ functional group.

Catabolism—The breakdown of metabolites and cell components to supply energy to the cell.

Catalyst—A substance that increases the rate of a reaction by lowering the activation energy, and is not consumed during the reaction.

Catenation—The ability of an element to bond to itself.

Cathode—In a cathode ray tube, the negative electrode.

—Electrode at which reduction occurs.

Cathode ray—Beam of electrons going from the negative electrode toward the positive electrode in a cathode ray tube.

Cathode ray tube—Closed glass tube containing a gas under low pressure, with electrodes near the ends and a luminescent screen at the end near the positive electrode; produces cathode rays when high voltage is applied.

Cathodic protection—Protection of a metal (making it a cathode) against corrosion by attaching it to a sacrificial anode of a more easily oxidized metal.

Cation—A positively charged atom or group of atoms.

—A positive ion: an atom or group of atoms that has lost one or more electrons.

Cell potential—Potential difference, E_{cell} between oxidation and reduction half-cells under nonstandard conditions.

Cellulose—A polymer of glucose joined by b-acetal bonds.

Central atom—An atom in a molecule or polyatomic ion that is bonded to more than one other atom.

Charles' Law—At constant pressure the volume occupied by a definite mass of a gas is directly proportional to its absolute temperature.

Chemical bonds—The attractive forces that hold atoms together in elements and compounds.

Chemical equation—Description of a chemical reaction by placing the formulas of reactants on the left and the formulas of products on the right of an arrow.

Chemical equilibrium—A state of dynamic balance in which the rates of forward and reverse reactions are equal; there is no net change in concentrations of reactants or products while a system is at equilibrium.

Chemical kinetics—The study of rates and mechanisms of chemical reactions and of the factors on which they depend.

Chemical periodicity—The variation in properties of elements with their positions in the periodic table.

Cis—A term indicating that a molecule has two identical groups on the same side of a double bond.

Citric acid cycle—(See Krebs cycle).

Coenzyme—A small molecule that combines with the apoenzyme to form the active enzyme.

Cohesive forces—All the forces of attraction among particles of a liquid.

Colligative properties—Physical properties of solutions that depend upon the number but not the kind of solute particles present.

Collision theory—Theory of reaction rates that states that effective collisions between reactant molecules must occur in order for reaction to occur.

Colloid—A suspension in which the particles will not settle due to the force of gravity. Particle size is between 2.6 and 1000 Å.

Combination reaction—Reaction in which two substances (elements or compounds) combine to form one compound.

Combustion reaction—Reaction of a substance with oxygen in a highly exothermic reaction, usually with a visible flame.

Common ion effect—Suppression of ionization of a weak electrolyte by the presence in the same solution of a strong electrolyte containing one of the same ions as the weak electrolyte.

Complex ions—Ions resulting from the formation of coordinate covalent bonds between simple ions and other ions or molecules.

Compound—A pure substance containing two or more elements combined chemically in a fixed ratio by weight.

Composition stoichiometry—Describes the quantitative (mass) relationships among elements in compounds.

Concentration—Amount of solute per unit volume or mass of solvent or of solution.

Condensation—Liquefaction of vapor.

Condensed phase—The liquid and solid phases; phases in which particles interact strongly.

Condensed states—The solid and liquid states.

Conduction band—A partially filled band or a band of vacant energy levels just higher in energy than a filled band; a band within which, or into which, electrons must be promoted to allow electrical conduction to occur in a solid.

Conjugate acid—The species formed by gain of H^+ by a base.

Conjugate acid-base pair—In Bronsted-Lowry terminology, a reactant and product that differ by a proton, H^+.

Conjugate base—The species formed by loss of H^+ by an acid.

Contact catalyst—See *Heterogeneous catalyst.*

Coordinate covalent bond—A single bond formed by the contribution of both shared electrons by one of the two bonded atoms.

—Covalent bond in which both shared electrons are furnished by the same species: bond between a Lewis acid and a Lewis base.

Coordination number—In describing crystals, the number of nearest neighbours of an atom or ion.

Conformations—Structures of a compound that differ by the extent of rotation about a single bond.

Conjugated double bonds—Double bonds that are separated from each other by one single bond, as in C=C—C=C.

Continuous spectrum—Spectrum that contains all wavelengths in a specified region of the lectromagnetic spectrum.

Corrosion—Oxidation of metals in the presence of air and moisture.

Coulomb—Unit of electrical charge.

Covalent bond—A bond in which a pair of electrons is shared by two atoms.

Covalent compounds (molecular compounds)—Substances in which the atoms are held together in clusters, called molecules, by a sharing of electrons.

Critical point—The combination of critical temperature and critical pressure of a substance.

Critical pressure—The pressure required to liquefy a gas (vapor) at its critical temperature.

Critical temperature—The temperature above which a gas cannot be liquefied; the temperature above which a substance cannot exhibit distinct gas and liquid phases.

Crystal lattice—Pattern of arrangement of particles in a crystal.

Crystal lattice energy—The energy change when one mole of formula units of a crystalline solid is formed from its ions, atoms, or molecules in the gas phase: always negative.

Crystalline solid—A solid characterized by a regular, ordered arrangement of particles.

Cyclic compound—A compound in which some of the bonded atoms form a ring.

Cycloalkanes—cyclic saturated hydrocarbons.

Cytochrome chain—(*See* Respiratory chain.)

Cytochromes—Iron-containing compounds involved in ATP production by aerobic organisms. (*See also* Respiratory chain *and* Electron transport system.)

Dalton's Law—See *Law of Partial Pressures.*

Debye—The unit used to express dipole moments.

Decarboxylation—The loss of CO_2 from a compound when it is heated.

Degenerate—Of the same energy.

Dehydration—The removal of water.

Dehydrogenation—The removal of hydrogen atoms.

Delocalization—Of electrons: refers to bonding electrons that are distributed among more than two atoms that are bonded together; occurs in species that exhibit resonance.

Denaturation—Alteration of a protein's secondary and tertiary structure.

Density—A ratio of a material's mass (weight) to its volume. D= M/V.

Deposition—The direct solidification of a vapor by cooling; the reverse of sublimation.

Detergent—A long alkyl chain with an —O—$SO_3^-Na^+$ group.

—A soap-like emulsifier that contains a sulphonate, —SOO_3^-, sulphate, —OSO_3^-, or phosphate group instead of a carboxylate, —COO^-, group.

Diabetes mellitus—A disease characterized by an inability to metabolize carbohydrates, particularly glucose, correctly.

Dialysis—The passage of small solute particles through a dialyzing membrane.

Diamagnetism—Weak repulsion by a magnetic field.

Differential thermometer—A thermometer used for accurate measurement of very small changes in temperature.

Digestion—The breakdown of a protein's primary structure by hydrolysis of peptide bonds.

Dilution—The addition of pure solvent to a solution to yield a larger volume of a less concentrated solution.

Dipole—Refers to the separation of charge between two covalently bonded atoms.

Dipole-dipole interactions—Attractive interactions between polar molecules, that is, between molecules with permanent dipoles.

Dipole-induced dipole interactions—See *London forces.*

Dipole moment—A measure of charge separation in a molecule resulting from the presence of polar bonds in a molecule.

Dipole moment (μ)—The product of the distance separating opposite charges of equal magnitude and the magnitude of the charge: a measure of the polarity of a bond or molecule; a measured dipole moment refers to the dipole moment of an entire molecule.

Disaccharide—A carbohydrate formed from two monosaccharides joined by an acetal or metal bond.

Dispersed phase—The solute-like species in a colloid.

Dispersing medium—The solvent-like phase in a colloid.

Displacement reactions—Reactions in which one element displaces another from a compound.

Disproportionation reactions—Redox reactions in which the oxidizing agent and the reducing agent are the same species.

Dissociation—The separation of a species into ions.

Dissociation constant—Equilibrium constant that applies to the dissociation of a complex ion into a simple ion and coordinating species (ligands).

Distillation—The separation of a liquid mixture into its components on the basis of differences in boiling points.

DNA—Deoxyribonucleic acid. The molecules carrying the genetic code.

Double bond—The sharing of two pairs of electrons by two atoms.

***d*-orbitals**—Beginning in the third energy level, a set of five degenerate orbitals per energy level, higher in energy than *s* and *p* orbitals of the same energy level.

Downs cell—Electrolytic cell for the commercial electrolysis of molten sodium chloride.

Dry cells—Ordinary batteries (voltaic cells) for flashlights, radios, and so on; many are Leclanche cells.

***d*-Transition elements (metals)**—B Group elements except IIB in the periodic table : sometimes called simply transition elements.

Dumas method—A method used to determine the molecular weights of volatile liquids. See Figure 10-5.

Dynamic equilibrium—A situation in which two (or more) processes occur at the same rate so that no net change occurs.

Effective collision—Collision between molecules resulting in reaction: one in which molecules collide with proper relative orientations and sufficient energy to react.

Effective molality—The sum of the molalities of all solute particles in solution.

Effective nuclear charge (Zeff)—The nuclear charge experienced by the outermost electrons of an atom; the actual nuclear charge minus the effects to shielding due to inner-shell electrons.

Electrical conductivity—Ability to conduct electricity.

Electrochemistry—Study of chemical changes produced by electrical current and production of electricity by chemical reactions.

Electrode potentials—Potentials, E, of half-reactions as reductions versus the standard hydrogen electrode.

Electrodes—Surfaces upon which oxidation and reduction half-reactions occur in electrochemical cells.

Electrolysis—Process that occurs in electrolytic cells.

Electrolyte—A substance whose aqueous solutions conduct electricity.

Electrolyte balance—The reabsorption or release of electrolytes from blood to ensure proper electrolyte concentration.

Electrolytes—Compounds that liberate ions when dissolved in water.

Electrolytic cells—Electrochemical cells in which electrical energy causes nonspontaneous redox reactions to occur.

Electrolytic conduction—Conduction of electrical current by ions through a solution or pure liquid.

Electromagnetic radiation—Energy that is propagated by means of electric and magnetic fields that oscillate in directions perpendicular to the direction of travel of the energy.

Electromotive series—The relative order of tendencies for elements and their simple ions to act as oxidizing or reducing agents; also called the activity series.

Electroplating—Plating a metal onto a (cathodic) surface by electrolysis.

Electron—The fundamental unit of negative charge.

—A subatomic particle having a mass of 0.00054858 *amu* and a charge of 1 –.

Electron affinity—The energy released when a gaseous atoms gains an electron.

—The amount of energy absorbed in the process in which an electron is added to a neutral isolated gaseous atom to form a gaseous ion with a 1– charge; has a negative value if energy is released.

Electron configuration—Specific distribution of electrons in atomic orbitals of atoms or ions.

Electron deficient compounds—Compounds that contain at lease one atom (other thân H) that shares fewer than eight electrons.

Electron dot formula—See *Lewis dot formula.*

Electronegativity—A measure of the relative tendency of an atom to attract electrons to itself when chemically combined with another atom.

Electronic geometry—The geometric arrangement of orbitals containing the shared and unshared electron pairs surrounding the central atom of a molecule or polyatomic ion.

Electron transition—The transfer of an electron from one energy level to another.

Electron transport system (ETS)—(*See* Respiratory chain).

Electron volt (eV)—Very small unit of energy; 1 eV = 3.8×10^{-20} cal.

Electronegativity—A measure of the ability of an atom in a molecule to attract electrons toward itself.

Element—A pure substance that cannot be decomposed to similar substances by normal chemical methods.

Emission spectrum—Spectrum associated with emission of electromagnetic radiation by atoms (or other species) resulting from electronic transitions from higher to lower energy states.

Empirical formula—See *Simplest formula.*

Emulsifier—See *Emulsifying agent.*

Emulsifying agent—A substance that coats the particles of a dispersed phase and prevents coagulation of colloidal particles; an emulsifier.

Emulsion—A colloidal suspension of an insoluble liquid in another liquid.

End point—The point at which an indicator changes color and a titration is stopped.

Endergonic process—Energy-consuming process.

Endocrine glands—Ductless glands that synthesize and secrete hormones.

Endothermicity—The absorption of heat by a system as a process occurs.

Endothermic reaction—A reaction during which energy is absorbed in the form of heat.

End point—The point at which an indicator changes colour and a titration is stopped.

Energy—The ability to do work. Present in two forms, potential (stored) energy and kinetic (moving) energy.

Energy level (shell)—A region of energy associated with electrons about a nucleus.

Enthalpy, *H*—The heat content of a specific amount of a substance; defined as $E - PV$.

Entropy, *S*— A thermodynamic state property that measures the degree of disorder or randomness of a system; $T\Delta S$ represents a form of energy.

Enzyme—A protein that acts as a catalyst in biological systems.

Equation of state—An equation that describes the behaviour of matter in a given state; the van der Waals equation describes the behaviour of the gaseous state.

Equations—The use of chemical formulas to indicate reaction between chemicals (reactants) to produce new chemicals (products).

Equilibrium—The condition of a system when the rates of opposing processes are equal.

Equilibrium constant, *K*—A quantity that characterizes the position of equilibrium for a reversible reaction; its magnitude is equal to the mass action expression at equilibrium. *K* varies with temperature.

Equilibrium or chemical equilibrium—A state of dynamic balance in which the rates of forward and reverse reactions are equal; the state of a system when neither the forward nor the reverse reaction is thermodynamically favoured.

Equivalence point—The point at which chemically equivalent amounts of reactants have reacted.

Equivalent—Moles of an ion multiplied by the charge on the ion.

Equivalent weight—The mass of an acid or base that furnishes or reacts with 6.022, 10^{23} H_3O^- or OH^- ions.

Erg—An energy unit smaller than a calorie: 1 erg = 2.4×10^{-8} cal.

Ester—A compound formed by the reaction of an alcohol and an oxygen-containing acid; functional group, $R—\overset{\overset{\displaystyle O}{\|}}{C}—O—R'$.

Ether—A molecule with a $—\overset{|}{\underset{|}{C}}—O—C—$ bonding arrangement.

—Compound in which an oxygen atom is bonded to two alkyl or two aryl groups, or one alkyl and one aryl group.

Eutrophication—The undersirable overgrowth of vegetation caused by high concentrations of plant nutrients in bodies of water.

Evaporation—Vaporization of al liquid below its boiling point.

Excited state—Any state other than the ground state of an atom or molecule.

Exergonic process—Energy-releasing process.

Exothermicity—The release of heat by a system as a process occurs.

Exothermic reaction—A reaction during which energy is given off in the form of heat.

Extensive property—A property that depends upon the amount of material in a sample.

Faraday—One faraday of electricity corresponds to the charge on 6.022 × 10^{23} electrons, or 96.487 coulombs.

Faraday's Law of Electrolysis—One equivalent weight of a substance is produced at each electrode during the passage of 96.487 coulombs of charge through an electrolytic cell.

Fat—Solid triester of glycerol and (mostly) saturated fatty acids.

Fatty acid—An aliphatic acid; many can be obtained from animal fats.

Fuel cells—Voltaic cells in which the reactants (usually gases) are supplied continuously.

***f*-orbitals**—Beginning in the fourth energy level, a set of seven degenerate orbitals per energy level, higher in energy than *s*, *p* and *d* orbitals of the same energy level.

***f*-Transition elements**—See *Inner transition elements.*

Fatty acids—A class of compounds with molecules consisting of long hydrocarbon chains (11 to 25 carbons) containing single and double bonds terminating in a carboxylic acid group.

Fatty acid spiral (beta oxidation)—The degradation of fatty acids to acetyl CoA fragments.

Fermentation—The metabolism of glucose by yeast and certain other microorganisms to produce ethyl alcohol.

First Law of Thermodynamics—The total amount of energy in the universe is constant (also known as the Law of Conservation of Energy); energy is neither created nor destroyed in ordinary chemical reactions and physical changes.

Fluids—Substances that flow freely; gases and liquids.

Foam—Colloidal suspension of a gas in a liquid.

Forbidden zone—A relatively large energy separation between an insulator's highest filled electron energy band and the next higher energy vacant band.

Formed elements—A collective term used to indicate red cells, white cells, and platelets in blood.

Formal charge—A method of counting electrons in a covalently bonded molecule or ion: counts bonding electrons as though they were equally shared between the two atoms.

Formula—The use of element symbols and subscripts to indicate the elements and the number of atoms of each element that have combined to form a compound. (*See also* Ionic formula *and* Molecular formula).

—Combination of symbols that indicates the chemical composition of a substance.

Formula unit—The smallest repeating unit of a substance, the molecule for nonionic substances.

Formula weight—The sum of the masses of all the atoms in the chemical formula of a compound.

—The mass of one formula unit of a substance in atomic mass units.

Fractional distillation—The process in which a fractionating column is used in a distillation apparatus to separate components of a liquid mixture that have different boiling points.

Fractional precipitation—Removal of some ions from solution by precipitation while leaving other ions with similar properties in solution.

Free energy, G (also known as Gibbs free energy)—The thermodynamic state function of a system that indicates the amount of energy available for the system to do useful work at constant T and P.

Free energy change, ΔG—The indicator of spontaneity of a process at constant T and P. If ΔG is negative, the process is spontaneous.

Freezing point depression—The decrease in the freezing point of a solvent caused by the presence of a solute.

Frequency—The number of repeating corresponding points on a wave that pass a given observation point per unit time.

Functional group—(a) The reactive portion of a molecule. (b) A characteristic grouping of atoms. For example, —CO_2H is a carboxylic acid group.

—A group of atoms that represents a potential reaction site in an organic molecule.

Galactosemia—A disease involving the inability to metabolize the monosaccharide galactose because of the lack of an enzyme and results in the buildup of galactose concentration in the blood.

Galvanic cells—See *Voltaic cells.*

Gamma radiation—Very penetrating, high energy radiation; carries no charge.

Gay-Lussac's Law—See *Law of Combining Volumes.*

Gel—Colloidal suspension of a solid dispersed in a liquid; a semirigid sol.

Geometrical isomers—Compounds with different arrangements of groups on eithe rside of a bond with restricted rotation, such as a double bond or a single bond in a ring; for example, cis-trans isomers of certain alkenes.

Gibbs free energy—See *Free energy, G.*

Glyceride—Triester of glycerol.

Glycogen—A highly branched polymer of glucose joined by α-acetal bonds produced by the liver and muscle tissue of animals.

Glycogenesis—The synthesis of glycogen from glucose.

Grahm's Law—The rates of effusion of gases are inversely proportional to the square roots of their molecular weights (or densities).

Ground state—The lowest energy state or most stable state of an atom, molecule, or ion.

Group—A vertical column of the periodic chart; contains chemically similar elements.

Group (family)—The elements in a vertical column of the periodic table.

Gram formula weight—The formula weight expressed in grams. This quantity is called a mole.

Half-cell—Compartment in which the oxidation or reduction half-reaction occurs in a voltaic cell.

Half-life—Time required for half of a radioactive sample to decay.

Half reaction—The equation showing either the oxidation portion or the reduction portion of a reaction involving oxidation and reduction.

Halogens—Elements of periodic Group VIIA.

Hard water—Water containing Fe^{3-}, Ca^{2-}, and/or Mg^{2+} ions, which form precipitates with soaps.

Heat—A form of energy that flows between two samples of matter because of their difference in temperature.

Heat capacity—The amount of heat required to raise the temperature of a body (of whatever mass) one degree Celsius.

Heat of condensation—The amount of heat that must be removed from one gram of a vapor at its condensation point to condense the vapor with no change in temperature.

Heat of crystallization—The amount of heat that must be removed from one gram of a liquid at its freezing point to freeze it with no change in temperature.

Heat of fusion—The amount of heat required to melt one gram of a solid at its melting point with no change in temperature. Usually experessed in J/g. The molar heat of fusion is the amount of heat required to melt one mole of a solid at its melting point with no change in temperature and is usually expressed in kJ/mol.

Heat of solution (molar)—Of an ion, the energy change accompanying the hydration of a mole of gaseous ions.

Heat of vaporization—The amount of heat required to vaporize one gram of a liquid at its boiling point with no change in temperature. Usually expressed in J/g. The molar heat of vaporization is the amount of heat required to vaporize one mole of a liquid at its boiling point with no change in temperature and is usually expressed in kJ/mol.

Heisenberg Uncertainty Principle—It is impossible to determine accurately both the momentum and position of an electron simultaneously.

Helix—The spiral coiling of proteins or RNA. The spiral may coil to the right (a-helix) or to the left. The right-handed helix is the naturally occuring one in protein and RNA.

Hemiacetal—A compound containing the

```
 OR
 |
 C—O—H
/ \
R   H
```

group.

Hemiketal—A compound containing the

```
    O—R
    |
 R—C—R
    |
    OH
```

group.

Hemolysis—Rupture of red blood cell membranes.

Henry's Law—The pressure of the gas above a solution is proportional to the concentration of the gas in the solution.

Hess' Law of heat summation—The enthalpy change for a reaction is the same whether it occurs in one step or a series of steps.

Hetrocyclic amine—Amine in which the nitrogen is part of a ring.

Heterogeneous catalyst—A catalyst that exists in a different phase (solid, liquid, or gas) from the reactants; a contact catalyst.

Heterogeneous equilibria—Equilibria involving species in more than one phase.

Heterogeneous mixture—A mixture that does not have uniform composition and properties throughout.

Heteronuclear—Consisting of different elements.

Homeostasis—A physiological term used to describe a relatively constant system.

Homogeneous catalyst—A catalyst that exists in the same phase (solid, liquid, or gas) from the reactants: a contact catalyst.

Homogeneous equilibria—Equilibria involving only species in a single phase, i.e., all gases, all liquids, or all solids.

Homogeneous mixture—A mixture that has uniform composition and properties throughout.

Homologous series—A series of compounds in which each member differs from the next by a specific number and kind of atoms.

Homonuclear—Consisting of only one element.

Hormones—Endocrine gland secretions that act as chemical messengers.

Hund's Rule—All orbitals of a given sublevel must be occupied by single electrons before pairing begins. See *Aufbau Principle*.

Hybridization—Mixing of a set of atomic orbitals to form a new set of atomic orbitals with same total electron capacity and with properties and energies intermediate between those of the original unhybridized orbitals.

Hydrate—A solid compound that contains a definite percentage of bound water.

Hydration—(a) Water molecules surrounding an ion or molecule. (b) The addition of water during a chemical reaction.

Hydration energy (molar)—Of an ion, the energy change accompanying the hydration of a mole of gaseous ions.

Hydride—A binary compound of hydrogen.

Hydrocarbonds—Class of compounds containing only carbon and hydrogen.

Hydrogenation—The addition of hydrogen atoms across a double bond.

Hydrogen bond—A fairly strong dipole-dipole interaction (but still considerably weaker than covalent or ionic bonds) between molecules containing hydrogen directly bonded to a small, highly electronegative atom, such as N, O, or F.

Hydrogen-oxygen fuel cell—Fuel cell in which hydrogen is the fuel (reducing agent) and oxygen is the oxidizing agent.

Hydrometer—A device used to measure the densities of liquids and solutions.

Hydrolysis—The breakdown of a bond by the action of water.

—Reaction of a substance with water.

Hydronium ion—H_3O^+, the combination of H+ and water.

Hydrophilic colloids—Colloidal particles that attract water molecules.

Hydrophobic colloids—Colloidal particles that repel water molecules.

Hypertonic fluid—A fluid that has a greater osmotic pressure than another solution.

Hypophysis—The pituitary gland.

Hypotonic fluid—A fluid that has a lower oxmotic pressure than another solution.

Ideal gas—A hypothetical gas that obeys exactly all postulates of the kinetic-molecular theory.

Ideal Gas Law—The product of the pressure and volume of an ideal gas is directly proportional to the number of moles of the gas and the absolute temperature.

Ideal solution—A solution that obeys Raoult's Law exactly.

Inhibitory catalyst—An inhibitor, a catalyst that decreases the rate of a reaction.

Inner transition elements—Elements 58 to 71 and 90 to 103; also called *f*-transition elements.

Indicators—A class of compounds that changes color with changes in pH.

—For acid-base titrations, organic compounds that exhibit different colours in solutions of different acidities: used to determine the point at which reaction between two solutes is complete.

Insulator—Poor electrical and heat conductor.

Insoluble compound—Actually a very slightly soluble compound.

Integrated rate equation—An equation giving the concentration of a reactant remaining after a specified time: has different mathematical form for different orders of reaction.

Intensive property—A property that does not depend upon the amount of material in a sample.

Intermolecular—Between two different molecules.

Intermolecular forces—Forces *between* individual particles (atoms, molecules, ions) of a substance.

Internal energy, *E*—All forms of energy associated with a specific amount of a substance.

Intramolecular—Within a single molecule.

Intramolecular forces—Forces between atoms (or ions) *within* molecules (or formula units).

Ion—An atom or group of atoms that carries an electrical charge.

Ionic bond—The electrical attraction between cations and anions in an ionic substance.

Ionic bonding—Chemical bonding resulting from the transfer of one or more electrons from one atom or group of atoms to another.

Ionic compounds—Substances composed of oppositely charged atoms (or groups of atoms).

Ionic formula—The simplest ratio of cations and anions that is electrically neutral.

Ionic geometry—The arrangement of atoms (not lone pairs of electrons) about the central atom of a polyatomic ion.

Ionic radius—Radius of an ion.

Ionization—The process of electron removal from atoms.

—In aqueous solution, the process in which a *molecular* compound reacts with water and forms ions.

—The breaking up of a compound into separate ions.

Ionization constant—Equilibrium constant for the ionization of a weak electrolyte.

Ionization energy—The minimum amount of energy required to remove the most loosely held electron of an isolated gaseous atom or ion

Ionization potential—The amount of energy required to remove an electron from a gaseous atom.

Ion product for water—Equilibrium constant for the ionization of water.

$$K_w = [H_3O^+][OH^-] = 1.00 \times 10^{-14} \text{ at } 25^\circ\text{C}.$$

Isoelectric point—pH at which an amino acid is in its zwitterion form,

$$\underset{\substack{|\\ {}^{(+)}NH_3}}{—CH}—COO_2^{(-)}.$$

Isoelectronic—Having the same electron configurations.

Isomers—Two or more molecules with the same molecular formula but different structural formulas.

Isomorphous—Refers to crystals having the same atomic arrangement.

Isotonic solutions—Solutions of equal osmotic pressure.

Isotopes—Nuclei having the same atomic numbers but different mass numbers. Joule—A unit of energy in the SI system. One joule is 1 kg · m^2/s^2, which is also 0.2390 calorie.

—Two or more forms of atoms of the same element with different masses; atoms containing the same number of protons but different numbers of neutrons.

Ketal—A compound containing the $\mathrm{R{-}C(OR)_2{-}R}$ group.

$$\begin{array}{c} \mathrm{O{-}R} \\ | \\ \mathrm{R{-}C{-}R} \\ | \\ \mathrm{O{-}R} \end{array}$$

Ketone—A compound containing the $\mathrm{{-}C{-}C({=}O){-}C{-}}$ group.

$$\begin{array}{ccccc} & & \mathrm{O} & & \\ & | & \| & | & \\ - & \mathrm{C} - & \mathrm{C} & - \mathrm{C} & - \\ & | & & | & \end{array}$$

—Compound in which a carbonyl group is bound to two alkyl or two aryl groups, or to one alkyl and one aryl group.

Kinetic-molecular theory—A theory, originally published by Clausius in 1857, that attempts to explain macroscopic observations on gases in microscopic or molecular terms.

Kinetic energy—Energy that matter possesses by virtue of its motion.

Kinetics—The study of how fast a reaction takes place and the factors that influence how fast the reaction occurs.

Krebs cycle—An exergonic sequence of reactions in which citric acid, a six-carbon acid, is converted to oxaloacetic acid (four carbons) and this, subsequently, reacts with acetyl CoA to produce citric acid (six carbons).

Lanthanides—Elements 58 to 71 (after lanthanum).

Law of Combining Volumes—At constant temperature and pressure, the volumes of reacting gases (and any gaseous products) can be expressed as ratios of small whole numbers; also known as Gay-Lussac's Law.

Law of Conservation of Energy—Energy cannot be created or destroyed: it may be changed from one form to another.

Law of Conservation of Matter—There is no detectable change in the quantity of matter during an ordinary chemical reaction.

Law of Conservation of Matter and Energy—The total amount of matter and energy available in the universe is fixed.

Law of Constant Composition—See *Law of Definite Proportions*.

Law of Definite Proportions—Statement that different samples of a pure compound always contain the same elements in the same porportions by mass; also called Law of Constant Composition.

Law of Partial Pressures—The total pressure exerted by a mixture of gases is the sum of the partial pressures of the individual gases; also called Dalton's Law.

Lead storage batter—Secondary voltaic cell used in most automobiles.

LeChatelier's Principle—States that a system at equilibrium, or striving to attain equilibrium, responds in such a way as to counteract any stress placed upon it.

Leclanche cell—A common type of dry cell.

Leveling effect—Effect by which all acids stronger than the acid that is characteristic of the solvent react with solvent to produce that acid; similar statement applies to bases. The strongest acid (base) that can exist in a given solvent is the acid (base) characteristic of that solvent.

Lewis acid—Substance that accepts a share in a pair of electrons from another species.

Lewis base—Substance that makes available a share in an electron pair.

Lewis dot formula—Representation of a molecule, ion, or formula unit by showing atomic symbols and only outer shell electrons; does not show shape.

Lipids—Substances of biological origin that are soluble in chloroform, ether, and other nonwater soluble polar solvents. Includes triglycerides, steroids, and lecithins.

Lipogenesis—The synthesis of lipids (primarily triglycerides) from acetyl coenzyme A.

Limiting reactant—Substance that stoichiometrically limits the amount of product(s) that can be formed.

Line spectrum—An atomic emission or absorption spectrum.

Liquid aerosol—Colloidal suspension of a liquid in gas.

London forces—Very weak and very short-range attractive forces between short-lived temporary (induced) dipoles: also called dispersion forces.

Lone pair—Pair of electrons residing on one atom and not shared by other atoms: unshared pair.

Magnetic quantum number (mL)—Quantum mechanicall solution to a wave equation that designates the particular orbital within a given set (*s*, *p*, *d*, *f*) in which an electron resides.

Manometer- -A two-armed barometer.

Mass—A measure of the amount of matter in an object. Mass is usually measured in grams or kilograms.

Mass action expression—For a reversible reaction,

$$aA + bB = cC + dD$$

the product of the concentrations of the products (species on the right), each raised to the power that corresponds to its coefficient in the balanced chemical equation, divided by the product of the concentrations of the reactants (species on the left), each raised to the power that corresponds to its coefficient in the balanced chemical equation. At equilibrium the mass action expression is equal to K; at other conditions, it is Q.

$$\frac{[C]^c[D]^d}{[A]^a[B]^b} = Q, \text{ or at equilibrium, } K$$

Mass number—The sum of protons and neutrons in an atom's nucleus.

—The sum of the numbers of protons and neutrons in an atom; an integer.

Mass spectrometer—An instrument that measures the charge-to-mass ratio of charged particles.

Matter—Anything that has mass (weight) and takes up space.

Mechanism—The sequence of steps by which reactants are converted into products.

Metabolite—A compound capable of undergoing biochemical degradation.

Metallic conduction—Conduction of electrical current through a metal or along a metallic surface.

Melting point—The temperature at which liquid and solid coexist in equilibrium; also the freezing point.

Meniscus—The shape assumed by the surface of a liquid in a cylindrical container.

Metal—An element below and to the left of the stepwise division (metalloids) in the upper right corner of the periodic table; about 80% of the known elements are metals.

Metallic bonding—Bonding within metals due to the electrical attraction of positively charged metal ions for mobile electrons that belong to the crystal as a whole.

Metalloids—Elements with properties intermediate between metals and nonmetals; B, Al, Si, Ge, As, Sb, Te, Po, and At.

Metathesis reactions—Reactions in which two compounds react to form two new compounds, with no changes in oxidation number. Reactions in which the ions of two compounds exchange partners.

Method of initial rates—Method of determining the rate-law expression by carrying out a reaction with different initial concentrations and analyzing the resulting changes in initial rates.

Mitochondria—Rodlike structures found in most living cells; the site of cellular ATP production.

Mixture—A substance containing two or more components in varying amounts.

Molality (m)—Concentration expressed as number of moles of solute per kilogram of solvent.

Molarity—A concentration unit giving the number of moles of solute per litre of solution.

Molar solubility—Number of moles of solute that dissolve to produce a liter of saturated solution.

Mole—6.022×10^{23} (Avogadro's number of) formula units of the substance under discussion.

—(See *Gram formula weight*).

Molecular equation—Equation for a chemical reaction in which all formulas are written as if all substances existed as molecules; only complete formulas are used.

Molecular formula—The combination of element symbols and subscripts indicating the number and kind of atoms that have combined to form a molecule.

—Formula that indicates the actual number of atoms present in a molecule of a molecular substance. Compare with *Simplest formula*.

Mole fraction—The number of moles of a component of a mixture divided by the total number of mole in the mixture.

Molecular geometry—The arrangement of atoms (*not* lone pairs of electrons) around a central atom of a molecule or polyatomic ion.

Molecular orbital (MO)—An orbital resulting from overlap and mixing og atomic orbitals on different atoms. An MO belongs to the molecule as a whole.

Molecular orbital theory—A theory of chemical bonding based upon the postulated existence of molecular orbitals.

Molecular weight—The mass of one molecule of a nonionic substance in atomic mass units.

Molecule—A cluster of atoms that behaves as a single particle and possesses its own unique chemical properties. The atoms are held together by covalent bonds.

Monoprotic acids—Acids that dissociate liberating one mole of H^+ per mole of acid.

—Acid that contains one ionizable hydrogen atom per formula unit.

—Acid that can form only one hydronium ion per molecule; may be strong or weak.

Monosaccharide—A carbohydrate with the general formula $C_nH_{2n}O_n$.

***m*RNA**—Ribonucleic acid. The messenger that carries the genetic code from the cell nucleus to the ribosomes where protein synthesis occurs.

Mutation—A change in DNA that is passed to new generations.

Nanometer (nm)—10^{-9} meter.

Natural radioactivity—Spontaneous decomposition of an atom.

Negative feedback—Regulation of a process by one of its own products.

Nernst equation—Corrects standard electrode potentials for nonstandard conditions.

Net ionic equation—Equation that results from canceling spectator ions and eliminating brackets from a total ionic equation.

Neutralization—Reaction between an acid and a base.

—The reaction of an acid with a base to form a salt and water, Usually, the reaction of hydrogen ions with hydroxide ions to form water molecules.

Neutron—Uncharged nuclear particle with mass similar to that of the proton.

—A neutral subatomic nuclear particle having a mass of 1.0087 amu.

Nickel-cadmium cell (nicad battery)—A dry cell in which the anode is Cd, the cathode is Nio_2, and the electrolyte is basic.

Noble gases—Elements of periodic Group 0; also called rare gases; formerly called inert gases.

Noble gas configuration—The stable electronic configuration of a noble gas.

Nodal plane (node)—A region in which the probability of finding an electron is zero.

Nonbonding orbital—A molecular orbital derived only from a atomic orbital of one atom; lends neither stability nor instability to a molecule or ion when populated with electrons.

Nonelectrolyte—A susbtance whose aqueous solutions do not conduct electricity.

Nonelectrolytes—Compounds that do not dissociate in water.

Nonmetals—Elements above and to the right of the metalloids in the periodic table.

Nonpolar bond—Covalent bond in which electron density is symmetrically distributed.

Normal boiling point—The temperature at which the vapor pressure of a liquid is equal to one atmosphere pressure.

Normality—The number of equivalent weights (or equivalents) of solute per liter of solution.

Normal melting point—The melting (freezing) point at one atmosphere pressure.

Normal oxide—A metal oxide containing the oxide ion, O^{2-} (oxygen in the –2 oxidation state).

Normal salt—A salt containing no ionizable H atoms or OH groups.

Nuclear shielding—See *Shielding effect.*

Nucleoside—The combination of one of the heterocyclic bases in DNA with deoxyribose.

Nucleotide—A compound containing a nitrogen base bonded to a sugar that is bonded to a phosphate group.

Nucleus—The very small, very dense, positively charged center of an atom containing protons and neutrons, as well as other subatomic particles.

Nuclide symbol—Symbol for an atom. $^{A}_{Z}E$, in which E is the symbol for an element, Z is its atomic number, and A is its mass number.

Octahedron—A polyhedron with eight equal-sized, equilateral triangular faces and six apices (corners).

Octane number—A number that indicates how smoothly a gasoline burns.

Octet rule—Many representative elements attain at least a share of eight electrons in their valence shells when they form molecular or ionic compounds; there are some limitations.

Oil—Liquid triester of glycerol and unsaturated fatty acids.

Olefins—(See Alkenes).

Open sextet—Refers to species that have only six electrons in the highest energy level of the central element (many Lewis acids).

Order—Of a reactant, the power to which the reactant's concentration is raised in the rate-law expression; of a reaction, the sum of the powers to which all concentrations are raised in the rate-law expression (also called overall order of a reaction).

Organic chemistry—The chemistry of substances that contain carbon-hydrogen bonds.

Ornithine cycle—A series of reactions producing urea from ammonia and carbon dioxide.

Osmolarity—A unit describing the concentration of particles in a solution

$$\text{osmolarity} = \text{molarity} \times \left(\frac{\text{moles of particle}}{\text{mole solute}}\right).$$

Osmosis—The flow of water through an osmotic membrane.

—The process by which solvent molecules pass through a semipermeable membrane from a dilute solution into a more concentrated solution.

Osmotic pressure—The pressure build up in opposition to the flow of water across an osmotic membrane.

Overlap—Of orbitals, the interaction of orbitals on different atoms in the same region of space.

Oxidation—(a) The loss of electrons. (b) The removal of hydrogen atoms from a molecule. (c) The addition of an oxygen atom to a molecule.

—An algebraic increase in oxidation number; may correspond to a loss of electrons.

Oxidation-reducing reactions—Reactions in which oxidation and reduction occur; also called redox reactions.

Oxidation states (numbers)—The charge assigned to an ion or an atom is a molecule indicating the number of electrons it has gained or lost.

—Arbitrary numbers that can be used as mechanical aids in writing formulas and balancing equations; for single-atom ions they correspond to the charge on the ion; more lectronegative atoms are assigned negative oxidation numbers.

Oxidative deamination—Oxidation of an amino acid to a keto acid with the loss of ammonia.

Oxidative phosphorylation—ATP formation using energy liberated during catabolism of metabolites.

Oxide—A binary compound of oxygen.

Oxidizing agent—The substance that oxidizes another substance and is reduced.

p[ ]—Negative logarithm of the concentration (mol/L) of the indicated species.

Pairing—Of electrons, favourable interaction of two electrons with opposite m_s values in the same orbital.

Paramagnetism—Attraction toward a magnetic field, stronger than diamagnetism, but still weak compared with ferromagnetism.

Partial pressure—The pressure exerted by one gas in a mixture of gases.

Pauli Exclusion Principle—No two electrons in the same atom may have identical sets of four quantum numbers.

Peptide bond—An amide linkage between two amino acids.

Percentage ionization—Of weak electrolytes, the percentage of the weak electrolyte that ionizes in a solution of given concentration.

Percent by mass—100% multiplied by the mass of a solute divided by the mass of the solution in which it is contained.

Percent Composition—The mass percent of each element in a compound.

Percent purity—The percent of a specified compound or element in an impure sample.

Percent solution (weight/volume)—The number of gramds of solute per 100 mL of solution.

Percent yield—100% times actual yield divided by theoretical yield.

Period—A horizontal row in the periodic chart.

Periodicity—Regular periodic variations of properties of elements with atomic number (and position in the periodic table).

Periodic law—The properties of the elements are periodic functions of their atomic numbers.

Periodic table—An arrangement of elements in order of increasing atomic number that also emphasizes periodicity.

Peroxide—A compound containing oxygen in the –1 oxidation state. Metal peroxides contain the peroxide ion. O_2^{2-}.

pH—A numerical scale indicating acidity, alkalinity, or neutrality of aqueous solutions.

—Negative logarithm of the concentration (mol/L) of the $H_3O^+[H^+]$ ion; scale is commonly used over the range 0 to 14.

Phase diagram—Diagram that shows equilibrium temperature-pressure relationships for different phases of a substance.

Phenol—Hydrocarbon derivative containing an —OH group bound to an aromatic ring.

Photoelectric effect—Emission of an electron from the surface of a metal caused by impinging electromagnetic radiation of certain minimum energy; current increases with increasing intensity of radiation.

Photon—A "packet" of light or electromagnetic radiation; also called a quantum of light.

Pi (π) bond—A bond resulting from the side-on overlap of atomic orbitals, in which the regions of electron sharing are on opposite sides of the imaginary line connecting the bonded atoms and parallel to this line.

Pi (π) orbital—Molecular orbital resulting from side-on overlap of atomic orbitals.

Plasma—The remaining component when the formed elements have been removed from blood; the liquid portion of circulating blood.

Polar bond—Covalent bond in which there is an unsymmetrical distribution of electron density.

Polar covalent bond—A covalent bond in which the shared pair or lectrons spends more time near one of the two bonded atoms creating a partial charge separation.

Polyamide—A polymeric amide.

Polarization—Of an electrode, buildup of a product of oxidation or reduction at an electrode, preventing further reaction.

Polyatomic ion—A cluster of covalently bonded atoms having a positive or negative charge (a charged molecule).

Polyene—A compound that contains more than one double bond per molecule.

Polyester—Polymeric ester.

Polyhydric alcohol—Alcohol containing more than one —OH group.

Polymerization—The combination of many small molecules to form large molecule.

Polymers—Large molecules formed by the combination of many small molecules.

Polymorphous—Refers to substances that crystallize in more than one crystalline arrangement.

Polyprotic acid—An acid that contains more than one ionizable hydrogen atom per formula unit.

—Acid that can form two or more hydronium ions per molecule; often at least one step of the ionization is weak.

Polysaccharide—A polymer of monosaccharides joined by acetal or ketal bonds.

Polyunsaturated fatty acids—Fatty acids containing more than one carbon-carbon double bond per molecule.

***p* orbitals**—Beginning with the second energy level, a set of three mutually perpendicular, equal-arm, dumbbell-shaped atomic orbitals per energy level.

Potential energy—Energy that matter possesses by virtue of its position, condition, or composition.

Precipitate—An insoluble solid that forms and separates from a solution. —Solid formed by mixing in solution the constituent ions of a slightly soluble compound.

Precipitation reactions—Reactions in which a precipitate forms.

Pressure—Force per unit area.

Pressure-volume work—Work done by a gas when it expands against an external pressure or work done on a system as gases are compressed or consumed in the presence of an external pressure.

Primary—Two hydrogen atoms on the nitrogen atom of an amine or amide or two hydrogen atoms on the carbon atom bearing the OH group of an alcohol.

Primary alcohol—An alcohol with no or one R group bonded to the carbon bearing the —OH group.

Primary amine—An amine in which one H atom of ammonia has been replaced by an organic group.

Primary standard—A substance of a known high degree of purity that undergoes one invariable reaction with the other reactant of interest.

Primary structure—The sequence of amino acids in a protein.

Primary voltaic cells—Voltaic cells that cannot be recharged; no further chemical reaction is possible once the reactants are consumed.

Principal quantum number (n)—Quantum mechanical solution to a wave equation that designates the major energy level or shell in which an electron resides.

Products—Substances produced in a chemical reaction.

Prostaglandins—Hormonelike derivates or essential fatty acids.

Proteins—Compounds consisting of large numbers of amino acid groups joined by peptide bonds.

Proton—Fundamental unit of positive charge : found in an atom's nucleus.

—A subatomic particle having a mass of 1.0073 amu and a charge of +1, found in the nuclei of atoms.

Protonic acid—An Arrhenius (classical) acid, or a Bronsted-Lowry acid.

Pseudobinary ionic compounds—Compounds that contain more than two elements but are named like binary compounds.

Qualitative description—A description telling the kinds of things present but not the amounts.

Quantitative description—A description giving the kind and amount of things present.

Quantized—Restricted to only certain values.

Quantum—A "packet" of energy.

Quantum mechanics—Mathematical method of treating particles on the basis of quantum theory, which assumes that energy (of small particles) is not infinitely divisible.

Quantum numbers—Numbers that describe the energies of electrons in atoms; derived from quantum mechanical treatment.

Rad—A unit of radiation equal to 100 ergs absorbed per gram of tissue.

Radiant energy—See *Electromagnetic radiation.*

Radical—A species containing one or more unpaired electrons; many radicals are very reactive.

Radioactivity—The spontaneous emission of radiant energy by unstable nuclei.

Raoult's Law—The vapor pressure of a solvent in an ideal solution decreases as its mole fraction decreases.

Rare earths—Inner transition elements.

Rare gases—See *Noble gases.*

Rate constant—The numerical constant in the rate law. The value of the constant specific for a particular reaction and the temperature at which it occurs.

Rate-determining step—The slowest step in a mechanism; the step that determines the overall rate of reaction.

Rate law—The mathematical equation describing the rate at which a reaction will occur.

Rate-law expression—Equation relating the rate of a reaction to the concentrations of the reactants and the specific rate constant.

Rate of reaction—Change in concentration of a reactant or product per unit time.

Reactants—Substances consumed in a chemical reaction.

Reaction coordinate—The progress along the pathway from reactants to products: sometimes called "progress of reaction".

Reaction intermediate—A species produced and then entirely consumed during a reaction: usually short-lived.

Reaction quotient, Q—The mass action expression under any set of conditions (not necessarily equilibrium); its magnitude relative to K determines the direction in which reaction must occur to establish equilibrium.

Reaction rate—The speed with which a reaction occurs.

Reaction ratio—The relative amounts of reactants and products involved in a reactions may be the ratio of moles, millimoles, or masses.

Reaction stoichiometry—Description of the qunatitative relationships among substances as they participate in chemical reactions.

Real gases—Gases that deviate from ideal gas behaviour; all real gases.

Redox reaction—An oxidation-reduction reaction.

Redox titration—The quantitative analysis of the amount or concentration of an oxidizing or reducing agent in a sample by observing its reaction with a known amount or concentration of a reducing or oxidizing agent.

Reducing agent—The substance that reduces another substance and is oxidized.

Reducing sugar—Any sugar that will react with Fehling's and Benedict's solutions, Tollen's reagent, or other weak oxidizing agents.

Reduction—(a) The gain of electrons, (b) The addition of hydrogen atoms to a molecule. (c) The removal of oxygen atoms from a molecule.

—An algebraic decrease in oxidation number; may correspond to a gain of electrons.

REM (roentgen equivalent man)—The amount of any absorbed radiation causing the effect equivalent to the absorption of one roentgen by man.

Replication—The synthesis of new DNA.

Representative elements—A group elements in the periodic table.

Resonance—Concept in which two or more equivalent dot formulas for the same arrangement of atoms (resonance structures) are necessary to describe the bonding in a molecule or ion.

Respiratory chain—A sequence of chemical reactions on mitochondria producing ATP.

Reverse osmosis—Forcing solvent molecules to flow through a semipermeable membrane from a concentrated solution into a dilute solution by application of greater hydrostatic pressure on the concentrated side than the osmotic pressure opposing it.

Reversible reactions—Reactions that occur in both directions; indicated by double arrows ($\rightleftharpoons$).

Roasting—Heating an ore of an element in the presence of air.

Roentgen—A unit of X-ray or gamma-ray radiation.

Rydberg equation—Empirical equation that relates wave-lengths in the hydrogen emission spectrum to integers.

s orbital—A spherically symmetrical atomic orbital; one per energy level.

Sacrificial anode—A more active metal that is attached to a less active metal to protect the less active metal cathode against corrosion.

Salt—A product of the reaction between an acid and base.

—A compound that contains a cation other than H^+ and an anion other than OH^- or O^{2-}.

Salt bridge—U-shaped tube containing electrolyte, which connects two half-cells of a voltaic cell.

Salt bridges—Attractions between protein side chain groupings of opposite charge.

Saponification—Hydrolysis of an ester using a strong base such as NaOH.

Saturated hydrocarbons—Hydrocarbons that contain only single bonds, They are also called alkanes or paraffin hydrocarbons.

Saturated solution—Solution in which no more solute will dissolve.

Secondary—One hydrogen atom on the nitrogen atoms of an amine or amide or one hydrogen atom on the carbon atom bearing the OH group of an alcohol.

Secondary alcohol—An alcohol with two R groups bonded to the carbon bearing the —OH group.

Secondary amine—An amine in which two H atoms of amonia have been replaced by organic groups.

Secondary standard—A solution that has been titrated against a primary standard. A standard solution is a secondary standard.

Secondary structure—Hydrogen bonding between peptide bond groupings in proteins.

Secondary voltaic cells—Voltaic cells that can be recharged; original reactants can be regenerated by reversing the direction of current flow.

Second Law of Thermodynamics—The universe tends toward a state of greater disorder in spontaneous processes.

Semipermeable membrane—A thin partition between two solutions through which certain molecules can pass but others cannot.

Semiconductor—A substance that does not conduct electricity at low temperatures but does so at higher temperatures.

Serum—The liquid remaining when fibrinogen is removed from blood plasma.

Shielding effect—Electrons in filled sets of s and p orbitals between the nucleus and outer shell electrons shield the outer shell electrons somewhat from the effect of protons in the nucleus; also called screening effect.

Sigma (σ) bond—A bond resulting from the head-on overlap of atomic orbitals, in which the region of electron sharing is along and (cylindrically) symmetrical to the imaginary line connecting the bonded atoms.

Sigma (σ) orbital—Molecular orbital resulting from head-on overlap of two atomic orbitals.

Simplest formula—The smallest whole-number ratio of atoms present in a compound; also called empirical formula. Compare with *Molecular formula.*

Single bond—The sharing of a pair of electrons by two atoms.

Soap—The sodium or potassium salt of a long-chain organic acid.

—An emulsifier that can disperse nonpolar substances in water; the sodium salt of a long chain organic acid; consists of a long hydrocarbon chain attached to a carboxylate group, —$CO_2{}^-Na^+$.

Sol—Colloidal suspension of a solid dispersed in a liquid.

Solid aerosol—Colloidal suspension of a solid in a gas.

Solid emulsion—Colloidal suspension of a liquid dispersed in a solid.

Solid foam—Colloidal suspension of a gas dispersed in a solid.

Solid sol—Colloidal suspension of a solid dispersed in a solid.

Solubility product constant—Equilibrium constant that applies to the dissolution of a slightly soluble compound.

Solubility product principle—The solubility product constant expression for a slightly soluble compound is the product of the concentrations of the constituent ions, each raised to the power that corresponds to the number of ions in one formula unit.

Solute—The substance being dissolved during the solution-making process.

Solution—A homogenous mixture, usually a solid or liquid dissolved in a liquid.

Solvation—Solvent molecules surrounding solute particles (ions or molecules).

Solvent—The substance that dissolves the solute as a solution is made.

Specific gravity—The ratio of the density of a substance to the density of water.

Specific heat—The amount of heat required to raise the temperature of one gram of a substance one degree Celsius.

Specific rate constant (also called **rate constant**)—An experimentally determined (proportionality) constant, which is different for different reactions and which changes only with temperature: k in the rate-law expression: Rate = $k[A]^x[B]^y$.

Spectator ions—Ions in solution that do not participate in a chemical reaction.

Spectral line—Any of a number of lines corresponding to definite wavelengths in an atomic emission or absorption spectrum; represents the energy difference between two energy levels.

Spectrum—Display of component wavelengths (colors) of electromagnetic radiation.

Spin quantum number (m_s)—Quantum mechanical solution to a wave equation that indicates the relative spins of electrons.

Spontaneity—Of a process, its property of being energetically favourable and therefore capable of proceeding in the forward direction (but not necessarily at an observable rate).

Square planar—A term used to describe molecules and polyatomic ions that have one atom in the center and four atoms at the corners of a square.

Standard cell potential—Potential difference, E^0_{cell}, between standard reduction and oxidation half-cells.

Standard conditions (STP or SC)—Standard temperature, 0°C, and standard pressure, one atmosphere, are standard conditions for gases.

Standard electrochemical conditions—1.0 *M* concentration for dissolved ions, 1.0 atm partial pressure for gases, and pure solids and liquids.

Standard electrode potential—By convention, potential, E^0, of a half-reaction as a reduction relative to the standard hydrogen electrode when all species are present at unit activity.

Standard electrodes—Half-cells in which the oxidized and reduced forms of a species are present at unit activity; 1.0 *M* solutions of dissolved ions, 1.0 atm partial pressure of gases, and pure solids and liquids.

Standard entropy, S^0 (of a substance)—The absolute entropy of a substance in its standard state at 298 K.

Standardization—The process by which the concentration of a solution is accurately determined by titrating it against an accurately known amount of a primary standard.

Standard molar enthalpy of formation. ΔH^0_f—The amount of heat absorbed in the formation of one mole of a substance in a specified state from its elements in their standard states.

Standard molar volume—The volume occupied by one mole of an ideal gas under standard conditions; 22.4 liters.

Standard reaction—A reaction in which the numbers of moles of reactants shown in the balanced equation, all in their standard states, are completely converted to the numbers of moles of products shown in the balanced equation, also all at their standard states.

Standard solution—A solution of accurately known concentration.

Starch—A polymer of glucose joined by α-acetal bonds produced by plants.

State function—A variable that defines the state of a system; a function that is independent of the pathway by which a process occurs.

Stoichiometry—Description of the quantitative relationships among elements and compounds as they undergo chemical changes.

Strong acids—Acids that undergo extensive dissociation in water.

Strong electrolyte—A substance that conducts electricity well in dilute aqueous solutions.

Structural isomers—Two or more molecules with the same molecular formula but the order of arrangement of the atoms within the molecule differs.

Sublimation—The direct vaporization of a solid by heating without passing through the liquid state.

Subsidiary quantum number (L)—Quantum mechanical solution to a wave equation that designates the sublevel or set of orbitals (*s, p, d, f*) within a given major energy level in which an electron resides.

Substitution reaction—A reaction in which an atom or a group of atoms is replaced by another atom or group of atoms.

Substrate—The molecule on which an enzyme acts.

Supercooled liquids—Liquids that, when cooled, apparently solidify but actually continue to flow very slowly under the influence of gravity.

Supercritical fluid—A substance at a temperature above its critical temperature.

Superoxide—A compound containing the superoxide ion. O^{2-} (oxygen in the –½ oxidation state).

Supersaturated solution—A (metastable) solution that contains a higher than saturation concentration of solute; slight disturbance or seeding causes crystallization of excess solute.

Surface tension—A measure of the inward intermolecular forces of attraction among liquid particles that must be overcome to expand the surface area.

Surroundings—Everything in the environment of the system.

Suspension—Particles with diameters greater than 1000 Å dispersed in a solvent.

—A heterogeneous mixture in which solute-like particles settle out of a solvent-like phase some time after their introduction.

System—The substances of interest in a process: the part of the universe under investigation.

Temperature—A measure of the intensity of heat, i.e., the hotness or coldness of a sample or object.

Ternary acid—A ternary compound containing H, O, and another element, often a nonmetal.

Ternary compound—A compound consisting of three elements; may be ionic or covalent.

Tertiary—The absence of hydrogen atoms on the nitrogen atom of an amine or amide or on the carbon atom bearing the OH group of an alcohol.

Tertiary amine—An amine in which three H atoms of ammonia have been replaced by organic groups.

Tertiary structure—Features such as disulfide bonds, salt bridges, and hydrogen bonding which give added organization to protein structure.

Tetrahedral—A term used to describe molecules and polyatomic ions that have one atom in the center and four atoms at the corners of a tetrahedron.

Tetrahedron—A solid four-sided figure with equilateral triangular faces.

Theoretical yield—Maximum amount of a specified product that could be obtained from specified amounts of reactants, assuming complete consumption of limiting reactant according to only one reaction and complete recovery of product. Compare with Actual yield.

Thermal cracking—Decomposition by heating a substance in the presence of a catalyst and in the absence of air.

Thermal pollution—Introduction of heated waste water into natural waters.

Thermodynamically favourable (spontaneous) reaction—Reaction that occurs with a net release of free energy, G; reaction for which ΔG is negative.

Thermodynamics—The study of the energy transfers accompanying physical and chemical processes.

Thermodynamic state of a system—A set of conditions that completely specifies all of the properties of the system.

Thermodynamic standard state of a substance—Its most stable state under one atmosphere pressure and at some specific temperature (usually 25°C unless otherwise specified).

Third Law of Thermodynamics—The entropy of a hypothetical pure, perfect, crystalline substance at absolute zero temperature is zero.

Titration—The process by which the unknown concentration of a scummer may be determined by allowing it to react completely with a measured volume of a solution of known concentration.

Titration curve—For acid-base titration, a plot of pH versus volume of acid or base solution added.

Total ionic equation—Equation for a chemical reaction written to show the predominant form of all species in aqueous solution or in contact with water.

Tracer—An internally administered radioactive sample whose presence can be determined using an appropriate radiation detector.

Trans—A term indicating that a molecule has two identical groups on opposite sides of a double bond.

Transamination—Transfer of ammonia from an amino acid to alkeno acid to produce a new amino acid (and a new keto acid).

Transcription—The transfer of genetic information from DNA to *m*RNA.

Transition state—Relatively high-energy state in which bonds in reactant molecules are partially broken and new ones are partially formed.

Transition state theory—Theory of reaction rates that states that reactants pass through high-energy transition states before forming products.

Translation—The recognition of the triplet of heterocyclic bases on *m*RNA by *t*RNA for the incorporation of a specific amino acid during protein synthesis.

Tricarboxylic acid cycle—(See Krebs cycle).

Triglyceride—A triester formed by the reaction of one mole of glycerol and three moles of fatty acid(s).

Trigonal bipyramid—A six-sided polyhedron with five apices (corners), consisting of two pyramids sharing a common triangular base.

Trigonal planar—A term used to describe molecules and polyatomic ions that have one atom in the center and three atoms at the corners of an equilateral triangle.

Triple bond—The sharing of three pairs of electrons by two atoms.

Triple point—Point on a phase diagram for a substance corresponding to the only pressure and temperature at which solid, liquid, and gas can coexist at equilibrium.

***t*RNA**—Transfer ribonucleic acid. Carries the amino acids to the ribosomes where protein synthesis occurs.

Turnover number—The number of substrate molecules a single enzyme molecule can act on per minute.

Tyndall effect—The scattering of light by colloidal particles.

Unit cell—Smallest repeating unit showing all the structural characteristics of a crystal.

Universal gas constant—*R*, the proportionality constant in the ideal gas equation, $PV = nRT$.

Universe—The system plus the surroundings.

Unsaturated compound—An organic compound with carbon-carbon multiple bonds.

Unsaturated hydrocarbons—Hydrocarbons that contain double or triple carbon-carbon bonds.

Unsaturated solution—A solution containing less than the maximum amount of solute that will dissolve in a given amount of solvent at a given temperature.

Unshared pair—See *Lone pair.*

Valence—The combining power power of a specific element.

Valence bond (VB) theory—Assumes that covalent bonds are formed when atomic orbitals on different atoms overlap and electrons are shared.

Valence electrons—Outermost electrons of atoms; usually those involved in bonding.

Valence shell electron pair repulsion (VSEPR) theory—Assumes that electron pairs are arranged around the central element of a molecule or polyatomic ion so that there is maximum separation (and minimum repulsion) among regions of high electron density.

van der Waals' equation—An equation of state that extends the ideal gas law to real gases by inclusion of two empirically determined parameters, which are different for different gases.

Vapor—A gas formed by boiling or evaporating a liquid.

Vapor pressure—The pressure exerted by a vapor in equilibrium with its mother liquid.

—The partial pressure of a vapor at the surface of its parent liquid.

van't Hoff equation—The relationship between ΔH^0 for a reaction and its equilibrium constants at two different temperatures.

Viscosity—The tendency of a liquid to resist flow : a measure of its fluidity.

Vitamin—A naturally occuring organic compound in foods that is not synthesized by the body but is necessary for normal body health.

Volatility—The ease with which a liquid vaporizes.

Voltage—Potential difference between two electrodes: a measure of the chemical potental for a redox reaction to occur.

Voltaic cells—Electrochemical cells in which spontaneous chemical reactions produce electricity; also called galvanic cells.

Vulcanization—The process in which sulfur is added to rubber and heated to 140°C.

Water equivalent—Of a calorimeter, the amount of water that would absorb the same amount of heat as the calorimeter per degree temperature increase.

Wavelength—The distance between two corresponding points of a wave.

Weak acids—Acids that are only slightly dissociated in water.

Weak electrolyte—A substance that conducts electricity poorly in dilute aqueous solution.

Weight—A measure of the gravitational attraction of the earth for a body.

Work—The application of a force through a distance; for physical changes or chemical reactions at constant external pressure, the work done on the system is $-P\Delta V$.

Zwitterion—An amino acid form in which the H+ from the carboxylic acid group has been transferred to the amine nitrogen, $\text{—}\underset{\underset{{}^{(+)}NH_3}{|}}{CH}\text{—}CO_2^{(-)}$.